✓ Emery-Pratt
12/31/84
36.00

Mechanisms of Morphological Evolution

Mechanisms of Morphological Evolution

A combined genetic, developmental and ecological approach

WALLACE ARTHUR

Senior Lecturer
Department of Biology
Sunderland Polytechnic

A Wiley–Interscience Publication

JOHN WILEY & SONS
Chichester · New York · Brisbane · Toronto · Singapore

Copyright © 1984 by John Wiley & Sons Ltd.

All rights reserved.

No part of this book may be reproduced by any means, nor transmitted, nor translated into a machine language without the written permission of the publisher.

Library of Congress Cataloging in Publication Data:

Arthur, Wallace.
 Mechanisms of morphological evolution.

 "A Wiley–Interscience publication."
 Bibliography: P.
 Includes index.
 1. Evolution. 2. Morphology. I. Title.
QH371.A76 1984   575   83-16993 83- 40291
ISBN 0 471 90347 7 (U.S.)

British Library Cataloguing in Publication Data:

Arthur, Wallace.
 Mechanisms of morphological evolution.

 1. Evolution
 I. Title
 575.01 QH366.2

 ISBN 0 471 90347 7

Typeset by Activity Ltd, Salisbury, Wilts.
Printed by St Edmundsbury Press, Suffolk.

"It is doubtful if anyone would have ever felt any need to resist the notion of evolution if all it implied was that the exact chemical constitution of haemoglobin gradually changed over the ages."

C. H. WADDINGTON, 1975.

Contents

Preface		xi
Acknowledgements		xiii
Chapter 1: Introduction		1
1.1	General introduction	1
1.2	An overview of evolution	5
1.3	Defining morphological evolution	8
1.4	Evolutionary mechanisms	10
1.5	Neo-darwinism	19
1.6	Summary	20
Chapter 2: The measurement of morphological characters		21
2.1	Individuals	22
2.2	Populations	27
2.3	Summary	29
Chapter 3: The genetic basis of quantitative variation		31
3.1	The connection between Mendelian and biometrical approaches	32
3.2	Components of variation	33
3.3	The measurement of heritability	37
3.4	Heritability and microevolution	41
3.5	Threshold characters	45
3.6	Summary	50
Chapter 4: Artificial selection experiments		51
4.1	Rationale and basic experimental design	51
4.2	Experiments involving directional selection	53
4.3	Experiments involving disruptive selection	62
4.4	Changes in gene frequency	64
4.5	Experimental selection	66
4.6	Summary	67

Chapter 5: Microevolution in natural populations — 68

5.1 The problem — 68
5.2 Shell shape in *Lymnaea* — 71
5.3 Shell size in *Cepaea* — 75
5.4 Beak size in *Geospiza* — 79
5.5 Ecotypes in *Potentilla* — 83
5.6 Summary — 87

Chapter 6: Speciation and interspecific differences — 88

6.1 Divergence of species and divergence of opinions — 88
6.2 The multitude of speciation models — 89
6.3 Comparisons of closely related species — 92
6.4 Partial reproductive isolation — 100
6.5 The Wallace effect and reproductive character displacement — 102
6.6 Summary — 103

Chapter 7: Further divergence after speciation — 105

7.1 Stages in the evolutionary divergence of species — 105
7.2 Competitive character displacement: an introduction — 108
7.3 The 'neo-classical' example — 109
7.4 Species ranges and species richness — 112
7.5 Summary — 114

Chapter 8: Long-term evolution at the species level — 116

8.1 Introduction: punctuated equilibrium versus phyletic gradualism — 116
8.2 Case studies of molluscs and men — 121
8.3 A critique of punctuated equilibrium — 128
8.4 Summary — 130

Chapter 9: Development and evolution — 131

9.1 Development, evolutionary theory and saltations — 131
9.2 Cell differentiation, pattern formation and morphogenesis — 135
9.3 The central problem of morphogenesis — 138
9.4 The morphogenetic tree: a hypothesis — 142
9.5 Summary — 145

Chapter 10: Developmental genetics — 147

10.1 Introduction — 147
10.2 Three categories of genes — 148
10.3 Maternal *D*-genes: shell coiling in *Lymnaea* and *Partula* — 151
10.4 Zygotic *D*-genes: insect homoeotics — 156
10.5 A general framework for *D*-gene action — 165

		ix
10.6	The molecular basis of *D*-genes	169
10.7	Summary	170

Chapter 11: The evolution of major genes and the origin of higher taxa ... 172

11.1	Introduction: two types of origin	172
11.2	Neo-darwinian origins	174
11.3	Saltational origins, and the roles of *w*- and *n*-selection	178
11.4	Examples of saltational evolution	195
11.5	Summary	203

Chapter 12: Variation in evolutionary rates ... 204

12.1	Introduction	204
12.2	Developmentally induced rate variation: genetic aspects	207
12.3	Developmentally induced rate variation: morphological aspects	212
12.4	The morphogenetic tree and the mega-evolutionary tree	216
12.5	Summary	219

Chapter 13: The evolution of morphological complexity ... 220

13.1	Introduction	220
13.2	What is complexity?	220
13.3	Alternative theories of the evolutionary increase in morphological complexity	222
13.4	*D*-genes, complexity and saltations	224
13.5	Four misconceptions to be avoided	226
13.6	Summary	227

Chapter 14: Limits to the evolutionary proliferation of species and body plans ... 228

14.1	Introduction: two types of limit	228
14.2	Species packing and ecological limits	230
14.3	Coadaptation and developmental limits	235
14.4	Summary	237

Chapter 15: Summary, conclusions and prospect ... 239

References ... 251

Index ... 266

Preface

There is increasing controversy among evolutionary biologists on the question of whether current evolutionary theory, largely formulated by population geneticists who usually study very short-term microevolutionary processes, provides an adequate explanation of long-term morphological evolution. This controversy is presently centred on variation in evolutionary rates at the species level, where reasonably accurate measurements are sometimes possible; but, in addition, other areas of doubt exist, such as whether current evolutionary theory is capable of explaining the origin of major taxonomic groups. Central to both these aspects of the controversy is the question of whether evolution sometimes proceeds by rapid, large-scale phenotypic changes — something that has become known as saltational evolution. Discussion of this possibility is often heated because proponents of saltational change are usually considered to be anti-Darwinian since Darwin emphasized the importance, in evolution, of slow change through the gradual accumulation of many, individually small, variations.

The aim of this book is to approach the problems of morphological evolution from a number of different angles — some conventional (chapters 3–8) and others relatively novel (chapters 9–14) — in the hope that a degree of synthesis of ideas can be achieved. Because the emphasis is on evolutionary *mechanisms* rather than on the elucidation of evolutionary pathways in particular taxonomic groups, the approach taken is largely neontological rather than palaeontological; and the book is essentially a synthesis of selected topics from quantitative and ecological genetics, evolutionary ecology and developmental biology that are relevant to morphological evolution.

No book of this kind, in which the approaches of different academic disciplines are juxtaposed and a synthesis attempted, can escape from being controversial and somewhat partisan. The particular synthesis that emerges reflects the convictions of the author; and it may be helpful to point out here what these are, in the case of the present book, in order to clarify its main theme. This book was written in the strong conviction that the so-called 'modern synthesis' of evolutionary theory, gradually formulated during the period from the re-discovery of Mendel's work in 1900 until about 1950, and the more recent refinements of it, involve only a partial synthesis and notably lack a developmental component connecting genotype and phenotype. I hold the view that we now have enough information — albeit fragmentary — about developmental genetics at least to begin the formulation of this missing developmental component; and that, until this has been done, current evolutionary theory in the form of neo-Darwinism is inadequate as an explanation of long-term evolution. I do not take the more extreme position of some recent authors that neo-Darwinism is incorrect; rather I argue that it is correct as far as it goes, and needs to be extended, not replaced. I put forward the view that saltational evolution at the morphological level is explicable in terms of natural selection acting on single genes of major developmental effect. This proposition involves acceptance of a lack of correspondence between 'developmental revolutions', which can be produced by single mutations, and 'genetic revolutions', which need not involve any large developmental change. Also, the form of selection proposed to be responsible for the establishment of some mutations of major morphogenetic effect is, as will become apparent, not a form envisaged by Darwin.

I have assumed the reader's familiarity with Mendelian genetics as well as with elementary population genetics and ecology. Only a minimal background in mathematics is required, mostly in simple algebra and biometrical techniques; knowledge of calculus, matrices and other advanced areas of mathematics is not needed. The book is aimed at (a) researchers in any of the areas of evolutionary biology dealt with here wishing to learn more of the others and to consider their possible interrelationships, (b) researchers in developmental biology interested in morphological evolution and its possible links with their own subject, and (c) senior undergraduates taking advanced courses in evolution, genetics, ecology or development who have already completed an introductory course in evolution.

It must be stressed that the book is not intended to be a complete review; the areas covered are much too diverse for this to be attempted. Throughout, the emphasis is on principles and hypotheses; and particular examples are used to illustrate specific points. It is not intended to imply in these cases that the examples chosen are the only, or even the best, ones that illustrate the point being made.

It is a pleasure to acknowledge my indebtedness to Professor Bryan Clarke, FRS, who helped me to start developing some of the ideas put forward herein by insisting, several years ago, that there was some fundamental difference between morphological and molecular evolution, and that this difference was not simply explicable in terms of one process being selective, the other neutral. Professor Clarke was also kind enough to read the entire manuscript and to suggest numerous improvements. As is usual in these situations, I must point out that any remaining errors, omissions and so on are entirely mine.

The production of the final script owes much to the help I have had from Susan Harrison, Jan Laverick and Julie Wilson, who did the typing, and Judith Middlecote, who prepared all of the original figures. I am also very grateful to the various authors, journals and publishers concerned for their permission to reproduce those figures and tables that are derived from other published works. Finally, I must thank the University of Newcastle upon Tyne for the use of its excellent library, in which establishment the bulk of the manuscript was written.

Acknowledgements

The author acknowledges with thanks the permission of various authors for the use of previously published material as indicated in the legends accompanying the relevant figures and tables.

He is also grateful to the following copyright holders for permission to reproduce the material listed.

Academic Press Inc.
> *Figure 10.6.* From: Poodry, C. A. (1980). Imaginal discs: morphology and development. In: *The Genetics and Biology of Drosophila*, Vol. 2d, eds. M. Ashburner and T. R. F. Wright. Copyright: Academic Press Inc. (London) Ltd.
> *Table 10.1.* From: Ouweneel, W. J. (1976). Developmental genetics of homoeosis. *Adv. Genet.*, **18**, 179–248. Copyright: Academic Press Inc., New York.

American Society of Zoologists
> *Figure 10.9.* From: Lewis, E. B. (1963). Genes and developmental pathways. *Am. Zool.*, **3**, 33–56. Copyright: American Society of Zoologists.

Cambridge University Press
> *Figure 2.1.* From: Thompson, D'A. W. (1942). *On Growth and Form*, second edition. Copyright: Cambridge University Press.
> *Figures 5.4, 5.5.* From: D. Lack (1947). *Darwin's Finches. An Essay on the General Biological Theory of Evolution*. Copyright: Cambridge University Press.
> *Figure 4.4.* From: Falconer, D. S. (1973). Replicated selection for body weight in mice. *Genet. Res.*, **22**, 291–321. Copyright, Cambridge University Press.
> *Figure 4.10.* From: Scharloo, W., den Boer, M., and Hoogmoed, M. S. (1967). Disruptive selection on sternopleural chaeta number in *Drosophila melanogaster*. *Genet. Res.*, **9**, 115–118. Copyright: Cambridge University Press.

Carnegie Institution of Washington
> *Figure 5.7.* From: Clausen, J., and Hiesey, W. M. (1958). Experimental studies on the nature of species. IV. Genetic structure of ecological races. *Carnegie Inst., Washington Publ.*, No. 615. Copyright: Carnegie Institution of Washington.

Columbia University Press
> *Figure 6.1.* From: Dobzhansky, T. (1951). *Genetics and the Origin of Species*, third edition. Copyright: Columbia University Press.

Figure 6.3, Table 6.4. From: Lewontin, R. C. (1974). *The Genetic Basis of Evolutionary Change.* Copyright: Columbia University Press.

Cook, L. M.
Figure 1.2. From: Cook, L. M. (1971). *Coefficients of Natural Selection.* Hutchinson, London. Copyright: L. M. Cook.

Ecological Society of America
Figure 5.6. From "Comparative ecology of Galapagos ground finches (*Geospiza* Gould): Evaluation of the importance of floristic diversity and interspecific competition." *Ecol. Monogr.,* **47**, 151–184. Copyright © 1977, The Ecological Society of America.

Excerpta Medica
Figure 10.10b. From: Garcia-Bellido, A. (1975). Genetic control of wing disc development in *Drosophila.* In: *Cell Patterning, CIBA Found. Symp.,* No. 29 (new series). Associated Scientific Publishers, Amsterdam. Copyright: Excerpta Medica.

W. H. Freeman & Co.
Table 6.1. From: White, M. J. D. (1978). *Modes of Speciation.* Copyright: W. H. Freeman and Company.
Figures 10.7, 10.8, 10.11. From Garcia-Bellido, A., Lawrence, P. A., and Morata, G. (1979). Compartments in animal development. *Sci. Am.,* **241** (1), 90–98. Copyright: W. H. Freeman and Company.

Falconer, D. S.
Figure 3.3. From: Falconer, D. S. (1960). *Introduction to Quantitative Genetics,* first edition. Longman, London. Copyright: D. S. Falconer.
Tables 3.2, 3.5; Figures 3.7, 3.8, 4.3, 4.11. From: Falconer, D. S. (1981). *Introduction to Quantitative Genetics,* second edition. Longman, London. Copyright: D. S. Falconer.

Genetical Society
Figure 5.1. From: Arthur, W. (1982). Control of shell shape in *Lymnaea stagnalis. Heredity,* **49**, 153–161. Copyright: Genetical Society.
Table 5.3. From Cook, L. M. (1967). The genetics of *Cepaea nemoralis. Heredity,* **22**, 397–410. Copyright: Genetical Society.
Figure 10.5. From Murray, J., and Clarke, B. (1976). Supergenes in polymorphic land snails. II. *Partula suturalis. Heredity,* **37**, 271–282. Copyright: Genetical Society.

Krebs, C. J.
Table 1.2. From: Krebs, C. J. (1972). *Ecology. The Experimental Analysis of Distribution and Abundance.* Harper and Row, New York. Copyright: C. J. Krebs.

Macmillan Journals Limited
Figure 4.9. Reprinted by permission from *Nature,* **193**, 1164–1166. Copyright (c) 1962 Macmillan Journals Limited.
Figure 5.2. Reprinted by permission from *Nature,* **263**, 496–497. Copyright (c) 1976 Macmillan Journals Limited.
Table 5.4. Reprinted by permission from *Nature,* **274**, 793–794. Copyright (c) 1978 Macmillan Journals Limited.

Figures 8.2, 8.3. Reprinted by permission from *Nature,* **293**, 437–443. Copyright (c) 1981 Macmillan Journals Limited.
Figures 8.4, 8.5. Reprinted by permission from *Nature,* **292**, 113–122. Copyright (c) 1981 Macmillan Journals Limited.
Figure 9.2. Reprinted by permission from *Nature,* **225**, 420–422. Copyright (c) 1970 Macmillan Journals Limited.
Figure 10.10a. Reprinted by permission from *Nature,* **265**, 211–216. Copyright (c) 1977 Macmillan Journals Limited.

Mather, K., and Jinks, J. L.
Figure 2.3. From: Mather, K., and Jinks, J. L. (1977). *Introduction to Biometrical Genetics.* Chapman and Hall, London. Copyright: K. Mather and J. L. Jinks.

Robertson, A.
Figure 4.7. From: Clayton, G. A., Morris, J. A., and Robertson, A. (1957). An experimental check on quantitative genetical theory. I. Short-term responses to selection. *J. Genet.,* **55**, 131–151.

Robertson, F. W.
Figure 4.6. From: Robertson, F. W. (1955). Selection response and the properties of genetic variation. *Cold Spring Harbor Symp. Quant. Biol.,* **20**, 166–177.

Royal Society
Figure 10.3. From: Boycott, A. E., Diver, C., Garstang, S. L., and Turner, F. M. (1930). The inheritance of sinistrality in *Limnaea peregra* (Mollusca, Pulmonata). *Phil. Trans. R. Soc. London, Ser. B,* **219**, 51–131. Copyright: The Royal Society.

Springer-Verlag
Figures 7.2, 7.3. From: Fenchel, T. (1975). Character displacement and coexistence in mud snails (Hydrobiidae). *Oecologia,* **20**, 19–32. Copyright: Springer-Verlag.

Valentine, J. W.
Figure 12.5. From: Ayala, F. J., and Valentine, J. W. (1979). *Evolving. The Theory and Processes of Organic Evolution,* Benjamin/Cummings.

Chapter 1

Introduction

"The current theory of evolution attributes this process to the natural selection of random variations. In a certain sense this phrase probably conveys the essential truth about the nature of the process concerned. It is, however, only too easy to accept it as an adequate answer to our questions, whereas it should really be taken as an introduction to a series of problems."

C. H. Waddington, 1957a

1.1 GENERAL INTRODUCTION

Of all biological processes, evolution is perhaps studied from a greater number of angles, and by the adherents of a greater variety of scientific disciplines, than any other. This is not surprising, since evolution involves the change, over time, not only of all organisms, but also of all genetically based characteristics of each. Thus not only may the study of evolution be fragmented on a taxonomic basis, but, in addition, it may be divided on the basis of levels of organization, such as molecular and morphological.

Special problems arise in the study of evolution because, unlike some other temporal biological processes, notably development, it cannot be studied directly in its entirety. Admittedly, certain direct studies of particular microevolutionary events, such as the rapid responses to human modification of the environment observed in industrial melanism in the Lepidoptera and heavy-metal tolerance in grasses, have yielded fascinating insights into the evolutionary process. However, these events are only tiny fragments of evolution, and their relevance to the mechanisms of long-term evolutionary change is increasingly being questioned. Two approaches, both in a sense alternatives to the direct study of small bursts of microevolution, are: firstly, a study of more extended evolutionary changes as preserved in the fossil record; and secondly, a comparative study of existing groups of organisms with a view to deducing by what means, and via which routes, the differences between them have evolved.

Within any one broad approach to the investigation of evolution, considerable differences in strategy may occur due to the emphasis of different problems. For example, both ecological geneticists and evolutionary ecologists are involved in the study of microevolution in natural populations. However, a preoccupation with the necessity of demonstrating the inheritance of the characters whose evolution is observed has led ecological geneticists to concentrate on discrete phenotypic variants with a simple genetic basis. Examples include allozymes, pigmentation in organisms such as *Biston* and *Cepaea*, chromosomal inversions in *Drosophila*, and various blood proteins in mammals, notably haemoglobin. On the other hand, an awareness of the need to understand the ecological importance of particular characters, and hence the likely forms of selection that may modfy them, has directed studies in evolutionary ecology to such characters as beak size in birds, feeding behaviour, and interspecific competitive ability. Neither of these approaches is without its problems and in fact the main problem of each is the strength of the other; for example, both the ecological significance of allozymes and the inheritance of a pattern of feeding behaviour may be, and indeed have been, strongly questioned. The particular success of evolutionary studies of pigmentation, including investigations of mimicry, derives largely from the fact that the system studied satisfies both the requirements of known inheritance and ecological importance, the latter usually relating to the activities of visually searching predators.

The existence of so many alternative approaches to the study of evolution has inevitable consequences for the coverage and emphasis of different parts of evolutionary biology in books on the subject. For the most part it has led to a fragmentation along the lines of fairly well established disciplines. A quick perusal of Crow and Kimura's (1970) *An Introduction to Population Genetics Theory*, Pianka's (1978) *Evolutionary Ecology* and Stanley's (1979) *Macroevolution. Pattern and Process* reveals little common ground. The result of this kind of fragmentation of the literature, which is reflected also in evolutionary journals, is often bewilderment and some difficulty in seeing the relevance of parts to the whole in the case of the beginning student of evolution; and enforced over-specialization for all but the most courageous of researchers. Clearly, neither of these consequences is desirable and it would be advantageous if something could be done to remedy the problem. The question is, what?

There appear to be two possible remedies, the first of which would involve the writing of a totally comprehensive, all-embracing book on modern evolutionary biology. Several factors militate against the success of such a venture if undertaken at anything but an introductory level. First, few potential authors are sufficiently well versed in the whole range of subjects, from molecular genetics to palaeontology, that would form the basis of a general evolutionary treatise; and a volume written by several authors often lacks the continuity of a book written by one. Second, the resulting tome would be of such a length that it would probably be most often used for reference, thus defeating the purpose for which it was written. Finally, there is some doubt at present that the same principles underlie all aspects of the evolutionary process. Thus while the role of natural selection in directing morphological evolution is rarely questioned, its importance in molecular evolution is doubted by a large group of population geneticists, the neutralist school, whose views have been summarized in highly readable form by Kimura (1979). Also, even if molecular evolution is driven largely by natural selection, it exhibits marked differences from morphological evolution, such as the greater constancy of its rate (though there are problems in obtaining comparable scales of rate measurement — see Arthur, 1982d). The relatively variable rate of morphological evolution may be attributable to an interaction between natural selection and the development of morphological characters (see chapter 12). Proteins do not 'develop' in the usual sense of the word so their evolution may be explicable, whether by selection or neutrality, without reference to developmental genetics; whereas it seems increasingly unlikely that the same is true of morphological evolution.

The second possible remedy to the problem of fragmentation of the evolutionary literature along conventional interdisciplinary borders is to treat one level of evolution from as many different disciplinary viewpoints as possible. As hinted above, I hold the view that the evolutionary mechanisms operating at molecular and morphological levels will turn out to be different, at least in terms of their relative importance. However, even if this is not the case, some current controversies such as phyletic gradualism versus punctuated equilbrium and selection versus neutrality, are restricted to one or other level (morphological and molecular, respectively, in the case of the named examples). Also, as mentioned earlier, the disciplines with which familiarity is required differ for an understanding of molecular and morphological evolution. For the former, subjects such as molecular genetics, enzyme kinetics and the theoretical population genetics of single and linked loci are important, as is a knowledge of recently discovered molecular entities like pseudogenes, eukaryotic transposable elements and various grades of repetitive DNA. For an understanding of morphological evolution, however, subjects such as quantitative genetics, biometry, developmental biology and evolutionary ecology are more important. Thus, both from the viewpoints of cohesion of the subject matter and of the areas of knowledge required by the author, there are strong arguments for treating molecular and morphological evolution separately. This does not, of course, mean that we should treat the evolution of morphological characters from a purely phenotypic angle; but rather that only those molecules involved in controlling development (see chapter 9), and the genes that produce them, need to be considered. Much of the data now available on molecular evolution relates to enzymes with no known role in development, and these data will not be discussed here.

Molecular evolution more or less exploded, as an area of research activity, from the papers of Hubby and Lewontin (1966), Lewontin and Hubby (1966) and Harris (1966), which described the use of gel electrophoresis in examining enzymic gene products in natural populations and revealed the extent of allozymic variation. There have been several subsequent books largely or wholly devoted to molecular evolution, including the multi-authored collection under that name edited by Ayala (1976) and Lewontin's (1974) *The Genetic Basis of Evolutionary Change*.

The situation as regards morphological evolution is rather different in two respects. Firstly, many of the books devoted to the subject are much older; for example, Goldschmidt (1940), Huxley (1942) and Simpson (1944). Secondly, while none of the texts on molecular evolution can be regarded as totally comprehensive, the division along disciplinary lines of material on or relevant to morphological evolution is even more marked. Particularly notable is the fact that quantitative genetics, which provides us with information about the inheritance of the ubiquitous continuous variation found in all morphological characters, is rarely dealt with in books on evolution. Instead, it is either treated in its own right (Mather and Jinks, 1971, 1977; Falconer, 1981) or as a subject linked to animal or plant breeding (Lerner, 1950). Sometimes (Bulmer, 1980) it is dealt with in a sufficiently mathematical form that it becomes inaccessible to a good many biologists.

The aim of the present book is to provide a multidisciplinary approach to the study of morphological evolution. Even within the morphological area, a totally comphrehensive text would be a mammoth task and would be subject to the problem of a lack of equal familiarity on the author's part with the various aspects of the material. This book is, like many, a compromise between this problem, which argues for a less general book, and the problem of fragmentation of the literature (which provided part of the motivation for writing it) which argues for a more general one. I will now point out which areas are covered and which receive the most emphasis.

First, as indicated by the title, this book is largely about evolutionary mechanisms and not about the reconstruction of particular evolutionary pathways which led to the origin of various taxonomic groups. The emphasis is thus on neontological approaches rather than on palaeontology. However, the book is not devoid of palaeontological material; it includes aspects of this discipline which bear most closely on evolutionary mechanisms. For example, the debate about the variability of palaeontological evolutionary rates at the species level is covered (chapter 8) because some proponents of the punctuated equilibrium view (Williamson, 1981a) argue that this pattern of evolution requires conventional views on speciation mechanisms to be altered.

Second, for no particularly good reason except by own leanings, the book is taxonomically biased in that examples are mostly drawn from animals rather than plants, and from certain groups of animals — notably molluscs, insects, and, to a lesser extent, vertebrates, at the expense of others. I hope, however, that these taxonomic biases will not disguise the generality of the principles.

Third, the book was written in the firm conviction that an adequate understanding of morphological evolution, particularly in the long term, requires a larger and clearer developmental component in evolutionary theory than exists at present, a view championed earlier by Waddington (1957a, 1975). The second 'half' of the book, chapters 9–14, is decidedly flavoured by this conviction. These chapters are thus permeated by a less conventional approach than the earlier ones, which represent an attempt at a synthesis of selected aspects of quantitative genetics (chapters 3, 4), ecological/population genetics (chapters 5, 6), evolutionary ecology (chapters 5, 7), and palaeontology (chapter 8).

Finally, the book is restricted to characters that are morphological in the strict sense of the word (which comes from the Greek μορφή(*morphē*) meaning form or shape) and is not concerned with other externally visible characteristics which are sometimes included in a 'broad' view of morphology. Notably, this means that pigmentation, which is sometimes treated as a 'morphological' character (for example, Gale, 1980; Jones *et al.*, 1980), will not be included. The rationale behind this omission, in addition to literary correctness, is that such characters are not the end-products of a complex sequence of developmental processes in quite the same way that the size, shape and structure of an organism, or part of it, are. (This point is expanded on in section 12.3.) They are thus unlikely to be of central concern in any future synthesis of evolutionary theory resulting from the incorporation of new concepts from developmental genetics.

Having now reached a stage where I hope the reader has a clear idea of why I decided to write this book, and what, at least roughly, is included in it, it is time to turn outwards and look at evolution itself.

1.2 AN OVERVIEW OF EVOLUTION

The earth is approximately 4.5 billion years old — about a third of the age of the universe — and as far as is known was totally unsuitable for, and devoid of, living organisms during its early stages, perhaps about the first 0.5 billion years. There is now widely accepted fossil evidence for the existence of prokaryotic organisms 3.5 billion years before the present (BYBP; Walter *et al.*, 1980) and thus it seems that the origin of cellular life-forms from pre-cellular organic material occurred sometime between about 4.0 and 3.5 BYBP. Whether such an origin occurred once or several times is uncertain and will likely always remain so, but whichever is the case it seems reasonable to assume that the first group of cellular organisms represented at most a handful of species, if indeed the term species as it is applied to living organisms (the biological species concept) is applicable to them at all. In contrast, there are now some two to three million recognized species of animal, plant and micro-organism in the biosphere with, doubtless, many more remaining to be discovered in all but the best-known taxa.

As far as the fossil record allows us to tell, the progression from a few species to a few million species, and the concurrent progress from a few body plans (mostly unicellular and filamentous) to many, has not occurred at a constant rate. If the present view of the origin of the eukaryotes at around 1.4 BYBP (Schopf and Oehler, 1976) is approximately correct, very little gross morphological change occurred for the first 60% of the history of life, whereas since then, and especially during the period 0.8 – 0.5 BYBP (see review by Olson, 1981), there has been considerable evolutionary change and radiation of morphology at the macroscopic level. Figure 1.1 illustrates the compression

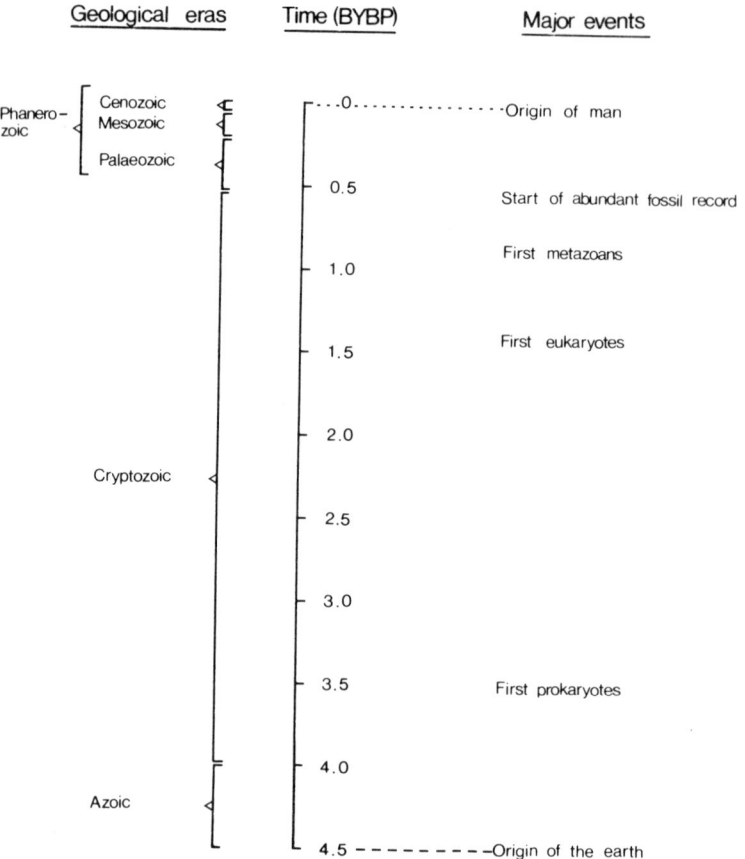

Figure 1.1. Evolutionary time-scale showing approximate times of some major biological and geological events.

of major evolutionary events into the more recent eras. Most of the fossil record relates, as indicated, to the Phanerozoic which began at 0.57 BYBP.

I have taken some licence in naming the time before the Palaeozoic. This has been done inconsistently in the past and there seems to be a need to fit the names to the fossil facts, as presently known. Some or all of the time period between 4.5 and 0.57 BYBP has on previous occasions been referred to as the Proterozoic, Archaeozoic, Cryptozoic, Pre-Palaeozoic and Azoic. I have used the term Cryptozoic, which has sometimes been employed as a label for all of the Pre-Palaeozoic (Gass *et al.*, 1971) to cover the period back to 4.0 BYBP, beyond which it seems unlikely that 'hidden life' (or, more literally, 'hidden animals') will be revealed through subsequent fossil finds. This leaves the period 4.5 – 4.0 BYBP for which the term Azoic seems appropriate. The often-used names for geological periods (Cambrian, Ordovician, etc.) do not appear in Figure 1.1 as it represents too macroscopic a view for them to be useful. These periods only apply, in fact, to the subdivision of the Phanerozoic, everything prior to 0.57 BYBP being often collectively referred to as the Pre-Cambrian.

One of the most long-standing evolutionary problems is the obscurity of the links between major taxonomic groups, within which evolutionary relationships are less difficult to see. This problem has been widely recognized and stated, but even moderate steps towards its solution have not been achieved. Thus, despite the increased knowledge of the fossil record that we now have compared to biologists working a century ago, the problem remains. Olson (1981) states, "At the level of phyla and higher categories, any information on transitions as far as the fossil record is concerned is essentially non-existent." The question of whether this lack of information is due to temporal patchiness of fossils, especially in early time periods, or to some form of saltational evolution, has given rise to heated controversy. For the most part, attempts to explain major evolutionary shifts in terms of macromutations (de Vries, 1905; Goldschmidt, 1940) have met with little acceptance. However, the time may be ripe for a re-examination of some of these views in the light of recent advances in developmental genetics; this will be done in chapters 11 and 12. A more recent but equivalent controversy at a lower taxonomic level — namely the debate over punctuated equilibrium at the level of the species — will be discussed in chapter 8. Unfortunately, much discussion of saltational evolution in the past has tended to be more heated than necessary, largely because proponents of this form of evolutionary change are seen as attacking Darwin. I will attempt not to follow this tradition and, at any rate, saltational evolution is not incompatible with natural selection, Darwin's main contribution to evolutionary theory.

Many biologists have noted that there appears to be something 'progressive' about evolution, and in earlier evolutionary works (for example, Huxley, 1942; Rensch, 1959) considerable space was often devoted to this issue. In more recent texts (for example, Dobzhansky *et al.*, 1977) the matter is sometimes discussed, but to a much lesser extent since evolutionary progress is not a concept which has been particularly fruitful. The idea of progress is linked to attempts to find something that always increases in evolution. There may in fact be no such thing; as will be seen in section 1.4, even fitness may not fall into this category. Many other characteristics which seem to exhibit a general, but by no means universal, increase in evolution can be included under the heading 'increasing phenotypic complexity'. Whether one compares prokaryotes with eukaryotes, unicellular organisms with multicellular ones, or coelenterates and sponges with vertebrates and insects, the later-evolved group is, on average, always more phenotypically complex than the earlier. The increased complexity can occur through one or more of a series of components. Thus comparing insects with sponges, for example, we find a more complex shape, an increased variety of cell types and a greater array of internal organs (sponges in fact have no real organs). However, the differentials in complexity only hold for comparisons involving the 'average' member of each of the groups concerned; it is possible to find certain species within later-evolved groups which are less complex or at least not noticeably more complex in overall shape than earlier groups. This would be the case, for example, if tunicates were compared with coelenterates. Although such examples show that there are instances of species or smallish groups of species evolving, at least in some respects, downwards on the scale of phenotypic complexity, the overall trend is certainly up. The possible reasons for this general feature of evolution will be discussed in chapter 13.

The origins of particular groups, the variability in the rate of evolution, and general trends such as increasing phenotypic complexity are all features of evolution which can to some extent be described or measured, hence revealing more detail about certain aspects of the evolutionary process. However, our primary concern here is with the *mechanisms* by which evolutionary change is brought about. Mechanisms are widely recognized to occupy a central position in evolutionary theory and indeed, to be acceptable, any theory of evolution must be based on a demonstrable mechanism. As we will see later, certain mechanisms do in fact lead to particular expectations, e.g. for evolutionary rates, and so measurement of rates can be used to provide a basis for inferring the action of certain mechanisms. This form of reasoning has played a major role in two recent attacks on established evolutionary theory — from the neutralists (who are largely concerned with molecular evolution) and from supporters of the theory of punctuated equilibrium (chapter 8). However, before examining this kind of reasoning it is necessary to examine the mechanisms themselves; and before we do this it is necessary to formulate a definition of morphological evolution since the form this takes influences the list of factors that can be considered as potential evolutionary mechanisms.

1.3 DEFINING MORPHOLOGICAL EVOLUTION

Evolution in general is often defined as a change in the gene frequency of a population (see, for example, Wilson and Bossert, 1971). This definition has the advantage of being short and clear; and it usefully excludes changes of genotype frequency where these are unaccompanied by changes in the frequencies of the alleles themselves. Thus, for example, the approach to Hardy-Weinberg equilibrium in a newly founded population is not taken to constitute evolutionary change. As well as possessing conciseness and clarity, a definition must also be workable. In the case of characters whose genetic basis is a single locus and whose variants can be easily identified, the above definition clearly is workable because the gene frequency can be calculated on a series of occasions in a single population or in several populations at a particular time. Thus it is possible to tell quickly whether evolutionary change has occurred and to move on to the more interesting business of attempting to elucidate the mechanism causing the change.

Turning to morphological evolution, little of the variation met with has such a simple genetic basis. The loci affecting a particular character cannot often be identified and thus changes in gene frequency cannot be calculated, though the fact that such changes have occurred can sometimes be inferred from particular kinds of phenotypic change. This sort of inference is most easily made in artificial selection experiments which are described in chapter 4. A second problem arises because the definition of evolution given above relates most easily to short-term changes; but in morphological evolution the existence of a fossil record — despite its deficiencies — enables and even urges us to consider long-term evolution as well. In the long term, instances, of gene duplication, chromosomal inversion and other genetic events not describable in terms of gene frequency occur. These two problems are in a sense complementary and any study is likely to encounter at least one of them. For example, if a study of phenotypic change in a series of fossils is based on a short enough time period that large-scale genetic rearrangements are unlikely, then the phenotypic changes involved will probably be small and hence within the range where it is difficult to distinguish between inherited and non-inherited variation. Inference of gene frequency changes from small phenotypic alterations is more dangerous in studies of fossil material than in studies of living organisms which can be experimentally manipulated. It is sometimes asserted that morphological changes occurring in geologically short time periods are genetic in origin because they are outside the range within which ecophenotypic variation is to be expected (Williamson, 1981a); however, this range is not always easy to delineate.

No formulation of a definition for morphological evolution can remove the difficulties outlined above, but it is clearly desirable to have a definition which at least includes some reference to the phenotype which is, after all, the usual source of quantitative information in morphological evolutionary studies. This need can be satisfied by adding to the more general definition, so that morphological evolution is said to occur when 'the mean value or variance of a morphological character in a population changes and this change is at least partially caused by changes in gene frequency at one or more of the loci contributing to that character or by a reorganization of the genetic material.' This will be adopted, here, as a working definition and consequently a few points of clarification are in order. First, the form given excludes any reference to skewness or kurtosis (see chapter 2). This is a minor problem since most studies of morphological evolutionary change are based on means and variances; and the definition could easily be expanded to take account of these other parameters if necessary. Second, morphological changes caused solely by changes in genotype frequency do not constitute evolution. Thus the increased variability in quantitative morphological characters seen in the F_2 generation of a cross between two inbred lines is not considered to be an evolutionary change; while increased morphological variability caused by changes in gene frequency in a disruptive selection experiment would be describable as evolution. Finally, the qualification that the morphological change need only be partially caused by genetic change is necessary since variation of quantitative characters both within and between populations usually has both genetic and environmental components. These will be described in detail in chapter 3.

1.4 EVOLUTIONARY MECHANISMS

Five mechanisms are recognized as being capable of causing evolutionary change. These are mutation, migration, meiotic drive, genetic drift and natural selection. In addition, there is the founder effect, which is a compound of migration and drift. Mechanisms outside the above five have been proposed from time to time, notably the Lamarckian mechanism of inheritance of acquired characters. There is no clear evidence for this proposed mechanism and recent attempts to revive it in the field of immunology (Gorczynski and Steele, 1980) have been strongly questioned on the basis of negative results obtained on repeating the experiments involved (Brent et al., 1982). Whether other evolutionary mechanisms outside the above-mentioned five will need to be acknowledged in the light of new findings in molecular genetics remains to be seen. A considerable amount of DNA in the Drosophila genome, and perhaps in most eukaryotes, occurs in the form of transposable elements whose evolution is not yet understood and may involve novel processes. However, a wide variety of molecular and chromosomal events can be included in the general category of mutation, as opposed to the more specific category of point mutation involving changes of individual nitrogenous bases, and it may thus be unnecessary to invoke additional categories of evolutionary mechanisms.

As stressed earlier, the definition of evolution that is adopted affects the list of agents which cause evolutionary change. For example, if changes in genotype frequency alone were accepted as evolutionary events, then assortative mating would constitute an evolutionary mechanism. Thus it is necessary to consider whether the definition of morphological evolution formulated in the previous section restricts the number of mechanisms that are effective in this component of evolution, because of the requirement that morphological change must occur in addition to genetic change.

In fact, the list of possible evolutionary mechanisms is unaltered for the following reasons. (1) Many mutations of large morphological effect are known, including the homoeotic mutations of *Drosophila* and other insects (Ouweneel, 1976) which will be discussed in later chapters. A single mutation of this kind will alter both the mean value and variance of several morphological characters in a population. Mutations of smaller effect will no doubt also affect mean values and variances but often by undetectable amounts. (2) Meiotic drive has been extensively studied at the t locus in the mouse *Mus musculus* by Dunn and colleagues (Dunn and Glueckson-Waelsch, 1953; Dunn and Suckling, 1956). Certain genotypes at this locus, including heterozygotes between alternative t alleles, are tailless, and others have shortened tails. Thus a genetic system involving meiotic drive is also capable of altering morphological characters. (3) Given that genetically and morphologically differentiated populations exist within many species, evidence for which is discussed in chapter 5, it is clear that migration between populations can cause the distribution of a morphological character to alter in a population which accepts immigrants and, if the migrants are a non-random sample, then in the donor population also. (4) As regards drift, this must be capable of causing morphological change in a population because, if a single individual dies due to some stochastic event, then, providing that individual's character value was not equal to the mean value in the population, both the mean and variance of the character concerned will alter as a result of the death. Finally, (5) the ability of selection to cause morphological changes in addition to changes of gene frequency is hardly in doubt; evidence for this is presented in chapters 4 and 5.

Given, then, that all five generally recognized evolutionary mechanisms are *capable* of altering the distributions of morphological characters as well as the gene frequencies in a population, what is their *relative importance* in morphological evolution? This is a more difficult question and one on which views are likely to differ, though perhaps not so widely as in the field of protein evolution where the neutralist school holds that drift is the major mechanism of evolutionary change (Kimura, 1968).

Despite the title of this book, I do not intend to spend the rest of it discussing all aspects of this question. Rather, I will dismiss in a fairly cursory way drift, drive and migration as primary agents of morphological evolutionary change, and will concentrate on mutation and selection. *Mechanisms* appears in plural form in the title because I believe that to obtain an adequate understanding of evolution at the morphological level it is necessary to distinguish between different forms of selectively driven evolutionary change, and in particular to contrast evolution caused by selection on mutations of individually large, and individually undetectable, effects on the phenotype.

There is in fact some basis, albeit limited, for excluding migration, drive and drift from further consideration. Migration may be important in introducing genetic variability into a population, but as regards the direction of evolutionary change of population parameters subsequent to such an introduction, it will have no effect. The extent of migration may be important in determining the spatial scale of evolutionary events. For example, restricted migration on island archipelagos may result in the production of subspecies and eventually new species in areas much smaller than required for the same events to occur in a continental land-mass (see section 5.4). However, there is nothing to suggest that the processes involved are fundamentally different. The case for excluding meiotic drive from consideration as a major agent of morphological evolution is simply its rarity. Loci exhibiting distorted segregation — which include the 'driving' locus and others on the 'driven' chromosome — appear to constitute a small fraction of all loci. Also, other examples of meiotic drive (e.g. *segregation distorter* in *Drosophila melanogaster*; see Sandler *et al.*, 1959) seem to be without morphological effect. As regards genetic drift, this is likely to be much less effective in morphological than in molecular evolution because sustained alteration of the mean value of a character with a polygenic basis will require parallel drifting of alleles at many different loci. While this may occasionally happen, it is likely to be a very rare event. Also, it seems unlikely that alterations manifested at the morphological level are often truly neutral, though this has been suggested for some characters (for example, bristle number in *Drosophila*; see Latter and Robertson, 1962).

We now return to mutation and selection, to which evolutionary mechanisms the rest of the book is devoted. Their treatment in later chapters varies from a consideration of evolution of phenotypic variation produced by mutations of individually undetectable effect, where selection is the prime focus (chapters, 4, 5), to a discussion of selection on mutations of large phenotypic effect, where the nature of the mutations themselves is of considerable interest (chapter 11). In chapter 10, the developmental nature of some large-effect mutations is considered. In the remainder of this section I will make a few introductory remarks about mutation and selection, taken separately.

Mutation

Any alteration in the sequence of DNA bases in the genome constitutes a mutation. This may occur either through loss, addition, substitution or rearrangement, and some or all of these types of change can occur at anything from the level of a single base to the level of whole chromosomes. If a mutation occurs in a somatic cell it will not be inherited and will consequently have no evolutionary potential; thus what follows relates to mutations in germ cells. (Mutations in germ-cell precursors are also evolutionarily important, but discussion of these will be postponed until chapter 11.)

Two comments about mutations are commonly made in introductory evolutionary texts: that they are important as a source of genetic variation (they are in fact the sole source) and that they are unimportant as agents of evolutionary change. The first of these statements can be accepted without reservation but the second requires some scrutiny. The main problem here is that the conclusion that mutation is unimportant as an evolutionary agent derives from the small effect of mutations on gene frequency, which is expanded on below in the context of a point mutation.

In diploid organisms the number of copies of a particular genetic locus is $2N$ except in the case of sex-linked loci which, in XX/XY systems of sex determination, will be represented by $(1 + f)N$ copies, where f is the proportion of the homogametic sex. Taking the autosomal case, the proportion of a mutant form of a gene is zero prior to its occurrence and $1/2N$ immediately afterwards. The frequency of the allele which mutates declines by $1/2N$. The value of Δq, the change in gene frequency caused by the mutation, is thus very small in all but the smallest populations. For example, with a fairly limited population size of $N = 1000$, $\Delta q = 0.0005$. Mutation rates are now known to be sufficiently low that it is unlikely for there to be many mutations at the same locus in a single generation. Thus mutation *per se* has a negligible influence on the gene frequency.

The fact that mutations have consistently small effects on gene frequency does not mean that they have consistently small effects on individuals; nor that their effect on the mean value of a morphological character in a population is always small. In fact, at the morphological level, the effects of mutations are extremely varied. For example, quantitative variation in morphological characters (see chapter 3) occurs — at least in part — because individuals differ in the alleles present at many loci; but the phenotypes of two individuals differing at only one of these loci may be identical, at least within the limits of our powers of observation. In contrast, individuals of *D. melanogaster* with a mutant form of a single locus in the bithorax series have a radically altered gross structure (Lewis, 1963). If body length is monitored as a quantitative character in a population of *D. melanogaster*, the occurrence of a mutation in the bithorax complex in one individual fly can, providing the mutation is dominant, result in a fly that is much longer than usual. The mean and variance of length in the population will be correspondingly affected, though the magnitude of these effects will of course depend on the population size.

In addition to the possibility of mutations of individual genes causing large morphological changes, the reverse situation of mutations that are large in genetic terms — for example chromosomal inversions — causing little or no morphological change is also found. For example, the seaweed fly *Coelopa frigida* has a naturally occurring multiple chromosomal inversion (see Day *et al.*, 1982) yet the different variants are morphologically indistinguishable except for a very slight difference in size (Butlin *et al.*, 1982). The many chromosomal inversions known in various species of *Drosophila* seem to have no morphological effects at all (Dobzhansky, 1970).

In conclusion, then, the effects of a mutation may be examined at genic, chromosomal, morphological and population levels and the magnitude of effect can differ enormously from one level to another. The supposedly minor role of mutation as an evolutionary mechanism must therefore be questioned. Even in cases where a large individual morphological effect is accompanied by a very small change at the population level because of large N, the new mutation may eventually be fixed in the population by selection and may then act as a basis for a whole range of polygenic modifications which were previously unavailable. The mutations which are important in this respect are those which produce distinct alterations to the pattern of development. Some of these will be discussed in detail in chapter 10.

Selection

While the origin of novel forms is of considerable interest, ultimately one morphological phenotype (or morphotype) completely replaces another in evolution, and mutation itself is clearly insufficient to explain this. As mentioned earlier, the primary mechanism of morphological evolution, working on material provided by mutation, is considered here to be natural selection; though drift, drive and migration may play minor roles. The basic idea of natural selection of course originated with Darwin and Wallace, and is lucidly explained by Darwin (1859) as well as in many current population genetics texts. There is thus no need to introduce the concept here except to state that it refers to the differential survival and/or reproduction of two or more variants within a species. However, I will discuss briefly three aspects of natural selection that are relevant to its role as an evolutionary agent.

1. The timing of reproduction and the measurement of 'absolute' fitness

Two genetic variants with identical life expectancies and leaving identical numbers of offspring may differ enormously in fitness in the sense that one of them may increase substantially in frequency relative to the other in a few generations. This can occur simply through differing mean ages of leaving progeny. Thus if two hypothetical variants both live for precisely two years and each gives rise to three progeny, but in one case the pattern of reproduction is one and two offspring at ages one and two, respectively, while in the other variant the pattern is reversed (2, 1 offspring at ages 1, 2), then five years after the start of a newborn cohort with a frequency of the latter variant of 0.5, its frequency will have risen to 0.75 (see Table 1.1).

Table 1.1. Increasing frequency of a variant (number 2) with a 2,1 pattern of leaving progeny, at the expense of another variant with a 1,2 pattern

	Number of organisms									
	Variant 1				Variant 2				Grand total	Frequency of variant 2
	Age group				Age group					
Year	0	1	2	Total	0	1	2	Total		
0	1	0	0	1	1	0	0	1	2	0.50
1	1	1	0	2	2	1	0	3	5	0.60
2	3	1	1	5	5	2	1	8	13	0.62
3	5	3	1	9	12	5	2	19	28	0.68
4	11	5	3	19	29	12	5	46	65	0.71
5	21	11	5	37	70	29	12	111	148	0.75

This effect of the timing of reproduction, together with the probability of survival and the overall number of offspring, can be conveniently combined into a single measure of fitness using the approach of life-and-fertility tables pioneered, in this field, by Fisher (1930). An example of a life-and-fertility table for the US human population is given in Table 1.2. In this instance, the net reproductive rate R_0, which can be considered as a measure of fitness, takes the value 1.212. In general, values above 1.0 mean a population is expanding; in a static population, $R_0 = 1$. The example in Table 1.2 refers to a case where R_0 is calculated for an entire, and genetically heterogeneous, population. However, in evolutionary studies,

Table 1.2. Life-and-fertility table for US women. Note that in dioecious organisms it is conventional to express these tables in terms of number of female offspring per female. In hermaphrodite or asexual organisms, number of progeny per individual is given. From Krebs (1972)

Age group (x)	Pivotal age	Probability of surviving to pivotal age (l_x)	No. female offspring per female aged x per time unit (5 years), (m_x)	Product of $l_x m_x$
0–9	5.0	0.9775	0.0	0.0
10–14	12.5	0.9752	0.0022	0.0021
15–19	17.5	0.9730	0.1656	0.1611
20–24	22.5	0.9698	0.4244	0.4116
25–29	27.5	0.9661	0.3478	0.3360
30–34	32.5	0.9613	0.1934	0.1859
35–39	37.5	0.9541	0.0939	0.0896
40–44	42.5	0.9434	0.0259	0.0244
45–49	47.5	0.9275	0.0017	0.0016
50–over	—	—	0.0	0.0

$R_0 = \Sigma_0^\infty \, l_x m_x = 1.212$

Source: National Center for Health Statistics, 1969.

the approach can be altered so that R_0 is calculated separately for different genotypes. Of course, R_0 for any variant or for a whole population must change over time, eventually becoming 1 if the population stabilizes.

2. The measurement of relative fitness and the possibility of phenotypic cycles

In practice the fitness of any genetic variant is usually measured relative to other variants with which it, at least temporarily, coexists. If we consider two variants X and Y, the fitness of Y relative to X can be measured by the cross-product ratio (CPR) for Y as follows:

$$\mathrm{CPR}_Y = \frac{y' \cdot x}{x' \cdot y}$$

where x and y represent the initial numbers of X and Y; and x' and y' their numbers after a specified period of time, usually a single generation. It is conventional to use a fitness value of 1 for the fitter type. (This can be thought of in terms of two groups of this same, 'fit' variant competing with each other, in which case the CPR value for either group is 1.0.) Any less fit variant then has a fitness value of between 0 and 1. Fitness, as measured by the CPR, is usually referred to as w, and the selective coefficient — a measure of the intensity of selection against a particular variant — as s. These are complementary, so that $w = 1 - s$. With this sort of notation, a lethal mutation has $w = 0$, $s = 1$; whereas the fittest form in a population has $w = 1$, $s = 0$. A new, fitter mutant arising later in such a population will have $w > 1$ and a negative value of s unless the system of symbols is altered so that the new variant is designated $w = 1$.

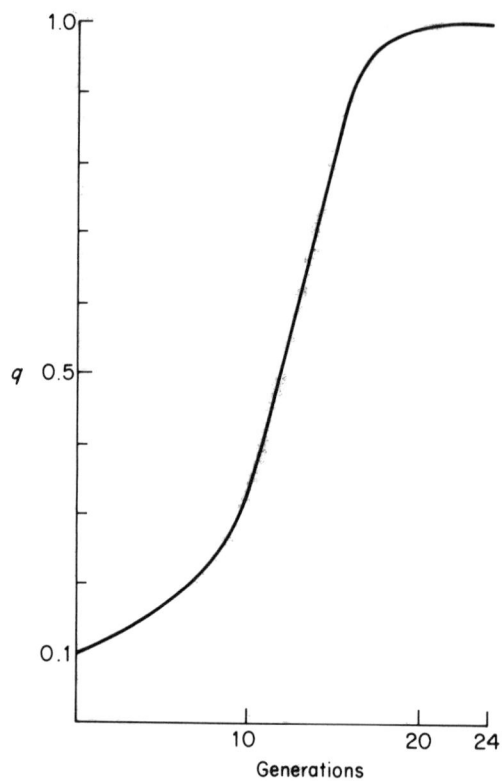

Figure 1.2. Increase in frequency (q) and eventual fixation of a selectively favoured recessive mutation. The curve requires nearly a hundred generations to move from 0.01 to 0.1. Modified from Cook (1971).

If a mutation occurs in a population of haploids and produces a variant of increased fitness, then unless its fitness is negatively related to its frequency (frequency-dependent selection: see review by Ayala and Campbell, 1974), and disregarding stochastic events, the mutant will eventually be fixed in the population. In diploids this will also occur, provided that the heterozygote is intermediate in fitness between the two homozygotes or equal in fitness to one of them (i.e. situations of complete, partial or no dominance in relation to fitness). Figure 1.2 illustrates a fixation occurring in the case of selection for a recessive mutant. The pattern shown in this figure, and its near equivalents for other dominance relationships, can be regarded as, in simplified form, an illustration of the fundamental process of evolutionary change under natural selection involving gradual change from one genetic state to another. However, in addition to fully recognized complexities to this picture, such as the determination of many characters by more than one locus, which means that Figure 1.2 shows only one component of the evolution of such a character, there is a less commonly cited complication to the seemingly simple picture of increase in frequency of the fitter type. This becomes apparent when a population is considered over a long time period.

Over many generations, it is possible that natural selection will be accompanied by, and indeed *cause*, a decline in fitness. This phenomenon could arise out of non-transitive fitness relationships among a series of variants; and may apply in some cases of fitness relationships between species (Gilpin, 1975; Jackson and Buss, 1975) as well as between genetic variants within species (Lewontin, 1968). Non-transitivity of fitness relationships among three variants A, B, C occurs when, for example, A > B and B > C but A < C. In a situation where this series of relationships applied, a mutant B could spread through a population of C and this could then give rise to subsequent invasions of A, C and B', where B' is a new mutation similar but probably not identical to B. Whether or not natural selection has led to a net increase or decrease in fitness depends on the fitness relationship between B and B' which is unspecified unless of course B and B' are identical, in which case the net change is zero. The series of evolutionary events from the original homogeneous population of B to the eventual fixation of B' can be referred to as a phenotypic cycle. (Corresponding genetic cycles are unlikely; it is more probable that the genetic basis of B' would be different from that of B regardless of how similar the two are phenotypically.)

How often non-transitive fitness relationships occur and lead to phenotypic cycles in practice will depend on: (1) how many variants are involved; (2) the order in which mutations occur; and (3) whether one fixation is complete before another mutation occurs. At present, it is difficult even to delimit a system within which to attempt to answer these questions, so the importance of phenotypic cycles must remain uncertain. As will be seen later (chapter 13) it is my view that the way in which natural selection acts on developmental systems will usually prohibit this sort of pattern in morphological evolution; and the fossil record seems to bear this out. However, the lack of developmental constraints in protein evolution may make phenotypic cycles more likely there — though so far this possibility has received little attention. Finally, it should be noted that each 'strategy' in a phenotypic cycle is unstable in an evolutionary sense, which makes such a system the opposite of the evolutionarily stable strategy (ESS) discussed by Maynard Smith (1972).

Before leaving this section, I would like to re-emphasize that we have two alternative measures of fitness: R_0 which measures absolute fitness; and w which measures relative fitness. Although w is predominant in population genetics theory, it may be that, in some unusual but important situations, fitness as measured by R_0 is more crucial. This possibility will be considered in chapter 11.

3. Levels of selection

The form of selection on which neo-Darwinian evolutionary theory is largely based is sometimes referred to as individual selection, since it favours some individuals at the expense of others. However, group selection, in which certain groups of individuals are favoured at the expense of other groups, has also been proposed, often in connection with patterns of behaviour which appear to be neutral or even detrimental to the individual but beneficial to the population. Behavioural mechanisms of population regulation have provided a focal point for the debate since the stabilizing of population numbers at a low enough level to avoid local extinction through over-exploitation of resources is clearly of benefit to the population itself; but if individuals contribute to this regulation by leaving fewer progeny when density is high, then this is clearly not beneficial to them in evolutionary terms, since individual fitness is defined largely in terms of number of progeny.

In fact, the two best-known behavioural mechanisms of population regulation are both explicable in ways which do not involve group selection. Territorial behaviour, which apparently helps to regulate many populations, particularly of birds (see Wynne-Edwards, 1962), may have evolved by individual selection since, although mean number of offspring per adult declines if the number of adults exceeds the necessary combined territory space for all of the population, this comes about by non-breeding of birds without territories. Individual selection may produce a characteristic, namely ability to take and defend a territory, which happens to benefit the population in the long term, but which also benefits the individuals exhibiting it in the shorter term.

The other example of population regulation occurring through decreased numbers of progeny at high density involves *all* members of a population leaving fewer offspring. This mechanism involves high density giving rise to smaller adults which in turn leave fewer offspring, and examples are known from many groups, including *Drosophila* (Bakker, 1961), *Tyria* (Dempster, 1982) and *Cepaea* (see chapter 5). There are two possible interpretations of this phenomenon which do not invoke group selection. First, it may be regarded as an inevitable effect in which semistarvation necessarily results in smaller, less healthy, individuals which are less capable of producing offspring. Alternatively, it may be thought of as having evolved through individual selection because individuals capable of attaining adulthood and reproductive maturity at smaller size, when resources are limiting, have a higher probability of surviving to leave offspring than those with a less flexible development, and this outweighs the fact that those small individuals that do survive are less prolific. Even if the latter (evolutionary) explanation is true, what has evolved is the ability to respond to high density with decreased adult size; the reduced size itself is not heritable.

Although many cases of apparently altruistic behaviour including those described above may be explicable in terms of individual selection, there is at least one example of an evolutionary situation which clearly could not have arisen through this kind of selection; namely, the existence of non-breeding 'worker' phenotypes amongst many social insects (largely in the order Hymenoptera). This phenomenon would appear to be explicable in terms of kin selection whereby genes which cause the individual possessing them to decline in fitness but to behave in such a way as to improve the fitness of related individuals increase in frequency when the genetic relationship between the donor and receptor of the altruistic behaviour is close enough (Maynard Smith, 1964). Since workers are morphologically distinct from other social castes, kin selection is clearly capable of causing morphological evolution. However, its role appears to be fairly restricted relative to selection acting at the level of the individual.

Finally, in addition to individual, group and kin selection, 'species selection' has been proposed by Stanley (1975). This phenomenon is only relevant to very long-term evolution, and so I will postpone discussion of it until chapter 12.

1.5 NEO-DARWINISM

Darwin clearly thought that selection was the primary agent of evolution. He says (1859, p. 6) "I am convinced that Natural Selection has been the main but not exclusive means of modification." What he had in mind as minor evolutionary mechanisms were Lamarckian processes and not meiotic drive or genetic drift since these phenomena, and even genetics itself, were not established when *On the Origin of Species* first appeared. However, many present-day evolutionary biologists still go along with Darwin's view, substituting the later-discovered mechanisms such as meiotic drive for the now discredited Lamarckian ones in the category of minor evolutionary mechanisms. This current view, together with the mathematical treatment of selection made possible by Mendelian genetics, is sometimes referred to as neo-Darwinism. However, a clear statement of *exactly* what is meant by the phrase 'neo-Darwinism' is rarely given, and this is problematical since much current controversy is centred on whether neo-Darwinism is correct and adequate as a theory of the mechanisms underlying evolution (see Gould, 1982).

That there are at least two senses in which neo-Darwinism is used is stressed by Waddington (1975, p. 201): "The term 'Neo-Darwinism' is often applied very loosely, to include almost any recent theorizing which has been influenced by Mendelian genetics ...... In its strict sense, it refers to the mathematical models of Sewall Wright, Haldane, and Fisher, and elaborated since then by many authors. This strict Neo-Darwinism does not involve any necessity to refer to the phenotype." Other authors discussing neo-Darwinism are rarely explicit about whether they are using the term in its broad or strict sence, but the answer is often apparent from the particular wording. Thus Maynard Smith (1972, p. 82) is clearly using the term in its broad sense when he says, "By Darwinism is meant the idea that evolution is the result of natural selection. Neo-Darwinism adds to this a theory of heredity."

Because of this dichotomy of usage (or it may in fact be a continuum if sufficient intermediate versions were collected from the literature) we cannot talk about the adequacy or otherwise of neo-Darwinism without making it very clear exactly which statement of neo-Darwinism is being referred to. I therefore intend to formulate a particular statement which can be used either in a general sense — 1 and 2 below — or in a more restricted sense which is applicable specifically to morphological evolution — all of 1, 2 and 3 below.

1. Natural selection is the major but not the sole means of evolutionary change.
2. The selection that causes evolutionary change acts on phenotypes which are wholly or partially based on allelic differences at one or more loci.
3. Morphological evolution usually occurs through selection acting on minor quantitative phenotypic variations of the type ubiquitously found in outbreeding natural populations.

I will return to this statement of neo-Darwinism towards the end of the book, after various case studies have been discussed, to examine the extent to which its claims are borne out. As will, I hope, become gradually apparent, the main problem lies in the adequacy of neo-Darwinism, not in its correctness. In particular, the tendency to concentrate on what 'usually' happens in evolution — that is, how most species evolve most of the time — may be misleading in that important events in evolution such as the origins of major taxa are by their nature rather unusual. Thus there is a danger that neo-Darwinism provides us with an adequate explanation of the bulk of evolution but not of those few evolutionary steps that have led to the main themes within the fascinating morphological diversity of the living world with which we are confronted.

1.6 SUMMARY

Having discussed the reasons for concentrating on morphology, given the briefest outline of the evolutionary process, and formulated a definition of morphological evolution, attention is focussed on evolutionary mechanisms. It is suggested that morphological evolution can be largely understood in terms of the action of mutation and selection, and that meiotic drive, genetic drift and migration play relatively minor roles. Some aspects of mutation are discussed and the lack of correspondence between the magnitudes of effect of a mutation at different levels of biological organization is stressed, together with the need to re-assess the importance of mutation as an evolutionary mechanism. Selection is also discussed and it is noted that this can be measured both in relative and in absolute terms. The different levels of selection are briefly considered and it is argued that the role of kin selection in morphological evolution is limited and that selection acting at the level of the individual is of paramount importance. A version of neo-Darwinism is formulated which serves to distil the central tenets of current evolutionary theory, especially as applied to morphological evolution. The adequacy of neo-Darwinism is questioned but its correctness is not.

Chapter 2

The Measurement of Morphological Characters

> "The mathematical definition of a 'form' has a quality of precision which was quite lacking in our earlier stage of mere description."
>
> D'Arcy Thompson, 1942.

Evolution involves the alteration, over time, of various characteristics of organisms; evolutionary change, therefore, is measured in terms of character units per time unit. The choice of time unit presents few problems, but unfortunately the same cannot be said in relation to the choice of units for the measurement of the evolving character. The problems are especially great in the case of organismic shape and structure, which are the prime concern here. Some molecular characters may be relatively easily quantified; for example, a difference in the primary structure of a protein which is represented by an identical number of amino acids in two closely related species is clearly measurable by the number of positions showing an interspecific difference in amino acid as a fraction of the total number of residues. However, when it comes to quantifying, say, the difference in morphology between two closely related species of insect, there is no single method that is clearly superior to all others.

In addition to solving the problem of how to measure a morphological character in a single individual, there is also a need to be able to summarize the distribution of the character in a population, since evolutionary changes are normally thought of (see section 1.3) as changes in the genetic and phenotypic structure of populations, not of individuals. The description of a population, with respect to some specified character, depends to a large extent on how that character is measured in the individuals within the population. This fact is reflected in the sequence followed in the current chapter: section 2.1 deals with measurements made on individuals, section 2.2 with the description of phenotypic distributions in populations.

2.1 INDIVIDUALS

The overall shape and size of a single organism of any species may be thought of as a scatter of points forming a surface in ordinary three-dimensional space. It would, indeed, be possible to build an exact replica of an unknown organism, given sufficient numerical co-ordinates relating to a three-dimensional grid. Morphological evolution represents genetically based change in this system of co-ordinates. Although this method of quantifying an organism's external structure is the most complete quantified description that can be achieved, it is impractical because (a) it is difficult to make sufficient measurements, and (b) the resulting overall description does not lend itself easily to any means of summarizing the information obtained for a whole population. These two difficulties both arise from the fact that the form of measurement is too complex; advocating the use of such systems would be rather like advocating the use of the landscape as the best possible map. What is needed is an abstraction which provides a simplified measurement yet retains the information necessary for the task at hand — in our case, quantification of evolutionary changes.

The most obvious simplification of our three-dimensional system of co-ordinates is to use the same approach but to consider only two dimensions. This of course loses considerable information, though the amount lost can at least be minimized by discarding the dimension into which the organism or structure under study has the smallest projection. Thus in organisms with dorsoventral flattening, such as isopods, a view from above or below is taken; whereas in cases of side-to-side flattening, such as amphipods, a side view retains more information. In the case of some structures, flattening is so extreme in one dimension as to make it particularly obvious which dimension to 'lose', one of the best examples of this being the leaves of deciduous trees. However, some organisms present considerable difficulty in the choice of appropriate dimensions, and indeed for any species the choice needs to be made on the basis of what is required of the description as well as the general shape of the organisms concerned.

It was of course D'Arcy Thompson (1942) who pioneered the use of two-dimensional Cartesian grid systems in examining morphological evolution. Particularly well known is his demonstration (1942, chapter 17) that the numerous small morphological differences separating two species can often be viewed as all resulting from a single transformation of the grid upon which the outline of an average member of one of the two species is plotted (see Figure 2.1 for an example). This approach, as Thompson points out, gives a clearer view of what Darwin (1859) referred to as 'correlation of characters'. Even in complex cases where a series of morphological differences cannot be accounted for by a single complete transformation, two or more partial transformations can often reduce the number of apparently separate characters that underlie the overall morphological difference.

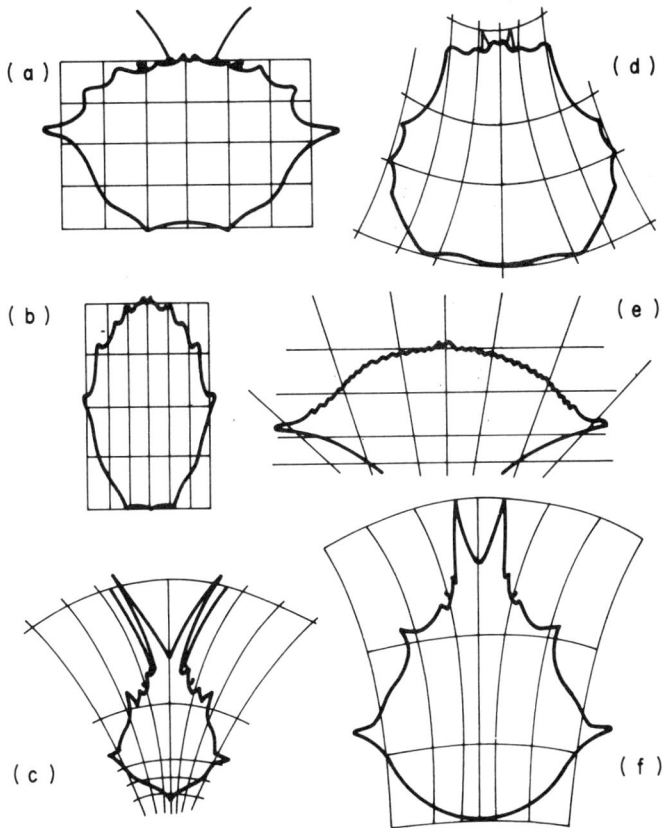

Figure 2.1. Two-dimensional outlines of the carapaces of several crabs (a, *Geryon*; b, *Corystes*; c, *Scyramathia*; d, *Paralomis*; e, *Lupa*; f, *Chorinus*) showing how appropriate transformations of the grid on which the outline of one genus is plotted give good approximations to the carapaces of other genera. Numerous, apparently unrelated, minor differences in shape are thus seen as part of the same overall transformation. From Thompson (1942).

Although many present-day evolutionary biologists regard Thompson's (1942) work as fascinating and inspired, his methods have not been pursued to any great extent. This is undoubtedly due, at least in part, to the difficulties of extending these methods to deal with populations rather than individuals. Disciplines involving the study of microevolution in populations have moved towards a more simplistic form of measurement of individual morphological characteristics which is more easily transposed from individual to population. This simpler approach has enabled, in quantitative genetics, the construction of a body of theory wherein the phenotypic value of a character in an individual organism and its distribution in the population are seen to be caused partly by the alleles present at a series of loci, partly by interactions between alleles and between loci, and partly to non-genetic factors and their interactions with genetic ones. Thus, on the basis of a simpler kind of individual measurement, an approach that is more orientated towards the study of cause and effect has been made possible, whereas Thompson (1942) was more preoccupied with mathematical elegance than with causation. However, although morphological microevolutionary processes have now been analysed in considerable detail (albeit largely in the laboratory) by quantitative geneticists, this discipline has made almost no impact on macroevolutionary problems. Also, Thompson's work may not be so far removed from cause and effect analyses as is sometimes thought, because both the transformations themselves, and the larger evolutionary differences to which Thompson realized his procedure of transformation could not be applied (see chapter 9), may ultimately be explicable in developmental terms.

The simplest of all morphological measurements, and a kind which has been frequently employed in quantitative genetics, is achieved by a further dimensional reduction, from two to one, coupled with the choice of a particular structure or substructure to be measured. In *Drosophila*, the commonly monitored characters of wing length, thorax length and total body length fall into this category, as do leg length and head width in mammals, the length and breadth of a blade of grass, and conchological measurements such as aperture height. The distributions of these characters in populations are readily described and hence they have been frequently used in microevolutionary studies. Other single measurements not expressible in units of length (for example, weight, number of bristles) have this same property. Little needs to be said about the actual measurement of such characters in individuals; their distribution in populations will be discussed in the next section.

The advantages of these simple measurements should not blind us to their several disadvantages, all of which arise from the enormous loss of information which accompanies the choice of a single measurement, linear or otherwise. Some of these problems have been discussed by Waddington (1957a). The discussion below relates largely but not exclusively to linear measurements rather than to weight or to the number of some substructure.

The first problem requiring consideration is that single measurements are often rather divorced from the functioning of the organism, and lack of awareness of this can result in somewhat naive speculation about the evolution of the characters concerned. For example, it is common in studies of character displacement to obtain linear measurements of structures involved in feeding, such as beak length or depth in birds, and to consider selection as acting, in areas of sympatry, to increase beak size in the species whose beak is already slightly larger and to decrease beak size in the other. The reasoning behind this proposed evolutionary process (which is discussed in much more detail in chapter 7) is that individuals least like the alternative species suffer least in interspecific competition. While this suggestion is not in itself unreasonable, the question arises of whether a larger (or smaller) beak is compatible with a given body size or, if the two characters are to be altered together, then

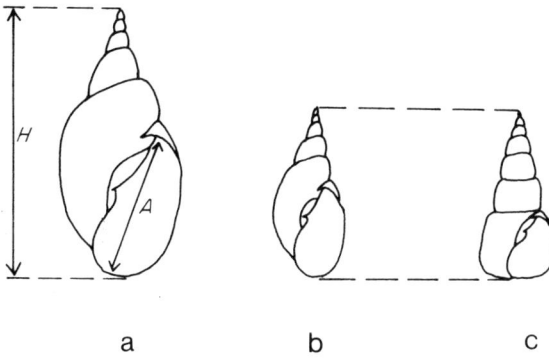

Figure 2.2. Illustration of gastropod shells differing in aperture height (A) for different reasons. Shells a and b are the same shape but differ in size, as measured by overall height (H). Shells b and c are identical for H but differ in shape. Shells a and c differ in size and shape.

whether there might not be some other form of selection tending to adjust body size in the opposing direction. The whole problem of compatibility of various characters with each other, and considerations about the operation of selection on character complexes, are automatically neglected if measurements on a single character only are made.

The second problem is that individual measurements are particularly uninformative on the developmental and biological basis of variation between individuals. As regards small variations, it is well known that the polygenes of quantitative genetics (see next chapter) are, developmentally, 'black boxes'. In the case of major variations, it is even more important to see *why* a character takes on an aberrant value, and this information is totally absent in single linear measurements. For example, to describe a *D.melanogaster* fly exhibiting bithorax and postbithorax phenotypes (in which case it has two mesothoraxes and four wings: Lewis, 1963) as having a body length of, say, 1.5 mm, would be singularly uninformative, while the fly's outline on a two-dimensional grid would be much more revealing.

The final problem of individual linear measurements is that they tend to confound size and shape. For example, taking the case of helicospiral gastropod shells (see Figure 2.2), a difference in aperture height between two shells can occur either because the shells are the same shape but different sizes (compare a and b) or for the converse reason (compare b and c). Shape is often a more satisfactory characteristic to work with, because it varies less with growth than does size, and in some cases (isometric growth) does not vary at all.

In order to examine shape rather than size it is possible to use two or more linear measurements to obtain a ratio or some more complex combined measurement. Thus, having broken down the overall form of an organism, as expressed by three-dimensional co-ordinates, into a series of linear measurements, it is possible to combine these in various ways and so to build up a more complex picture again, but one which is very different from the original system of co-ordinates, and which may be easier to handle statistically, when whole populations are considered.

Simple ratios are widely used for describing shape. For example, a component of the shape of the shells illustrated in Figure 2.2 is measured by the ratio H/A, where H is the overall height of the shell from the lower rim of the lip to the apex and A is the aperture height. Examples of the use of this ratio are provided by Piaget (1929a) and Arthur (1982c) for *Lymnaea* and Crothers (1977, 1980) for *Nucella*, amongst many others. More complex ratios may in some cases be more biologically meaningful than simple ones consisting of two linear measurements. For instance, the rate of desiccation of a snail, either a terrestrial one or an aquatic one during unimmersed periods, will be influenced considerably by the size of the aperture (Machin, 1967). One crucial feature is the area available for rapid water loss per unit of body size. Thus either aperture area/body weight or aperture area/shell volume would be more informative than the simple H/A ratio in this respect. Some other measurements on molluscan shells which may be combined into ratios or used singly are provided by Raup (1966).

An alternative to using two or more linear measurements to describe certain components of shape is to combine the various measurements using a multivariate statistical technique. For example, in principal component analysis (PCA), individuals become typified by particular values taken by 'new' variables — usually the first and second principal components — after being originally measured for a larger number of 'old' variables. An example of the use of PCA, as well as canonical variate analysis, is provided by Williamson (1981a), and the whole field of the application of multivariate techniques to morphology has been dealt with by Blackith and Reyment (1971). The problem with these techniques (and with other, non-statistical, ways of combining many separate measurements: see Westoll, 1949) is that it is often very difficult to interpret the transformed data biologically.

Morphological characters have been treated in a rather static way in this section, with only adults, and thus only three dimensions, being explicitly considered. In fact, the structure of an individual organism is a four-dimensional phenomenon (Bonner, 1974; Arthur, 1982d). For example, the character 'wing length' is just as meaningless in the case of a present-day *Drosophila* larva as it is in the case of an ancient wingless insect. However, discussion of the fourth dimension of the phenotype — its development —will be postponed until chapter 9. This is because microevolution has been analysed with some success in the absence of developmental considerations. Very long-term evolution is another matter, and the discussion of development is left until chapters 9 and 10 so as to place it immediately prior to that part of the book dealing with mega-evolution.

2.2 POPULATIONS

Any character which can be measured, and which exhibits a particular numerical value in each individual organism, will also have a distribution of values in the population or, for that matter, in any sample taken from it. Morphological evolution, as defined earlier, involves changes in this distribution, and it is thus necessary to devote some space to its statistical description. As mentioned at the beginning of this chapter, these distributions are affected by the nature of the measurements made on individual organisms, and some methods of describing individuals lead to difficulties at the population level. For example, it is difficult to imagine, or at least to measure the characteristics of, a distribution of two-dimensional co-ordinate systems. However, distributions of individual linear measurements, or of simple ratios, are much easier to work with, and consequently have been frequently employed by quantitative geneticists and others working with populations. The discussion below is largely restricted to the simplest of all characters, single measurements — both linear and others. This discussion provides a basis for the genetic analysis of such measurements which is considered in chapter 3.

There are three types of variation and, correspondingly, three fundamentally different sorts of distribution a character may exhibit in a population. If individual values fall into a few disjunctive classes, variation is said to be discrete. This is common in the case of pigmentation, an example being the yellow, pink and brown shells of the land snail *Cepaea nemoralis*. It is rare, but not entirely absent, in the case of morphological characters; an example is the direction of coiling in some gastropods, with individuals falling into two categories, sinistral and dextral. Discrete variation in morphological characters often has a clear link with development, and so discussion of this phenomenon will be deferred until chapters 9 and 10. Suffice it here to say that characterization of the distribution is relatively easy in the case of discrete variation. Such distributions are usually described by the frequency of a particular variant or, pictorially, by bar or pie diagrams.

The second kind of variation is much more common among morphological characters and is referred to as quantitative or continuous. It occurs when a variable can take any of a series of values differing by infinitesimal amounts. The distribution of values for a character varying in this manner is often plotted graphically as a histogram and described by the arithmetic mean together with a group of statistics related to the so-called central moments. A set of data of this kind is shown in Figure 2.3, and the formulae for the central moments and related statistics are given in Table 2.1. These formulae are all based on the idea of a character represented by a single measurement $x_1, x_2 \ldots x_n$ in each of n individuals.

Since skew and kurtosis *per se* do not feature in the definition of morphological evolution given earlier, and since relatively few evolutionary

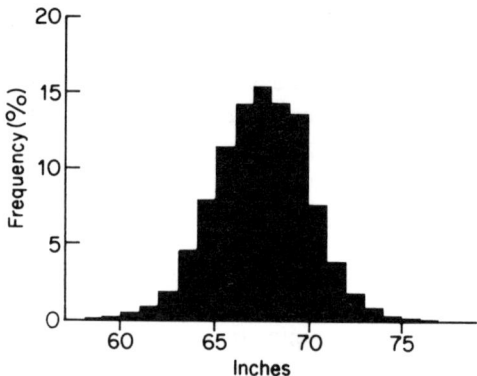

Figure 2.3. Distribution of a continuously variable character: heights of 8585 men. From Mather and Jinks (1977).

Table 2.1. Statistics commonly used to measure various properties of the distribution of a continuously variable character. n is the number of individuals in the sample, s is the standard deviation (positive square-root of the variance). In practice, $n - 1$ is often used instead of n as the denominator of the variance

Central moment	Related statistic	Property measured
$2\text{nd} = \dfrac{\Sigma(x - \bar{x})^2}{n}$	$\text{Variance} = \dfrac{\Sigma(x - \bar{x})^2}{n}$	Dispersion
$3\text{rd} = \dfrac{\Sigma(x - \bar{x})^3}{n}$	$\text{Coefficient of skewness} = \dfrac{\Sigma(x - \bar{x})^3}{ns^3}$	Symmetry
$4\text{th} = \dfrac{\Sigma(x - \bar{x})^4}{n}$	$\text{Coefficient of kurtosis} = \dfrac{\Sigma(x - \bar{x})^4}{ns^4} - 3$	'Peakedness'

studies have been preoccupied with these statistics, I will say little more about them. The only point that should be made is that their exclusion should be regarded as a temporary measure and it may be that in some evolutionary processes, measures of skewness and/or kurtosis will be important. For example, in character displacement, one might expect that there would be a phase during which the distributions of the character concerned in the two displacing species would be skewed away from each other.

The variance is often used as a measure of the variation within a group of data items, but, because of two potential problems, derivatives of it are sometimes used instead of the variance itself. The first problem is that the units of the variance are squared units of the original measurements; if the latter is in cm, the variance is expressed in cm². In cases where it is more appropriate to have the units of variation identical to the units of the original measurement, the standard deviation (the positive square root of the variance) is preferable. The second problem is that both the variance and the standard deviation of a measurement are likely to be related to the mean. We would expect leg length in giraffes to have a higher standard deviation (expressed in cm or some other unit of length) than leg length in mice. This problem is overcome, when two distributions of different means are to be compared for variability, by measuring the latter in the form of the coefficient of variation:

$$CV = \frac{s}{\bar{x}} \cdot 100$$

Thus variation is expressed in terms of the standard deviation (s) as a percentage of the mean ($\bar{x}$).

The distribution of a morphological character exhibiting quantitative variation often approximates to the normal distribution. This is convenient because, amongst other things, it allows the use, without prior transformation, of parametric statistical tests rather than their less efficient non-parametric equivalents, in the analysis of morphological data. I will not describe the features of the normal distribution here, since they are more than adequately covered in a wide range of statistical and biometrical texts (for example, Sokal and Rohlf, 1969). However, it should be noted that for many characters in many species, normal distributions are only obtained by excluding juveniles and all the members of one sex. Juveniles tend to skew the distribution, sexual dimorphism to render it platykurtic or even bimodal.

The third type of variation is intermediate between discrete and continuous types, and is referred to as meristic, quantal or quasi-continuous. Here, individuals fall into a number of disjunct categories (a characteristic of discrete variation) but the categories are both ordered and numerous, so that the distribution in the population resembles that of a continuously variable character. One classic example of a meristic character is the number of bristles on particular body segments of *Drosophila*. Meristic characters are also genetically similar to continuously variable characters in that their values are underlain by a large number of loci, as will be discussed in the next chapter.

The recognition of three categories of variation is a convenience adopted for the purposes of investigation. In fact, a single character in a single population may vary simultaneously in more than one way. For example, the freshwater snail *Lymnaea peregra* has some natural populations that are polymorphic for the direction of coiling (Boycott et al., 1930). If the character 'degrees turned through from point of origin' was measured in each individual in a polymorphic population, there would be one distribution of values centred on $+360W_1°$ and another centred on $-360W_2°$ where W_1 and W_2 are the average numbers of whorls in the two groups. The large gap separating these two distributions represents the discrete element of the variation while the two distributions themselves represent the continuous element.

2.3 SUMMARY

The rate of evolution is measured as character units per time unit. The choice of a unit of time presents no real problems but the measurement of morphological characters does. Several possible forms of measurement are discussed, all of which have their advantages and disadvantages. Single linear measurements provide easily described distributions in populations but lose considerable information; while outlines on two-dimensional grids retain much more information but are not so easily extended from individual to population. Ultimately, the choice of a particular kind of measurement depends on the task at hand. It is noted that a reasonable degree of understanding of microevolutionary processes has been achieved on the basis of single linear measurements but that such measurements tell us much less about long-term evolution.

Chapter 3

The Genetic Basis of Quantitative Variation

"Any variation which is not inherited is unimportant to us."

Charles Darwin, 1859.

One of the main problems met with in investigating microevolution of morphological characters is that of assessing the degree of inheritance of the continuous and meristic variation exhibited by such characters. The approach of ecological genetics, which includes demonstration of Mendelian segregation patterns in breeding experiments, cannot be successfully employed. Many early workers simply ignored this problem and drew evolutionary inferences from studies of morphological variation, having assumed at the outset that the variation was completely heritable. A classic example of this is the work of Mozley (1935, 1939) on H/A ratios in *Lymnaea* (see section 5.2). Although this approach is now less common, it is by no means extinct, and many recent studies within the field of evolutionary ecology have been based on assumptions, rather than tests, of the inheritance of the variation being investigated.

The main alternative to making such assumptions, apart from the rather negative one of ceasing to investigate the evolution of characters whose variation is continuous, is to undertake a study of inheritance using the methods of quantitative genetics. While the use of these methods if commonplace in the area of animal and plant breeding, it is not so widespread in evolutionary studies, because of the difficulties involved in applying quantitative genetic analyses to natural populations, to long periods of time, and to variation between rather than within populations. These difficulties will be discussed in section 3.4. The account of the methods themselves (sections 3.2, 3.3, 3.5) is brief and centred particularly on the concept of heritability. This is perhaps the most important quantitative genetic concept required for an understanding of morphological microevolution. Other aspects of quantitative genetics not included here, notably the theory of inbreeding, can be found in Falconer (1981).

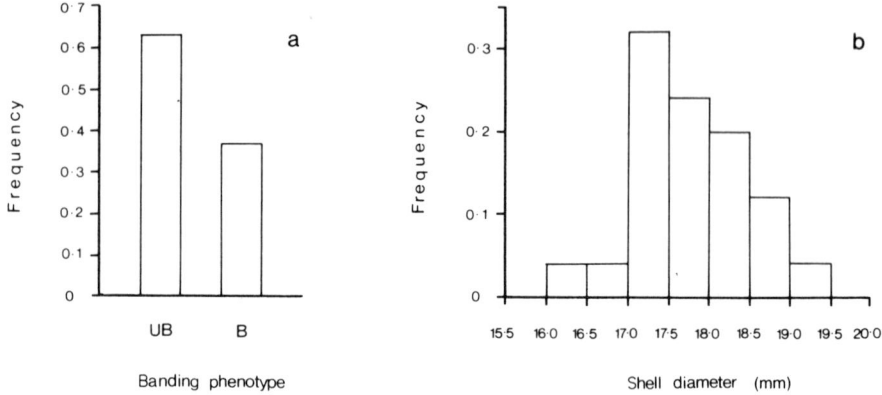

Figure 3.1. Examples of discrete and continuous variation. (a) Frequencies of banded (B) and unbanded (UB) phenotypes in a sample of adult *Cepaea hortensis*. (b) Distribution of maximum shell diameter in the same sample.
In (a) the banded category consists of a single genotype; the unbanded category includes both unbanded homozygotes and heterozygotes, since unbanded is dominant to banded.

3.1 THE CONNECTION BETWEEN MENDELIAN AND BIOMETRICAL APPROACHES

If continuous and discrete phenotypic distributions are plotted graphically (Figure 3.1), the results are strikingly different. The phenotypic categories within the system of discrete variation have a simple relation to genotypes, the correspondence being 1:1 in cases where there is no dominance; with the number of phenotypic categories being reduced below the number of genotypes where complete dominance occurs. In contrast, the 'categories' of phenotypes in Figure 3.1b have no obvious relationship to genotypes, and indeed such categories are completely arbitrary except in the case of meristic variation.

This lack of an apparent connection between continuous phenotypic variation and the genes led some early workers to believe that quantitative variation was always entirely environmental. There was considerable argument, soon after the turn of the century, between followers of the Mendelian and biometrical schools of thought, about the relative importance of, and the relationship between, the two types of variation. Their relationship was subsequently revealed by Fisher (1918) who showed that a large number of loci each having a small effect on a particular phenotypic character would give rise to a continuous distribution of values in a population. This is sometimes referred to as multiple factor inheritance or polygenic inheritance, and the numerous genes of small individual effect involved in it as polygenes. An example of a distribution that would result from a simplified scheme of this kind is shown in Figure 3.2, the corresponding numerical information being given in Table 3.1.

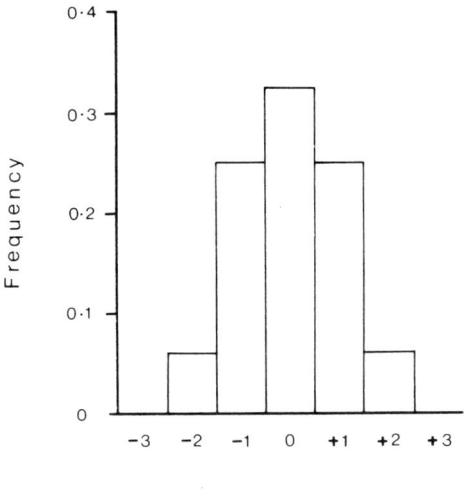

Figure 3.2. Distribution of values for a hypothetical morphological character whose genetic determination is described in Table 3.1.

Table 3.1. Frequencies of character values determined by two loci, A and B, at each of which a single allele substitution alters the value by one unit. Gene frequency of $p = q = 0.50$ and Hardy-Weinberg equilibrium are assumed

Genotype	Character value	Frequency	
		of genotype	of character value
$A_1A_1B_1B_1$	+2	0.0625	0.0625
$A_1A_1B_1B_2$	+1	0.125	0.25
$A_1A_2B_1B_1$	+1	0.125	
$A_1A_2B_1B_2$	0	0.25	
$A_1A_1B_2B_2$	0	0.0625	0.325
$A_2A_2B_1B_1$	0	0.0625	
$A_1A_2B_2B_2$	−1	0.125	0.25
$A_2A_2B_1B_2$	−1	0.125	
$A_2A_2B_2B_2$	−2	0.0625	0.0625

3.2 COMPONENTS OF VARIATION

The scheme outlined in Table 3.1 is simplified in that it excludes direct environmental effects as well as the effects of interactions between alleles at a locus (dominance), interactions between loci (epistasis) and interactions between genotype and environment. In fact, all of these may occur and may help to determine the value of a character in any specified individual and hence the variation of the character in a population. This section and the next are concerned with the breakdown of character values and their variation into components attributable to these different sources, and with the use of ratios measuring the relative magnitude of genetic contributions to the overall phenotypic variation. The treatment of the material here largely follows that of

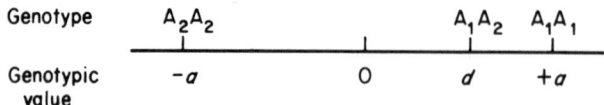

Figure 3.3. Genotypic values assigned to a hypothetical locus A at which allele A_1 increases the value of a morphological character while allele A_2 reduces it. From Falconer (1981).

Falconer (1981), though there are some differences of emphasis. In order to enable readers to follow up the discussion in more detail in Falconer's book, I employ the same range of symbols, though I have altered some of the subscript notation to avoid ambiguity.

Before discussing variation in the population, it is necessary first to consider the value of a particular quantitative morphological character in an individual. The breakdown of this value — the phenotypic value — into components can then be extended to the variation in the population. The simplest breakdown of the phenotypic value (P) into components is the split into genetic and environmental contributions. Thus, for any individual, we may write

$$P = G + E \tag{1}$$

While the phenotypic value of a morphological character can be measured in any given individual (in cm, g, etc.), the values of G and E cannot. Since G and E usually remain abstract entities they may be visualized in a number of different ways. It is conventional to regard G, the genotypic value, as the phenotypic value occurring under 'average' environmental conditions; and E as an environmental deviation superimposed on G, which takes a value of zero in average conditions and any one of a certain range of positive and negative values in other conditions. This is not the only way to picture G and E, though it is the conventional one. We could, for example, consider G to be the phenotypic value in an optimum environment and all environmental deviations to be either zero or negative. However, the conventional view has certain useful aspects. It leads, for example, to the equalities $\bar{E} = 0$, and hence, $\bar{P} = \bar{G}$. That is, the population's mean phenotypic value and its mean genotypic value coincide. It is common, because of this coincidence, to refer simply to the 'population mean' which is usually designated M. A practice sometimes adopted is to measure individual phenotypic values in relation to M. For example, if we have the absolute values $M = 10$ cm and, for some specified individual, $P = 12$ cm, an alternative representation is $M = 0$, $P = +2$ cm.

In order to stress the connection between the bulk of this section, which is biometrical, and the genes themselves, I will digress temporarily from the main theme of breaking the phenotypic value down into its components, to demonstrate the connection between M and the gene frequency. If we consider, first of all, one of the polygenes which affects a particular character, the genotypic values can be assigned as in Figure 3.3. In the absence of

Table 3.2. Calculation of population mean from genotypic values and gene frequencies. From Falconer (1981)

Genotype	Frequency	Value	Frequency × value
A_1A_1	p^2	$+a$	p^2a
A_1A_2	$2pq$	d	$2pqd$
A_2A_2	q^2	$-a$	$-q^2a$
		Sum =	$a(p-q) + 2dpq$

dominance, $d = 0$; in cases of complete dominance, $d = -a$ or $+a$. The mean value, M, of a character affected only by locus A in a population where the gene frequency is p of A_1 and q of A_2 can be determined as shown in Table 3.2. Note that M is obtained by summing the frequency × value column without dividing by N (the population size) because division by N has already taken place in calculating frequencies from absolute numbers. For the simplified situation described above we can write

$$M = a(p-q) + 2dpq \qquad (2)$$

Where many polygenes simultaneously affect a character, which is the situation commonly found in practice, then, providing their action is additive,

$$M = \sum a(p-q) + 2 \sum dpq \qquad (3)$$

where the summation is over all the loci concerned.

Returning to the main theme, expression (1) is only correct when there is no interaction between genotype and environment. When there is, that is when the deviation produced by a particular environment depends on the genotypic value, then instead

$$P = G + E + I_{GE} \qquad (4)$$

Throughout this section, interactions of this kind will be assumed to be negligible, which simplifies the theory. However, the lack of such interactions cannot be relied upon in practice.

When there is no interaction between genotype and environment, and when in addition there is no genotype–environment correlation (see below), the variance of phenotypic values in a population may be partitioned in the same way as the values themselves. Thus

$$V_P = V_G + V_E \qquad (5)$$

where V stands for variance — a practice in quantitative genetics which contrasts with the statistical practice of using σ^2 or s^2. Genotype–environment correlation is distinct from genotype–environment interaction, and occurs when individuals with different genotypic values select different environments. An example would be where individuals with higher genotypic values select environments causing higher positive deviations. Again, for simplicity, this sort of effect will be taken, here, to be negligible.

From an evolutionary viewpoint, the important component of the variation is clearly V_G, because any selection that occurs on variation between individuals that is due to direct environmental effects will not have its results passed on to subsequent generations. The relative contribution of genetic factors to the total phenotypic variation is known as the degree or coefficient of genetic determination (Falconer, 1981) or the broad heritability (Mather and Jinks, 1971). Thus

$$DGD = V_G / V_P \tag{6}$$

The values of this statistic must of course fall in the inclusive range from zero, indicating no genetic component to the variation, to 1, indicating no environmental component.

The split so far achieved of values and variance into 'genetic' and 'environmental' components is in fact rather uninformative since each of these components is heterogeneous. The DGD is not as useful, from an evolutionary viewpoint, as it might at first seem, because it tells us only the proportion of the variation that is due to all genetic sources, some of which, as will shortly become apparent, are not passed on directly from parent to offspring. Also, the DGD measures this combined genetic contribution relative to the proportion of the variation due to all non-genetic sources, which are not all environmental in the usual sense of the word. I will now expand on both of these complications.

First, as regards the genetic component of a phenotypic value, this may itself be split into three parts — the additive value, the dominance deviation and the interaction (epistatic) deviation — as follows

$$G = A + D + I \tag{7}$$

The meaning of these three components is illustrated in Table 3.3, which contrasts cases with and without dominance and epistatic effects.

Second, the 'environmental' component of values and variance is not entirely environmental because it includes, as well as the effects of factors such as temperature and food, effects due to stochastic developmental processes. An observation which illustrates this was made by Reeve and Robertson (1954). These authors, working on bristle number in *Drosophila*, showed that the overall 'environmental' variation between individuals, which can be measured by the phenotypic variation in isogenic lines, exceeds by very little the variation between sides of the body 'within' individuals.

This second complication has no effect on the construction of an index of the genetic contribution to the variance because it is clearly important to exclude stochastic developmental effects from the genetic component, and this is already done in the DGD. However, acknowledgement that the genotypic value has three components (A, D, I) is important in this respect, because in any index of genetic contribution to the variance that is to be of use in studies of selection, we want to consider only those genetic properties passed on directly from parents to offspring. Since gametes are haploid, dominance deviations

Table 3.3. Examples of additive and non-additive combination of genes. *Top*: Values of a quantitative character (arbitrary units) produced by different genotypes at a single locus, showing the effect of dominance. *Bottom*: Values produced by different two-locus genotypes, illustrating an epistatic effect. If both dominance and epistasis occur together, more complex patterns will occur. The dominance deviation is zero for the first column of figures in the top half of the table; and the interaction deviation is zero in the first column of the lower half. The other columns show non-zero deviations but these cannot be specified numerically as they depend on the gene frequency

Genotype	Dominance		
	None	Partial	Complete
A_1A_1	15	15	15
A_1A_2	10	13	15
A_2A_2	5	5	5

Genotype	Epistasis	
	Absent	Present
$C_1C_1D_1D_1$	15	15
$C_1C_1D_2D_2$	10	6
$C_2C_2D_1D_1$	10	6
$C_2C_2D_2D_2$	5	5

are clearly not transmitted; and although an entire complement of loci is transmitted, the interaction deviation of the offspring depends in a rather unpredictable way on its combined (diploid) genotype, so this also is not passed on directly in families. Thus in arriving at an index of genetic contribution to the variance which is useful in evolutionary studies, it is first necessary to partition further the variance in the same way as the genotypic value was partitioned in expression (7):

$$V_G = V_A + V_D + V_I \tag{8}$$

Substituting this into expression (5), we have

$$V_P = V_A + V_D + V_I + V_E \tag{9}$$

Having partitioned the phenotypic variance more fully, it is now possible to express the fraction of it due to additive genetic effects; that is, those effects due to the alleles themselves and so passed on, through the haploid stage of the life-cycle, to the offspring. This fraction is known as the heritability (Falconer, 1981) or the narrow heritability (Mather and Jinks, 1971) and is given the symbol h^2 (not h). Thus

$$h^2 = V_A / V_P \tag{10}$$

and again, as with the degree of genetic determination, we have a possible range of values from 0 through 1. Note that because V_A is a subset of V_G, each h^2 value must always be equal to or less than its corresponding DGD value.

3.3 THE MEASUREMENT OF HERITABILITY

Expression (10) cannot be used to estimate the heritability of a character because, although V_P can be directly measured, V_A cannot. Therefore, some

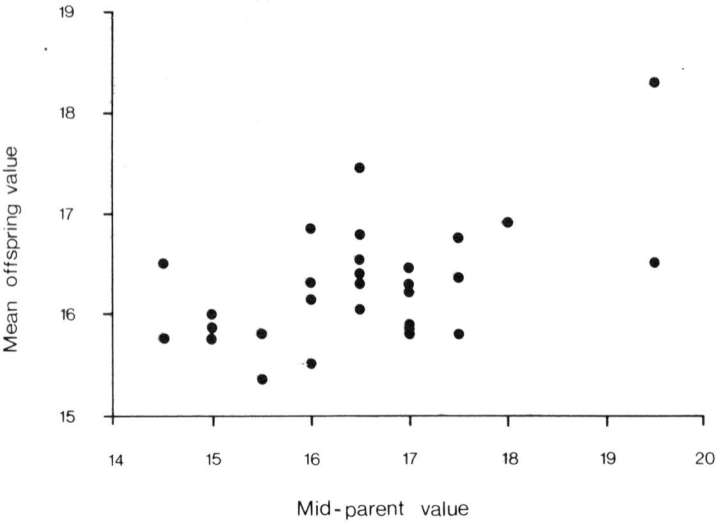

Figure 3.4. Plot of mid-parent v. mean offspring values for abdominal bristle number in a wild-type strain of *Drosophila melanogaster*.

other means must be found by which to determine heritability in practice. There are several ways of estimating heritability, most of which involve measuring the degree of phenotypic resemblance between related individuals. Since heritability represents the proportion of the overall variance that is due to factors passed on directly from parents to offspring, it follows that the higher the heritability of a character, the greater the resemblance between parents and their offspring as regards that character. Comparisons between close relatives other than parents and their progeny also show increasing likeness with increasing heritability. This occurs because close relatives share a high and fairly consistent proportion of genes with each other. In the case of two full sibs, for example, the average proportion of shared genes is 0.25, compared with 0.5 for parent and offspring.

The phenotypic resemblance between relatives for a group of families, i.e. for a population, can be illustrated graphically by plotting the mean value of one group of relatives against that of the other group. Figure 3.4 shows a plot of this kind for a meristic character, abdominal bristle number, in *D. melanogaster*, and the resemblance shown here is between parents and offspring. This particular comparison has the most direct relationship to the passing on of genes from one generation to the next, and is widely used in the estimation of heritability. In fact, the numerical value of the regression coefficient for offspring/parent data is also the value of the heritability. That is,

$$b_{\overline{OP}} = h^2 \tag{11}$$

It is of course clear that there should be some positive relationship between

$b_{\overline{OP}}$ and h^2, but their identity is not readily apparent. It is thus necessary to demonstrate why this identity exists; that is, to prove it.

First, it should be noted that for any two variables x and y, the regression coefficient of y on x can be obtained by dividing the covariance of y and x by the variance of x. Thus for parental and offspring phenotypic values

$$b_{\overline{OP}} = \frac{\text{cov}_{\overline{OP}}}{V_{\overline{P}}} \tag{12}$$

Thus what needs to be proved is that

$$\frac{V_A}{V_P} = \frac{\text{cov}_{\overline{OP}}}{V_{\overline{P}}} \tag{13}$$

Since we now have P-subscripts referring to 'phenotypic' and to 'parental', which are each in fact examples of two series of subscripts referring to components of variation and groups of individuals respectively, I have adopted, from expression (13) onwards, the procedure of italicizing subscripts relating to components of variation, which should help to avoid ambiguity. In cases where no italicized subscript is given, as in the right-hand side of expression (13), the variance or covariance referred to is phenotypic.

Since the variance of the mid-points of a series of pairs of values is half of the variance of the values themselves, then

$$V_{\overline{P}} = \tfrac{1}{2} V_P \tag{14}$$

Now since V_P of expression (13) will be approximately the same for parents and offspring, and since V_P of expression (14) is, according to the convention adopted, V_{PP}, it follows that

$$V_{\overline{P}} = \tfrac{1}{2} V_{PP} \tag{15}$$

Thus, if it can be shown that the covariance of mean offspring and mid-parent values is equal to $\tfrac{1}{2} V_A$, then the identity of $b_{\overline{OP}}$ and h^2 will be proved. This is done below.

Because mean phenotypic and genotypic values correspond, $\text{cov}_{\overline{OP}}$ is also the covariance of mean genotypic values of parents and offspring. Also, the mean genotypic value of a group of offspring produced by random mating of one parent is equal to half the additive genetic component of that parent's phenotypic value, expressed as a deviation from the population mean. Thus

$$\text{cov}_{\overline{OP}} = \text{cov}_{G, 1/2A} \tag{16}$$

However, assuming negligible epistatic interactions ($G = A + D$), then

$$\text{cov}_{\overline{OP}} = \text{cov}_{A+D, 1/2A} \tag{17}$$

Since A and D are expressed as deviations,

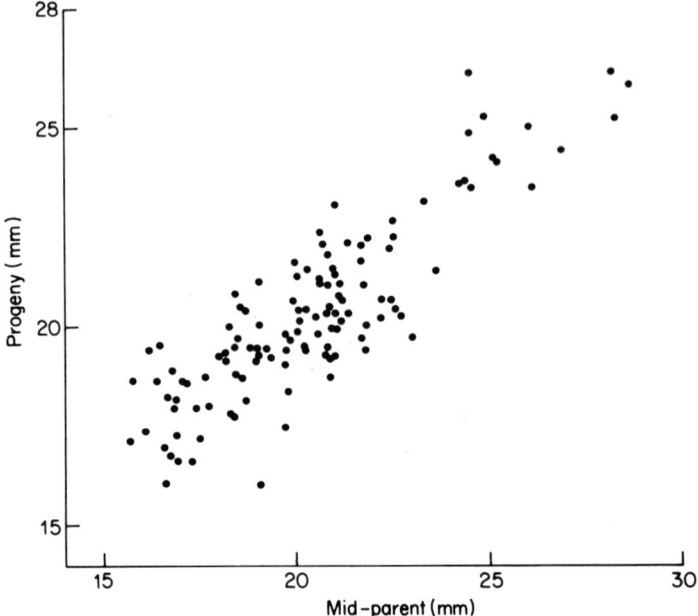

Figure 3.5. Scattergram showing parent–offspring shell size relationship in *Arianta arbustorum*. Progeny values are means for 5 individuals in each family. Regression coefficient, $b = 0.697$. From Cook (1965).

$$\text{cov}_{A+D,\,1/2A} = [\sum \tfrac{1}{2} A\,(A + D)]/(n - 1)$$
$$= (\tfrac{1}{2} \sum A^2 + \tfrac{1}{2} \sum AD)/(n - 1)$$
$$= \frac{\tfrac{1}{2}\sum A^2}{n - 1} + \frac{\tfrac{1}{2}\sum AD}{n - 1},$$
$$= \tfrac{1}{2} V_A + \tfrac{1}{2} \text{cov}_{AD} \tag{18}$$

It can be shown (Falconer, 1981, chapter 8) that cov_{AD} is equal to zero, and thus from expressions (16), (17) and (18)

$$\text{cov}_{\overline{OP}} = \tfrac{1}{2} V_A \tag{19}$$

It follows, then, from expressions (15) and (19) that

$$\frac{\text{cov}_{\overline{OP}}}{V_{\overline{P}}} = \frac{\tfrac{1}{2} V_A}{\tfrac{1}{2} V_P} = \frac{V_A}{V_P} \tag{20}$$

and thus the identity of the heritability of a character and the regression coefficient of mean offspring phenotypic values on mid-parent values is confirmed.

One example of the use of $b_{\overline{OP}}$ as a measure of heritability is given in Figure 3.5. This shows the results of a study by Cook (1965) of the heritability of shell diameter in the land snail *Arianta arbustorum*. As can be seen from Figure 3.5, the heritability is rather high (0.7). In general, it seems that heritability values for morphological characters are often higher than those for ecological/behavioural ones, but within each of these categories there is considerable heterogeneity (see Table 3.4).

Table 3.4. Heritability estimates for a number of morphological and non-morphological characters

Character	Species	h^2	Reference
Morphological			
Shell diameter	*Partula suturalis*	0.53	Murray and Clarke (1968a)
Shell diameter	*Arianta arbustorum*	0.70	Cook (1965)
Beak length	*Geospiza fortis*	0.97	Boag and Grant (1978)
Height	*Homo sapiens*	0.65	Roberts et al. (1978)
Abdominal bristle no.	*Drosophila melanogaster*	0.52	Clayton et al. (1957a)
Non-morphological			
Phototaxis	*Drosophila pseudoobscura*	0.1	Dobzhansky and Spassky (1969)
Egg production	*Drosophila melanogaster*	0.20	Robertson (1957)
Milk yield	Cattle	0.35	Barker and Robertson (1966)

In some cases it is impossible to estimate b_{OP}^- and in others it is impossible to use it as a measure of heritability. For example, when sampling from natural populations of invertebrates, it is usually impossible to delineate family groups; though in a few vertebrates this may be possible. Thus b_{OP}^- is more often used in cases where breeding experiments are performed in the laboratory, which in itself is problematical in that environmental variance among organisms reared in the laboratory is likely to be less than among individuals reared in nature. One situation where b_{OP}^- can be calculated but cannot readily be used as an estimate of heritability is when there is sexual dimorphism for the character whose heritability is being investigated. In such cases, h^2 may be estimated from resemblances between relations of the same sex. In fact, a variety of methods for estimating heritability are available, each based on the resemblance between a different pair of relations. These are summarized in Table 3.5. In each case, it is possible to demonstrate the identity of the 'measurement formula' and the ratio of additive genetic to total phenotypic variance; these proofs are given by Falconer (1981).

3.4 HERITABILITY AND MICROEVOLUTION

While it is often relatively easy to estimate the heritability of a character in a particular population, the use of such estimates in elucidating evolutionary processes is much more difficult. Here, students of the microevolution of morphological characters are less fortunate than students of the microevolution of characters such as allozymes which are often determined by a single locus. Although in both cases an absence of genetic information means that any study of geographic variation in the character concerned in inconclusive, the obtaining of a value of h^2 is much less effective in eliminating this inconclusiveness than the obtaining of, say, a 1:2:1 F_2 phenotypic ratio. In

Table 3.5. Estimation of heritability from regression or correlation coefficients for various pairs of relatives. From Falconer (1981)

Relatives	Regression (b) or correlation (r)
Offspring and one parent	$b = \tfrac{1}{2}h^2$
Offspring and mid-parent	$b = h^2$
Half sibs	$r = \tfrac{1}{4}h^2$
Full sibs	$r \geq \tfrac{1}{2}h^2$

this section I will outline the potential use of heritability estimates in evolutionary studies, together with the problems involved.

Any microevolutionary study of a morphological character is based on an investigation of variation between individuals in the phenotypic value of that character. Usually both variation within and between populations is examined, and in most cases an attempt is made to reconstruct evolutionary processes occurring over time from observed differences between populations in space. Occasionally, we may be lucky enough to observe temporal change in a character directly, but this is very unusual.

All purely phenotypic observations of variation leave open the question of whether there is an inherited component and, if so, how large it is. This is where heritability estimates come in, but they may be used in an attempt to answer two rather separate questions. These are:

1. Is there any additive genetic variation in the character concerned in the population under study? Or, to put it another way, Has the population the potential to respond to natural selection and to evolve, as regards the character concerned?

2. Is a particular pattern of variation between populations the result of divergent evolution in the past or of direct environmental influences occurring at present?

Estimates of heritability, as outlined in the previous section, are more directly applicable to the first of these questions than the second, and indeed any significantly non-zero heritability value means that the answer to that question must be 'yes'. However, there is a problem in making long-term statements about evolutionary potential — and it is evolution over many generations that evolutionists are usually interested in — because the heritability itself may change as a result of selection. Clearly, if all the polygenes affecting a character are monomorphic in a particular population, then $V_A = 0$ and so $h^2 = 0$. Thus a non-zero heritability implies that one or more of the relevant polygenes are polymorphic. However, the effect of directional selection at the morphological level will, except in cases of over-dominance, be to render the polygenes monomorphic for the alleles which cause the character to vary in the favoured direction. In practice, the effect of directional selection may not be quite so drastic, because of pleiotropic effects of some polygenes, which may prevent their fixation. But while heritability will not necessarily decline to zero, it will certainly decline, and under strong selection will do so much faster than it can be 'reflated' by new mutations. Thus a high heritability is certainly indicative of a potential for evolutionary change in the short term, but it does not tell us much about long-term evolutionary potential.

It is much more common, in evolutionary studies, to aim at an explanation of why, and whether, a particular pattern of interpopulation variation has evolved in the past, rather than to attempt a prediction of the future evolution of a character, which is generally agreed to be impossible except in very artificial situations. Thus the emphasis is usually on question 2 above, rather than on question 1. Unfortunately, the use of heritability estimates to help answer this second question is more difficult. There are two main reasons for this, as follows. First, the heritability of variation in a character within one population may not be the same as the heritability of the same character in another population. An estimate of heritability is restricted, in the information it provides, to the population in which the estimate was made. Here, as before, we find greater difficulty working with quantitative characters as compared to characters of single-locus inheritance where the 'heritability' is always 1 and the segregation ratios are unlikely to differ significantly between populations.

The second problem is that even where the heritability within each of a series of populations is high and fairly constant, there is no reason to believe that differences in mean phenotypic value between the populations are inherited to the same degree as variation among individuals within each. (See Jones and MacDonald (1976) for a colourful discussion of this problem.) The methods of calculating heritability given in the previous section relate specifically to variation within populations, and are not directly applicable to variation between them.

The difficulties involved in attempting to explain the inheritance or otherwise of a pattern of morphological variation between populations by using h^2 values relating to variation within them, can be emphasized by considering the following two contradictory points of view. The first of these is as follows. If within-population heritabilities are high, then this was probably the case in the fairly recent past, and there is therefore a good chance that the interpopulation variation now observed is a product of evolutionary change. Opposing this view, we have the following line of reasoning. If a character has evolved divergently in two populations due to selection favouring high values in one and low in the other, the end-result of such divergent selection will be fixation of the polygenes concerned for different alleles in the two populations, and consequently low values of h^2. Thus high values suggest that the interpopulation difference is largely environmental in origin.

It is clear, then, that to answer question 2 it is necessary to find some way of assessing the heritability of interpopulation differences themselves. There are two possibilities here. First, we may assess the heritability of an interpopulation difference indirectly, without analysis of family resemblances. Alternatively, we may adapt the methods discussed in sections 3.2 and 3.3 in order to obtain a direct numerical estimate of the heritability of the difference. Both of these alternatives are possible, but both have their problems.

A commonly used indirect method of approaching the heritability of a difference between populations is the transplant experiment. If two populations P1 and P2 inhabit different environments E1 and E2, then a sample of P1 can be cultured, through a complete life-cycle, in environment E2. If the difference between P1 and P2 was entirely due to genetic factors, it will persist; if due to direct environmental effects, it will disappear. If the difference was partially due to genetic factors and partially to environmental ones, then the transplanted population (P1) will converge towards, but not coincide with, the mean value exhibited by P2. Experiments of this kind may be conducted unilaterally or reciprocally, and the transplanted population may be cultured either in the appropriate natural environment or in a comparable laboratory one. Clarke et al. (1978) discuss several cases of the use of laboratory transplants in the study of morphological variation of the shells of pulmonate molluscs. The results of one such transplant are discussed in section 5.2.

Transplants are a rather crude way of examining the degree of genetic causation of interpopulation differences. They are useful if the difference turns out to be entirely, or very largely, due to genetic or to environmental factors; but in intermediate cases, transplants do not enable any partitioning of the variation which connects easily with the theory of quantitative genetics. It is possible, of course, to examine how close the mean phenotypic value of the transplanted population is to (a) its donor population, and (b) the population into whose environment it has been transplanted, but the interpretation of such information is hindered by a number of problems. Not least of these is the possibility that parasites, any effect of which is clearly environmental rather then genetic, may be unwittingly transplanted along with their host. A tendency for the transplanted population to retain its original morphology might be attributed to genetic factors yet in fact be due to the parasite. In this context, it is of interest to note that Oldham (1931) demonstrated a rather dramatic effect of parasitic mites on shell shape in the helicid land snail *Arianta arbustorum*: the mites made the shells of their hosts develop in a scalariform manner.

The alternative to transplant experiments, for the investigation of the heritability of interpopulation differences, is to carry out a breeding programme using members of two or more populations, and to subject the resultant family data to an analysis comparable to that described in section 3.3 for variation within a population, in order to separate genetic and environmental components. The simplest approach would be to use the offspring/parent regression technique with a representative pair of individuals from each of the n populations under study, the individuals being chosen so that all have phenotypic values close to their own population's mean. We are then effectively dispensing with variation within each population. All variation between $\bar{P}$ values in such a breeding programme would be variation between populations. The value of $b_{\overline{OP}}$ would reflect the heritability of the difference between population means. Such a programme would, of course, have to be carried out in the laboratory, so that individuals from all populations experienced the same environment.

The above approach suffers from the complication that by selecting individuals near the mean within each population, we may be causing assortative mating, which is known to affect h^2 values (Falconer, 1981); so, at best, this method can only give a rough estimate of heritability. More refined methods for the analysis of differences between populations have been relatively slow to develop. A variety of techniques are available for analysing differences between inbred lines, but most of these are not directly applicable to differences between natural populations which, unlike lines, are genetically heterogeneous. However, Lande (1981) has shown that Wright's (1952, 1968) formulae for analysing differences between inbred lines can be extended to deal with natural populations. In the procedure described by Lande, crosses are made between the populations, rather than within them as in the simple scheme outlined above, and backcrosses between the hybrid and parental populations are also made. The main use of Lande's method, however, is to estimate the number of genes (or the minimum number, to be more exact) contributing to the original interpopulation difference, rather than to estimate the heritability of this difference. There is considerable scope for future work aimed specifically at devising methods for separating genetic, environmental, and genotype–environment interaction components of morphological variation between natural populations.

3.5 THRESHOLD CHARACTERS

Because one of the major issues in the latter half of this book is the degree to which long-term evolution takes place in a saltational manner, it is necessary to discuss, at least briefly, the genetic basis of a form of discrete variation known as threshold variation. The existence of this class of variation needs to be acknowledged because otherwise the impression might be given that discrete variation necessarily implies inheritance by a single Mendelian major gene.

The nature of threshold characters can perhaps be most easily introduced in terms of the number of progeny an individual has, although this is in fact a physiological/behavioural characteristic, not a morphological one. It is clear that there is considerable variation within and between species in the character 'number of offspring per breeding season'. Whether the variation within a species is describable as meristic or discrete depends on the species concerned. For example, in many species of tree, the number of seedlings produced by a single individual runs into the thousands, and within a population the range of possible values an individual can have for this character is likely to span at least several hundred. Turning to birds, for many species the average number of eggs per pair per season is in single figures, and the range of possible values within a population is therefore also in single figures. Thus here we have a contrast where the same character is seen to vary in a meristic manner in one taxonomic group and in a discrete way in another. In fact, there is no clear line separating these two categories of variation, but a working definition of meristic variation as occurring when there are ten or more categories may be adopted. (Also, in meristic variation, the categories are always ordered whereas in discrete variation they may or may not be.)

An explanation of the existence of several hundred seedling-number categories in terms of a single-locus genetic basis is clearly ridiculous; the number of alleles required would far exceed the numbers of alleles found at those loci in natural populations where an exhaustive search has been conducted by employing a combination of different techniques of protein separation (for example, Singh *et al.* (1976) working on the *Xdh* locus of *D. pseudoobscura*). It is universally accepted that characters such as 'seedling number' have a quantitative genetic basis, with an underlying, continuously variable tendency to produce offspring (fecundity) being manifested in a meristic form simply because it is not possible to have fractions of offspring. However, in organisms such as birds, the number of categories of the character 'egg number' is within the range where a simple Mendelian genetic basis would be possible. There are two good reasons for dismissing such an explanation: first, we are dealing with a character which in other groups is recognized to be polygenic; and second, any single locus with such a large effect on fitness would be very rapidly made monomorphic by selection (unless there is some strong balancing mechanism at work), and if this were the case then we would not observe intraspecific variation in number of offspring.

The above example serves to illustrate that there are characters which vary in a discrete manner with a small number of categories (as low as two in some cases) which are nevertheless underlain by a less tangible quantitative character which has a polygenic basis. The question now arises of whether there are morphological examples of this phenomenon, and indeed several of these are known. One example concerns the number of pre-sacral vertebrae in the mouse *Mus musculus*, which can take any of the values 25, 26, 27. Falconer (1981), using data taken from Green (1962), calculates that for this character, in the strains used, DGD = 0.68.

The method used to obtain the above estimate involved crosses between inbred lines where the variability in the F_1 is entirely V_E while variation in the F_2 is composed of $V_G + V_E$. This allows calculation of the degree of genetic determination, but not of h^2. In fact, the concept of heritability (as well as the DGD) is directly applicable to threshold characters, but the methods of estimation necessarily differ because the underlying quantitative character cannot be measured in cases of threshold variation. It is thus necessary to examine briefly the method of analysing threshold characters that enables heritability estimates to be made. As in sections 3.2 and 3.3, I will largely adhere to the symbols used by Falconer (1981).

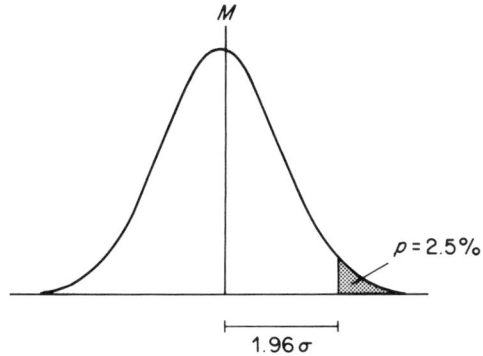

Figure 3.6. Normally distributed liability underlying a threshold character which is manifested in 2.5% of a population. The population's mean liability, M, is 1.96 standard deviations below the threshold.

The connection between a threshold character and its underlying continuous variable, which may be referred to as the *liability*, is shown in Figure 3.6. If the liability is, or can be transformed to be, normally distributed, then its mean phenotypic value can be estimated from the percentage of the population showing the 'affected' as opposed to the 'standard' phenotype, which is referred to as the *incidence*. The estimation of M from incidence follows directly from the known shape of the normal distribution. For example, we know that 5% of data items fall outside the range $M \pm 1.96\,\sigma$, with 2.5% in each tail. Thus, when the incidence of a threshold character is 2.5%, $M = -1.96\sigma$ relative to the location of the threshold.

Having discussed how M, which cannot be measured, is estimated from the incidence, which can, it is necessary to digress, before using this information to calculate the heritability of a threshold character, to consider a general method for calculating h^2 which has not yet been dealt with here. This estimate is sometimes referred to as the realized heritability and the reasoning behind it is as follows.

Consider a population in which a quantitative character is being monitored, and in which individuals mate at random. Over a single generation, a group of families can be represented by a group of points (one per family) as shown in Figure 3.7. If we suppose that parents with above a certain value of the character were used in an artificial selection experiment, then we can measure the mean values of the character in: (a) the whole population prior to selection, (b) the group of selected parents, and (c) the offspring resulting from those parents. These will be designated M_1, M_2 and M_3, respectively. Now it is possible to define the selection differential (S) and response to selection (R) in the following manner:

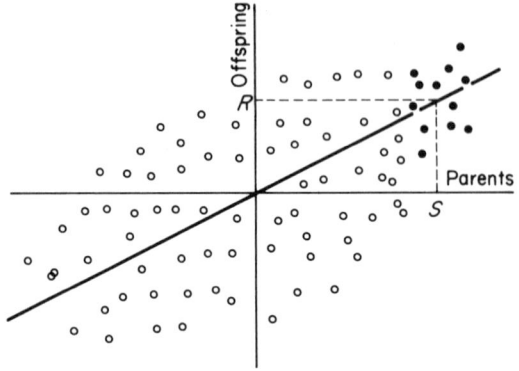

Figure 3.7. Scattergram in which each point represents a single family. The co-ordinates of a point are the mid-parent and mean offspring phenotypic values of the morphological character under consideration. From Falconer (1981).

$$S = M_2 - M_1$$
$$R = M_3 - M_1$$
(21)

Since the point whose co-ordinates are R, S will lie on the regression line through the scatter of points as shown in Figure 3.7, then

$$R = b_{\overline{OP}} S \qquad (22)$$

However, since we already know that $b_{\overline{OP}} = h^2$, then it follows that

$$R = h^2 S \qquad (23)$$

Thus the heritability may be calculated simply as the ratio R/S.

Returning to the specific problem of determining the heritability of threshold characters or, more correctly, the heritability of their liability, the procedure is to estimate mean values from incidences, to use the mean values to calculate R and S, and hence to determine the heritability from expression (23). An example of this procedure will now be discussed, with reference to Figure 3.8, which depicts the results of a selection experiment in which the affected individuals of the population (top curve, shaded) are used as parents, and the resulting progeny (lower curve) show an increased incidence. Assuming that the two curves have equal standard deviations (σ), the means of upper and lower curves (M_1, M_3) are -1.6σ and -0.8σ, respectively, and so $M_3 - M_1 = +0.8\sigma = R$. The mean of all affected individuals in the parental generation (M_2) can be obtained from a knowledge of the shape of the normal distribution, and the deviation $M_2 - M_1$ is the selective differential; in this example, $S = +2.1\sigma$. So in this case, the heritability is $0.8/0.21 = 0.38$.

The above example relates to the simplest possible case of two phenotypic categories separated by a single threshold. It also involves certain assumptions, namely normality of the distribution of liability and equality of variances in

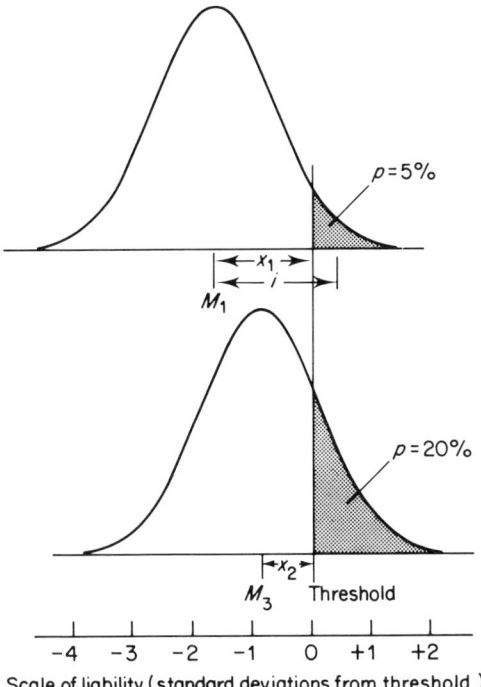

Figure 3.8. The response, over one generation, of selection on a threshold character, showing increased frequency of affected individuals (from 5% to 20%). x indicates distance from threshold to mean liability; i is the intensity of selection (see section 4.1). From Falconer (1981).

parental and offspring generations. More complex cases, and the consequences of violating assumptions, are discussed by Falconer (1981).

Before leaving the subject of threshold characters, it is necessary to introduce Waddington's concept of genetic assimilation. This has been discussed in a series of papers (e.g. Waddington, 1953) and in *The Strategy of the Genes* (Waddington, 1957a). Basically, the phenomenon involved is selection for higher liability underlying a threshold character, with the effect that a phenotype which initially only occurred in some individuals subjected to an environmental 'treatment' begins to appear in more of the treated individuals and eventually appears spontaneously, without aid from the environment. Several experiments have demonstrated that this can actually occur. The best known of these, which will be discussed in the next chapter, involved genetic assimilation of 'cross-veinlessness' in *Drosophila*, where the lack of a wing vein, originally caused by heat-shock, appeared spontaneously after selection. The important point about this and similar experiments is that they provide a conventional selective explanation for a phenomenon which might, in the absence of such an explanation, be considered Lamarckian.

3.6 SUMMARY

After describing how continuous variation is based on the combined action of many loci with individually small effects on the phenotype, a more detailed analysis of the phenotypic values of individuals, and the phenotypic variation in populations, is presented. Discussion is centred on the concept of heritability, which is particularly relevant to evolutionary studies. The questions that microveolutionists often ask about inheritance are noted, as are the difficulties involved in using heritability estimates to answer them. Because of later discussion (chapters 11, 12) of saltational evolution, special consideration is given to threshold characters, which have a polygenic basis despite their discrete variation at the phenotypic level. Waddington's concept of genetic assimilation is briefly introduced as the result of selection acting on threshold characters.

Chapter 4

Artificial Selection Experiments

"The great power of this principle of selection is not hypothetical."

Charles Darwin, 1859.

Having outlined the genetic basis for the ubiquitous quantitative variation found in morphological characters, it is now possible to begin discussion of microevolutionary change. The discussion is split between the present chapter, which deals with artificially induced microevolution in the laboratory, and the next, which deals with microevolution in nature. Both of these are important, though in different ways: natural microevolution because this is ultimately what we wish to understand; and artificially induced microevolution because experiments performed in the laboratory have elucidated several important facts about the genetic structure of populations and its progressive alteration through selection, at least some of which are general facts and undoubtedly apply also in the wild.

4.1 RATIONALE AND BASIC EXPERIMENTAL DESIGN

Artificial selection experiments have been conducted for two rather separate reasons: first, to improve stocks of domestic animals or crop plants as regards some aspect of their yield to man; and second, to examine the effect of selection on the genetic and phenotypic properties of populations with a view to improving our understanding of microevolutionary processes and of population structure in general. Different taxonomic groups have been used in experiments addressing these two different issues, since the groups from which most information can be obtained most quickly and the groups that are of agricultural importance overlap very little. Thus while many experiments of an 'applied' nature have been conducted on cows, sheep, pigs and poultry, the bulk of 'fundamental' experiments has been conducted on *Mus musculus* and on several species of *Drosophila*, notably *D. melanogaster*. While the two categories of experiment are not completely separate, I will concentrate here on experiments of the latter kind, since the goal, in the present context, is an understanding of evolution rather than of animal and plant breeding. An important difference between these experiments and the studies of natural populations discussed in chapter 5 is that the former investigate the genetic and phenotypic effects of selection, given that it has occurred, while the latter must also enquire about whether selection has actually taken place, and must attempt to distinguish between its effects and the effects of other evolutionary agents, as well as direct environmental effects, whose action in the laboratory can usually be either prevented or monitored.

The basic design of an artificial selection experiment is as follows:

1. A character exhibiting quantitative variation is chosen for study.
2. A particular population of the species being investigated is chosen, avoiding highly inbred populations which have been kept in the laboratory at low N or allowed to go through a 'bottleneck'.
3. The form of selection to be effected is decided upon. Most experiments have involved attempts to induce directional selection, which is the most relevant to microevolutionary change, and thus a common method of selection is to choose individuals with the largest, or the smallest, values of the character concerned within each generation to act as the parents of the next.
4. The mean phenotypic value of the character is monitored regularly, usually once per generation, over a number of generations — in *Drosophila*, often 10–30 (5–15 months).
5. In some cases, artificial selection is discontinued after a number of generations; this allows examination of whether the character value produced by artificial selection is stable or whether it reverts, under 'natural' selection, to its original, or to some other, value. An alternative to removal of any form of artificial selection is to attempt 'back-selection' at some stage.

The kind of selection exercised in such an experiment differs from the way natural selection is generally thought to operate in that it occurs by truncation; that is, all individuals of the ith generation that are chosen as parents of the $(i + 1)$th generation have character values above some predetermined cut-off point. In this sort of regime, the probability of leaving offspring switches suddenly from 0 to 1 (or, in reality, to nearly 1) as the character value increases. In the case of directional natural selection, the situation usually envisaged is a gradually, but not necessarily linearly, increasing probability of leaving offspring as we consider individuals with character values approaching the favoured end of the range.

The fact that artificial selection experiments act by truncation means that it is a relatively easy matter to calculate S, the selective differential. Since the group to be used as parents of the next generation is specified unambiguously by the character values, it is possible to calculate the mean value of the group so specified, and hence to obtain S, as well as R, the response to selection, after also measuring the overall means for parental and progeny generations. The numerical values of R and

S obtained can be used in a number of ways. First, if no other estimate of heritability is available, the ratio R/S provides such an estimate, as discussed in the previous chapter. Second, if an estimate of h^2 is already available, for example one obtained from measurements on siblings (see Table 3.5) which can be determined prior to running an experiment over a complete generation, then the ratio R/S provides a check on the other estimate, and hence a check on the assumptions on which it is based. Finally, having outlined a scheme of selection for several generations, and assuming that the variability available continues to allow the selective differential to be imposed, then using either R for the first generation and S, or h^2 and S, a prediction can be made for the value of the character after a number of generations of selection. Such predictions assume that change in h^2 over the course of the experiment is negligible, which is often true for at least five or ten generations (Falconer, 1981).

A statistic sometimes used in connection with artificial selection experiments is i, the intensity of selection. This is simply the selective differential expressed in units of phenotypic standard deviation. S itself is measured in units of the character concerned, and so may be in cm, g, etc. Since $i = S/\sigma_P$, and since both S and σ_P are in units of the original measurements, these latter units disappear when i is determined.

Long-term artificial selection experiments provide both qualitative information, for example demonstration of the effectiveness of selection, and quantitative information, such as confirmation or otherwise of a mean character-value predicted for (say) generation 5 from information on S and h^2. Particular examples of these experiments will be discussed in the next two sections, with emphasis on the more qualitative aspects. The experimental results help to illustrate not only general principles, but also a number of practical points which have, for simplicity, been omitted from the above discussion.

4.2 EXPERIMENTS INVOLVING DIRECTIONAL SELECTION

In this section I have separated out a series of general points which have emerged from artificial selection experiments, and have illustrated each with a particular example. The choice of the experiment used to exemplify each point was fairly arbitrary, since very many experiments involving directional selection have been conducted, often with broadly similar conclusions.

1. Selection is effective in causing significant microevolutionary change

MacArthur (1949) conducted an experiment in which he monitored the character '60-day body weight' in the mouse, *Mus musculus*. In separate but contemporary experiments, selection for high and for low values of this character was effected, so that in both experiments the selection was

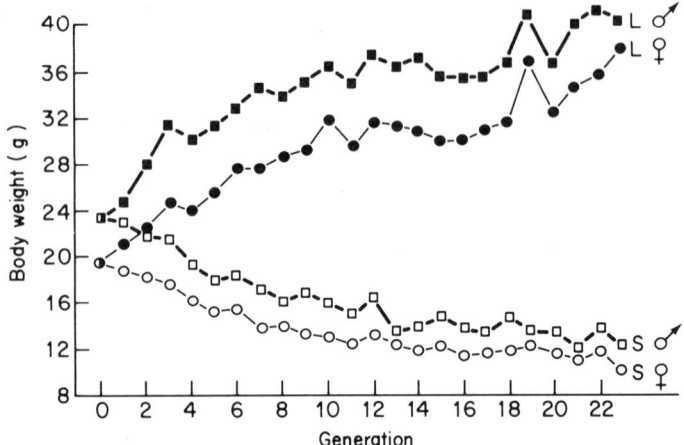

Figure 4.1. Response of 60-day body weight in mice to selection for large (L) and small (S) individuals. From MacArthur (1949).

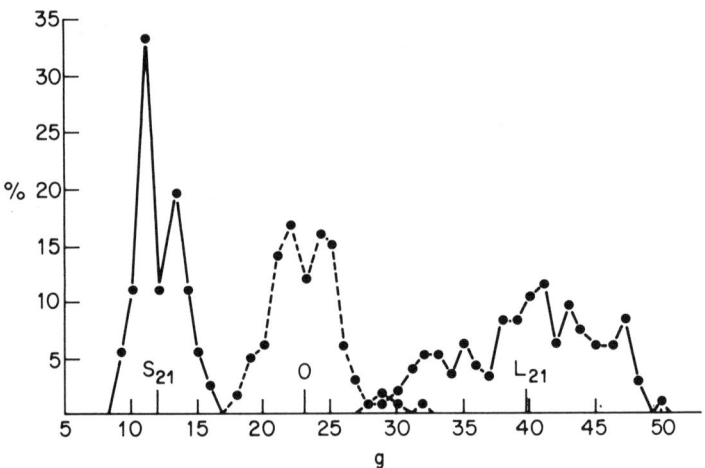

Figure 4.2. Body-weight distributions in large (L) and small (S) selection lines at generation 21 and in the base population (0). The body weights of the mice were determined at an age of 60 days. From MacArthur (1949).

directional. The mice used were from a stock created by interbreeding several inbred strains of laboratory mice in order to 'replace' the genetic variability lost in prior inbreeding (MacArthur, 1944). This stock was subjected to artificial selection for 23 generations and the results, expressed as generation means, are illustrated in Figure 4.1, with males and females in large (L) and small (S) lines shown separately since 60-day weight is sexually dimorphic. The variability within the selected lines at generation 21 is compared with the variation in the original stock in Figure 4.2. It can be seen that (a) selection was effective in both directions and in both sexes, and (b) by the 21st generation the ranges of variation of the two selected lines hardly overlap at all with the range of values in the foundation stock. In fact, the changes produced by selection over this rather short period of time were such that the mean value was almost halved in the down-selected line and almost doubled in the up-selected line.

2. Prediction of the quantitative response to selection varies in accuracy

An experiment in which an observed response agreed well with that predicted from prior measurement of h^2 and S was performed by Clayton *et al.* (1957a) on abdominal bristle number (fourth and fifth sternites) in *Drosophila melanogaster*. Heritability was estimated by three techniques, all based on resemblance between relatives, including parent–offspring regression as described in the previous chapter. The values obtained were combined to give an overall estimate of 0.52. Bristle number was then subjected to selection in both directions, the data on up-selection being re-analysed by Falconer (1981) to show that with $S = 5.3$ and $h^2 = 0.52$, giving an expected response of $R = 2.8$, the observed response was in fact 2.6 — very close to the prediction. However, Clayton *et al.* (1957a) point out that the response to selection was considerably less than predicted in the down-selected line, which they attribute to rapid depletion of genetic variation in this line. This particular sort of heterogeneity of response is discussed further below.

3. The response to 'up'- and 'down'-selection often differs

A difference between the responses to selection in opposite directions is in fact a rather widespread phenomenon, and it is not consistent in terms of whether the upward or downward response is the greater. For example, the results of MacArthur (1949) shown in Figure 4.1 illustrate a more rapid upward response, while those of Falconer (1953) show the reverse phenomenon despite being based on the same character and species (see Figure 4.3). There are many possible causes of such asymmetry (see Falconer (1981) for a discussion) and it is often difficult to identify which is operative in an individual experiment. One of these, which causes down-selected lines to respond less, is the correlation between the variance and the mean, discussed in chapter 2, which leads to a smaller variance in a line whose mean has been reduced by selection. This effect can be seen in Figure 4.2 where the variance of the up-selected line at generation 21 greatly exceeds that of the down-selected line. Thus the effectiveness of selection towards increasingly high character values is enhanced by the increased variance available, while selection for lower values is inhibited.

4. Between-replicate heterogeneity occurs in magnitude but not direction of response

The results of artificial selection experiments have been discussed so far without any reference to the problem of whether replicate samples of a population,

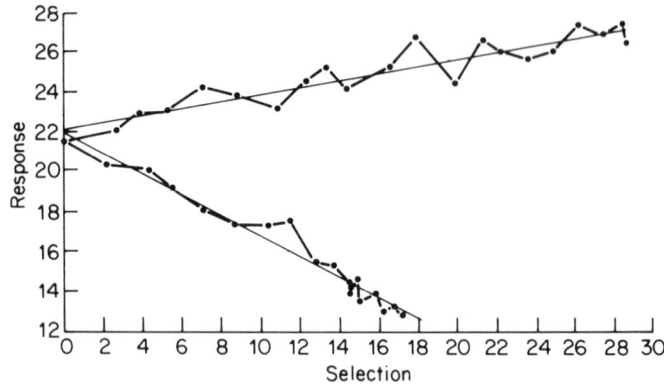

Figure 4.3. Responses to up- and down-selection for six-week body weight in mice, showing more marked response to selection in down-line. Note, however, that the x-axis takes the form of cumulative selective differential rather than time, so this figure is not directly comparable to Figure 4.1. From Falconer (1981).

subjected to the same selective regime, actually respond in the same way. This question has been examined by Falconer (1973), using six-week body weight in mice (see Figure 4.4). It is clear that there is variation in response between replicates within generations. However, the effect of such variation on the cumulative response to selection after 23 generations is clearly small compared with the variation between treatments. Thus, as this set of results indicates, although the precise response over a single generation may vary, general conclusions such as the elementary but important one that selection is indeed capable of altering the character value in a specified direction in the long term are not altered by this heterogeneity between replicates.

5. *If artificial selection is discontinued, partial or complete reversion occurs*

What happens to the mean value of a selected character when the selection ceases depends on two things: first, why prior to selection the character took the value it did; and second, whether the genetic variation originally present in the population has been exhausted by the selective process of repeatedly breeding from the most extreme individuals. If the original character value was a result of strong optimizing selection and the genetic variation is not depleted, then, on discontinuance of artificial selection, the character will tend to return to its original value under natural selection. On the other hand, if the original value represented a neutral equilibrium and/or the genetic variability was exhausted, then the character will fluctuate around its new artificially produced value. In fact, a tendency to return to the original value is commonly found, even in experiments where selection has been continued for many generations and the response has almost ceased. This indicates that cessation of response under long-continued artificial selection is sometimes due to the reaching of a selection limit caused by something other than exhaustion of genetic variation.

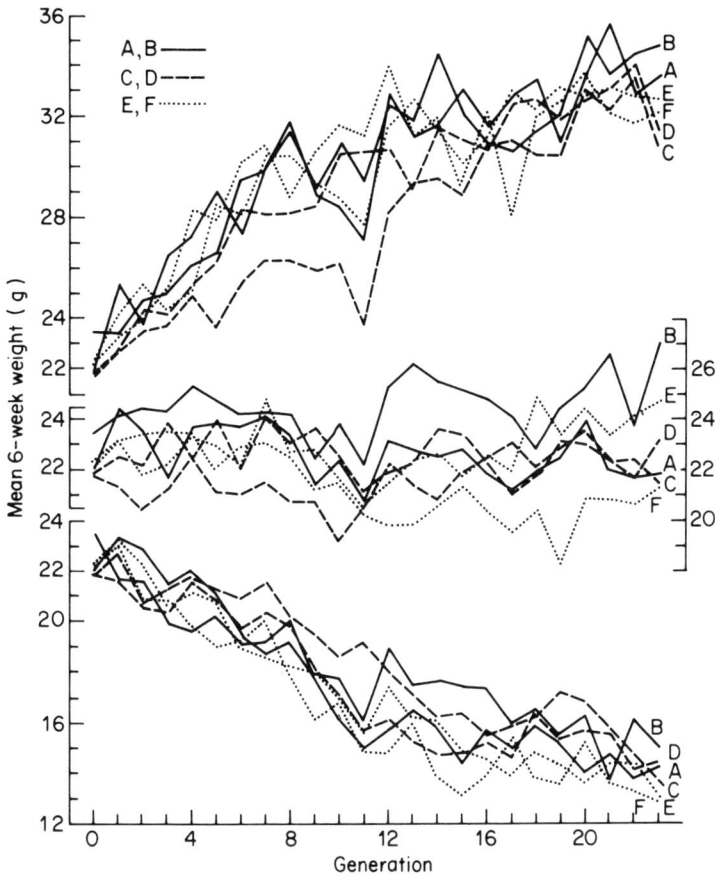

Figure 4.4. Results of selection on 6-week body weight in mice, with generation means being given separately for each of the replicates A–F in control (centre) and selected lines. The construction of the base population was such that replicates A and B were genetically similar to each other as were C and D, and also E and F. The three pairs of replicates were less genetically similar to each other than were the two replicates within each pair. From Falconer (1973).

An example of the reversion of a character to its original value or thereabouts is shown in Figure 4.5. This example involves a behavioural character (phototaxis in *Drosophila pseudoobscura*: Dobzhansky and Spassky, 1969) but is included because the reversion is particularly clear. In other cases a less strong tendency towards reversion has been observed; for example, Robertson (1955) — see below.

6. *If artificial selection is continued indefinitely, the response ceases to occur*

The results of an experiment by Robertson (1955) on thorax length in *D. melanogaster* are shown in Figure 4.6. It can be seen that, in all three stocks (the

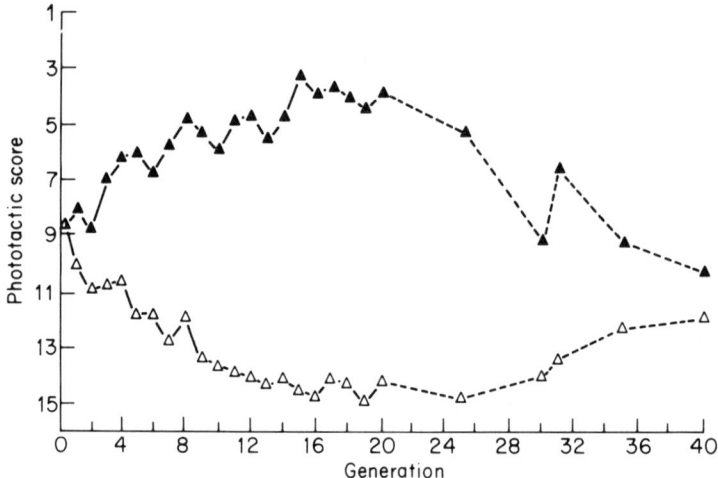

Figure 4.5. Results of 20 generations of directional selection for photopositive (▲) and photonegative (△) behaviour in *D. pseudoobscura* followed by 20 generations during which artificial selection was suspended (-----). Low values of the phototactic index used indicate photopositive behaviour, high values photonegative. A score of 8.5 corresponds to photoneutrality. From Dobzhansky and Spassky (1969).

names of which correspond to the Italian and Scottish localities from which they were collected), the response to selection was positive up to about generation 10 for up-selection and generation 15 for down-selection. After these points little further progress was made. The results are remarkably consistent for the three stocks, suggesting that the pattern of response to selection in this character is a rather general one. As mentioned above, it can also be seen from Figure 4.6 that discontinuance ('relaxation') of selection resulted in a much less marked return to the original character values than in the case of selection on phototaxis in *D. pseudoobscura*. Lack of pronounced reversion is not necessarily due to exhaustion of genetic variation. This is illustrated by Robertson's (1955) results which show that back-selection was effective in bringing the up-selected lines back to their original character values, even when it was not initiated until between generations 15 and 20. (Note though that back-selection was not effective when started at a late stage in the down-lines, suggesting that monomorphism of genes affecting thorax length was eventually attained in these lines.)

7. Other characters often show correlated responses

The response of a population to selection is never entirely restricted to the character whose value is used to determine which individuals are employed for breeding; rather, a whole series of characters will be modified in addition to the principal one under investigation. The responses of these additional characters to artificial selection, that is, 'correlated responses', have been studied by

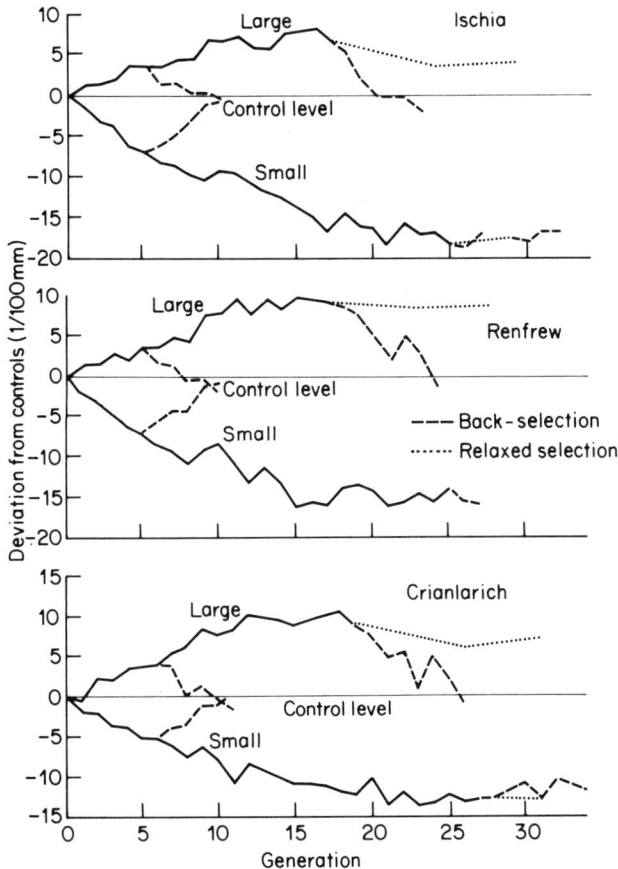

Figure 4.6. Response of thorax length in three strains of *D. melanogaster* to up- and down-selection, as well as to relaxation of selection and to back-selection. From Robertson (1955).

Reeve and Robertson (1953; thorax length and wing length in *D. melanogaster*) and Clayton *et al.* (1957b; sternital and sternopleural bristles in *D. melanogaster*), among others. The patterns of correlated response observed have often been complex, one particular complexity being that correlated responses to up-and down-selection in the 'primary' character tend to differ. Such asymmetry is illustrated in Figure 4.7 where a significant correlated response of sternopleural bristles is apparent in the high lines, that is those selected upwards for number of sternital bristles, whereas there is little if any correlated response to down-selection (Clayton *et al.*, 1957b). A theoretical analysis of asymmetry in correlated responses was carried out by Bohren *et al.* (1966), and one of these authors' main conclusions was that asymmetrical correlated responses are likely to be the rule rather than the exception. This conclusion stems largely from the fact that the genetic covariance of two characters, which

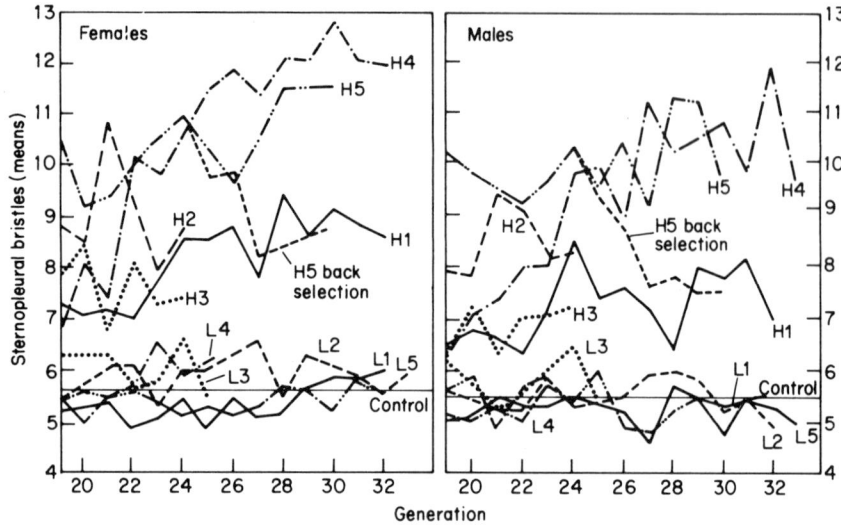

Figure 4.7. Correlated responses of sternopleural bristles in lines of *D. melanogaster* selected for high (H) and low (L) numbers of abdominal sternital bristles. From Clayton et al. (1957b).

underlies correlated responses, is affected by changes in gene frequency which of course occur in different directions in up- and down-selection.

Genetic covariance of two characters can arise in two distinct ways. First, the groups of genes controlling the two characters may be the same, or overlap, or alternatively one group may be a subset of the other. Second, close linkage between some or all of the genes in the two groups may cause covariance of the characters, even if none of the genes affect both of them. This second source of covariance is likely to decay over time, however, due to recombination, whereas the other source of covariance is effectively permanent. It is of course conceivable that a correlated response will occur due to epistatic effects without any genetic covariance of the characters involved. That is, up-selection in charcater A may cause 'natural' selection for higher values of character B than those that were optimal when character A's mean value was lower. Thus there are several distinct reasons why selection for one character may progressively alter another, and no doubt their relative importance varies in different situations.

8. Genetic assimilation can be produced by artificial selection

The examples cited so far all relate to continuously variable or meristic characters. However, we have seen (section 3.5) that there are certain threshold characters whose discontinuous phenotypic variation is underlain by a polygenic system similar to those of more 'conventional' quantitative characters. It should, therefore, be possible to conduct artificial selection

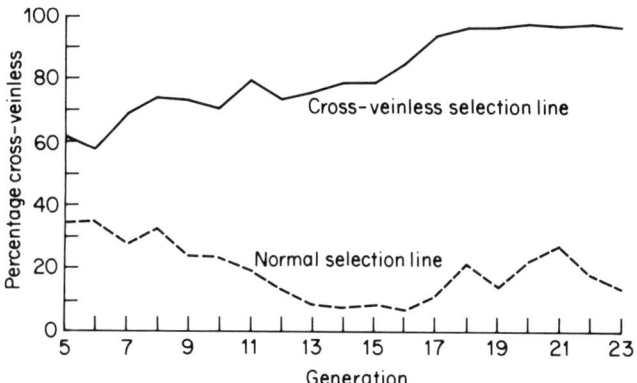

Figure 4.8. Results of selection for CVL and for normal wings (i.e. cross-veins present) in *D. melanogaster*. From Waddington (1953).

experiments on threshold characters, and indeed several such experiments have been carried out. One group of these is of particular importance in an evolutionary context, namely the experiments demonstrating genetic assimilation (Waddington, 1942, 1953). In these experiments, a characteristic originally appearing only as a direct response to an unusual environmental regime is acted upon by selection until it eventually appears as an inherited trait that does not require any environmental stimulus for its production.

A clear demonstration of genetic assimilation was given by Waddington (1953) in an experiment involving selection on the character 'cross-veinlessness' (CVL) in *Drosophila melanogaster*. If individuals of this species are subjected to a short period of high temperature (40 °C) during pupation, many of the adult flies that eclode lack a posterior vein in the wing which normally connects two of the main veins. In some cases, these flies also lack an anterior cross-vein. Waddington (1953) took a stock of *D. melanogaster* in which no flies showed the CVL characteristic without prior heat-shock and by producing CVL initially in about one third of the flies by using heat-shock treatment, selected in up and down directions for this character. The results, from generation 5 onwards, are shown in Figure 4.8, expressed as percentage CVL (when heat-shock is applied) changing over time. The observed pattern of change is not unlike the results of many conventional artificial selection experiments on continuously variable characters, though the response to selection in the down-line was not very marked. The interesting thing about Waddington's (1953) result is that in generation 14 individuals appeared that showed the CVL characteristic without a heat-shock; and further selection in the absence of heat-shock produced stocks in which up to 100% of the flies were cross-veinless when allowed to develop at low temperatures (18 or 25 °C). Thus a character which was initially produced only as a direct response to an abnormal enviroment was, by the end of the experiment, genetically fixed.

Waddington (1953) investigated the genetic basis of CVL and found it to be polygenic; thus CVL is a threshold character in the sense described in chapter

3. What Waddington was presumably doing, as pointed out by Falconer (1981), was selecting for high liability values for CVL, this being possible, in the absence of direct measurement of liability, because of the heat-shock technique effectively unmasking differences in liability to CVL which would otherwise remain hidden in the base population.

Although CVL is a fairly minor morphological character, Waddington (1956) has also demonstrated genetic assimilation of the bithorax phenotype, in which the metathorax is partially converted into a second mesothorax. This phenotype, and variations of it, can be produced by mutations of a series of closely linked loci on chromosome 3 (Lewis, 1963) which will be described in more detail in chapter 10. However, bithorax phenocopies can be produced by exposure of eggs to ether vapour. Using this method of producing bithorax in a normally wild-type stock, Waddington (1956) selected in both directions for percentage bithorax phenocopies. This eventually gave rise to spontaneous production of bithorax without ether treatment. Thus, again, genetic assimilation was successfully accomplished, this time for a rather extreme morphological variant. Genetic analysis showed that the assimilated form was produced by a number of genes, spread throughout the chromosomes, and including a recessive X-linked allele, with maternal action, capable of acting alone to produce bithorax in the offspring of a female homozygous for it (Waddington, 1957b). This, it should be noted, is very different from the normal genetic basis of the bithorax phenotype (see section 10.4).

4.3 EXPERIMENTS INVOLVING DISRUPTIVE SELECTION

Of the three possible kinds of selection on a quantitative character — directional, disruptive and stabilizing — only the first two are capable of causing microevolutionary change of the sort that can conceivably lead to the production of a new species. So far, I have discussed only directional selection and, although the present section is devoted to disruptive selection, the discussion here will be much less extensive. This reflects both the predominance of studies on directional selection and also a view which I share with most, but perhaps not all, evolutionists, that directional selection in different directions in geographically separate populations is the commonest form of selection leading ultimately to speciation in nature. However, it is theoretically possible for microevolutionary divergence and sympatric speciation to occur through disruptive selection operating in a single population (Maynard Smith, 1962) and so it is important not to ignore this kind of selection.

One well-known experiment involving disruptive selection on a morphological character was conducted by Thoday and Gibson (1962) on sternopleural chaeta number in *Drosophila melanogaster*. This experiment involved selecting the 8 males and 8 females with the highest number of bristles, as well as those (again 8 + 8) with the lowest, allowing these 32 flies to mate in mass culture, and repeating the procedure for a total of 12 generations. In each generation, fertilized 'high' and 'low' females were separated so that their

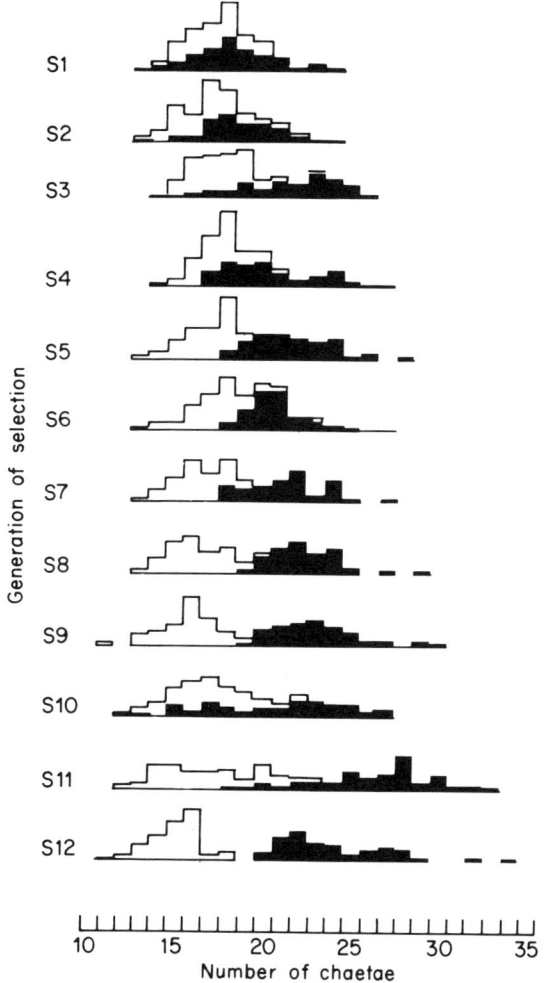

Figure 4.9. Divergence of sternopleural chaeta number in progeny of high females (shaded) and low females during 12 generations of disruptive selection. From Thoday and Gibson (1962).

offspring could be scored separately. The result of this experiment was that the two groups, i.e. the progeny of high and of low females, rapidly diverged in chaeta number to the extent that by generation 12 the two distributions did not overlap (Figure 4.9). Thoday and Gibson (1962) attribute their result to reproductive isolation, either through behavioural isolation or through hybrid inviability.

Because of the potential importance of this experiment in providing evidence that sympatric speciation may be possible, several workers have attempted to repeat it (Scharloo et al., 1967; Chabora, 1968; Barker and Cummins, 1969; Robertson, 1970). The results have been largely negative in

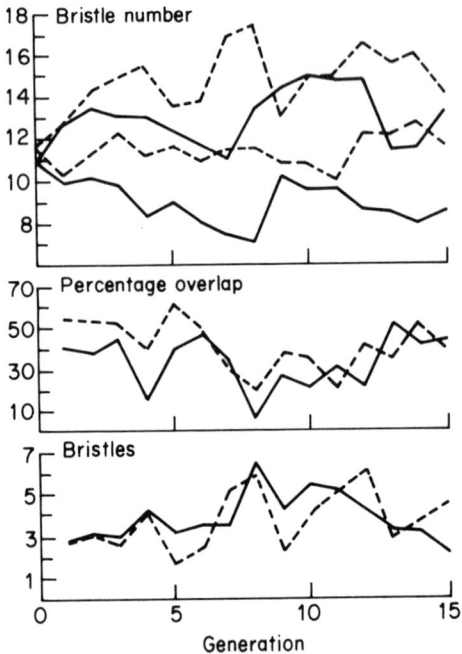

Figure 4.10. Responses to disruptive selection on sternopleural bristle number in two strains of *D. melanogaster*: Kaduna (—), and Pacific (---). The top panel shows mean values for progeny of 'low' and 'high' females. The percentage overlap of the bristle-number distributions is shown in the middle panel, while in the bottom one the difference between mean values of low and high groups is given. From Scharloo *et al.* (1967).

that, although some divergent responses were obtained, clear separation of two groups with respect to bristle number only occurred in one of the later experiments (discussed by Thoday, 1972). The results of Scharloo *et al.* (1967) are shown in Figure 4.10, and it should be noted that these show approximately 30% overlap of high and low bristle distributions at generation 12, compared with the 0% overlap of Thoday and Gibson (1962). Thus although these experiments do indeed show a response to selection, the response is limited and, as Chabora (1968) concludes, disruptive selection on bristle number must only in very rare cases lead to reproductive isolation.

4.4 CHANGES IN GENE FREQUENCY

So far, experiments on the effects of artificial selection have been described largely in phenotypic terms, with little reference to the underlying genetic changes. However, although these genetic changes usually cannot be observed directly in terms of alterations in the gene frequencies at specified loci, the general principles of the connection between gene frequencies and the phenotypic mean are understood, as outlined in chapter 3. It is thus possible to

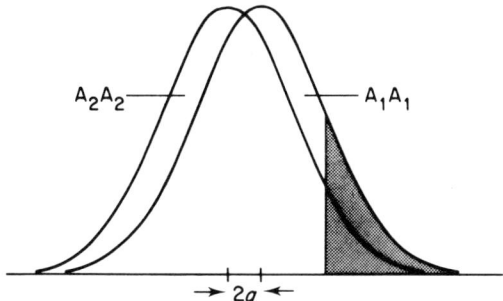

Figure 4.11. Distributions of phenotypic values for random samples of genotypes A_1A_1 and A_2A_2. The intersection of the vertical truncation line with the distribution curves separates those individuals to be used as parents in an up-selection experiment (shaded) from the rest. From Falconer (1981).

discuss, at least in general terms, the genetic changes which are produced in artificial selection experiments and which underlie the systematic trends in phenotypic values that are actually observed.

If we take two random samples of the hypothetical genotypes A_1A_1 and A_2A_2 whose genotypic values were given in Figure 3.3, and consider their distributions of phenotypic values, then we obtain a picture as illustrated in Figure 4.11. The normal distribution of values within each genotypic category results from variation at other loci and from direct environmental effects; and the gap separating the means of the two distributions is $2a$, since the distributions for A_1A_1 and A_2A_2 will be centred on $+a$ and $-a$, respectively (Figure 3.3).

In an artificial selection experiment involving up-selection, all individuals used for breeding have character values above some point of truncation, as discussed in section 4.1. The truncation point is indicated by a solid vertical line in Figure 4.11. It is readily apparent from the intersection of this line with the two curves that the effect of up-selection will be to increase the frequency of allele A_1. If we start with equal numbers of the two genotypes, for example, then more than 50% of the individuals chosen as parents will be A_1A_1. This trend towards increasing frequency of A_1 in up-selection will not be qualitatively altered by the degree of dominance (providing there is no over-dominance) though exactly how much the frequency of A_1 rises with a given point of truncation will depend on dominance. Thus with any starting frequency and any degree of dominance from none to complete (in either direction), one generation of up-selection will increase the frequency of allele A_1 and continued selection will lead to eventual fixation of that allele. Clearly, down-selection will have the opposite effect. So we can see the qualitative relationship between changes in population mean and changes in the gene frequency of a locus contributing to it. The quantitative relationship between the two, and also between selective coefficients at individual loci (s) and selection intensity (i) on the phenotypic character are given by Falconer (1981).

There is still little known about which loci actually cause quantitative variation in morphological characters, or, to put it another way, what the polygenes actually produce. At the molecular level we can recognize genes that produce enzymes, blood proteins, structural proteins, tRNA and rRNA among others. Some examples of these groups of genes have been investigated in various ways, such as by examining the mean phenotypic values of alternative genotypes or by comparing gene frequencies in up- and down-selected lines, and the results have been largely negative. For example, body weight in cattle was not affected by the genotype at any of a number of blood-group loci (Niemann-Sørensen and Robertson, 1961); and selection on the same character in mice did not produce changes of gene frequency at the five polymorphic enzyme loci monitored by Garnett and Falconer (1975). One positive result, arising from a study by Frankham et al. (1978), is that bristle number in *D. melanogaster* appears to be affected by a locus producing rRNA. Whether some polygenes are 'regulatory' genes is uncertain and indeed the mechanisms of gene regulation in eukaryotes are still poorly understood. However, whether or not some polygenes are regulatory in a genetic sense, they are *all* regulatory in a developmental sense in that they manifestly do affect morphogenesis, albeit in a very small way. The distinction between genetic and developmental regulation will be discussed in chapters 9 and 10.

4.5 EXPERIMENTAL SELECTION

In all the experiments discussed so far in this chapter, the predominant form of selection exercised was artificial; that is, individuals were selected because of a morphological criterion established by the experimenter. I say predominant because other forms of selection, outside the control of the experimenter, also occur in these experiments. One example of this, discussed in section 4.2, is the selection that frequently causes up- and down-selected character values to return towards their initial values when artificial selection is suspended. A related phenomenon is that, even while artificial selection proceeds, its effects may be modified by fertility differences within that group selected for breeding. This often takes the form of reduced fertility of those individuals with the most extreme character values. A practice of allowing for such effects by calculating an 'effective selection differential' is sometimes adopted, which involves weighting the various selected individuals according to the number of offspring they produce.

All such forms of selection occurring alongside, or after relaxation of, artificial selection are often referred to by quantitative geneticists as 'natural selection'. This is a perfectly acceptable label to distinguish the additional, uncontrolled forms of selection from the artificial selection effected by the experimenter, but it is unsatisfactory from the viewpoint of a microevolutionist, who wants to distinguish genetic changes occurring in the laboratory from those occurring in nature under 'truly natural' selection. I have chosen to call this 'intermediate' form of selection experimental selection, since it is observed during population experiments. It is in fact often found in experiments not involving artificial selection but set up to investigate some other population process. For example, if two species are maintained in mixed culture for several generations to investigate competitive interactions, the species–frequency trajectory sometimes shows an abrupt change at some point during the experiment, suggesting evolution of one of the species in its interspecific competitive ability. There changes, which have been observed in *Drosophila* (Ayala, 1966, 1969) and in mixed populations of houseflies and blowflies (Pimentel *et al.*, 1965), are often such that a non-genetic explanation cannot be excluded (Arthur, 1982a). In a few experiments of this sort, conclusive evidence of evolutionary change is indeed available (Arthur and Middlecote, 1984), though most such examples do not involve *morphological* change (but see de Souza *et al.*, 1970).

4.6 SUMMARY

Artificial selection experiments have provided considerable information on the genetic and phenotypic effects of selection, much of which is relevant to microevolution in nature as well as in the laboratory. The basic design of artificial selection experiments is outlined, and some of the main generalizations that these experiments have allowed us to make are discussed. Most experiments have confirmed the effectiveness of directional selection in causing microevolutionary change. The results of the much smaller number of experiments that have been conducted on disruptive selection are somewhat heterogeneous; again the effectiveness of selection is confirmed, but an early claim that divergence under disruptive selection proceeded as far as partial reproductive isolation has been disputed. After discussion of the phenotypic changes brought about by selection, the connection between these and changes in gene frequency is briefly outlined, and our lack of knowledge of the molecular identity of polygenes noted. Finally, attention is drawn to the fact that in many population experiments an uncontrolled form of selection is acting either on its own or in addition to the artificial selection promoted by the experimenter.

Chapter 5

Microevolution in Natural Populations

"Each local population is the product of a continuing selection process."

E. Mayr, 1963.

5.1 THE PROBLEM

It is clear from the results of the experiments described in the previous chapter that selection is capable of producing morphological change, and indeed that it can produce, in a handful of generations, variant forms which are outside the range of variation exhibited by their ancestral stocks. These facts are hardly surprising because, given a system with certain attributes, selection is not merely a possibility but a logical necessity. The only attributes required, in fact, are reproduction, variation and inheritance (Maynard Smith, 1972) though mortality is of course also necessary if any form is to be eliminated. All species possess all these four attributes and hence must evolve, however slowly, through natural selection in the wild; and the increased speed of evolutionary change observed in aritificial selection experiments is easily explicable in terms of the closer linking of reproduction with variation than is commonly found in nature.

Although microevolutionary change through artificial selection is easily demonstrated, and although it is known that natural selection of some sort is a logical necessity in the wild as well as in the laboratory, the demonstration of natural selection actually causing morphological evolutionary change in field populations is by no means a simple matter. Indeed, such demonstrations are fraught with several problems which have been inadequately dealt with in many case studies and which have consequently rendered the results inconclusive. We now briefly consider the standard approach taken in studies of microevolution in nature, and examine what these problems are.

It is generally, though not universally, accepted among evolutionary biologists that *most* new species, at least in the animal kingdom, arise out of reproductive isolation of geographically or microgeographically separated populations which have diverged genetically. This is the allopatric model of speciation which, along with alternative models, is discussed in more detail in chapter 6. If allopatric speciation is indeed the usual method by which new species are formed, then it should be possible to find cases in which a species exhibits geographical variation in genetically based characters. A species showing such variation can be viewed as one stage (number 3 in the scheme below) of an evolutionary cycle which, in its simplest form, consists of four stages:

1. A newly formed species consisting of a single population or a group of neighbouring populations.
2. A widespread species consisting of many genetically similar populations capable of hybridization.
3. A widespread species consisting of many genetically different populations still capable of hybridization.
4. A widespread 'superspecies' whose genetically distinct populations or 'semispecies' are developing reproductive isolation from each other and becoming new species in their own right.

The processes responsible for the three transitions in such a series of stages are migration, directional selection (together with drift, drive, etc.), and various isolating mechanisms, respectively. It should be noted that the existence of stage 2 depends on migration initially outpacing local selection, so this stage may be rare in groups of low mobility such as land snails. Viewing the total transition from stage 1 to stage 4 as a cycle is simplistic in that the first two processes, migration and selection, are reversible; however, the cyclical concept is useful providing that the possibility of such reversals is not forgotten. Also, there may be additional complexities in the pattern of evolution in some taxonomic groups, particularly plants, due to the frequent formation of new species by hybridization between 'old' ones (see Grant, 1981).

The time required for a complete evolutionary cycle of this kind must be very long by ecological standards. Even in a case where a species gives rise to two or more daughter species through steps 1–4 without any back-tracking due, for example, to renewed migration diluting or negating the effects of prior selection, the complete cycle may take a million years. The encountering, by one of the diverging populations, of a particularly extreme environment may speed up the process, as suggested by the apparent beginnings of reproductive isolation in heavy-metal tolerant populations of the grass *Agrostis tenuis* (McNeilly and Antonovics, 1968). However, even in such cases of extremely rapid evolution, the complete cycle of events is much too long to be observed, as it takes place, by a single investigator.

Since the overall cycle cannot be followed through in its entirety, the usual approach taken is to attempt to verify the individual processes of migration, selection and reproductive isolation through shorter-term studies, often involving an investigation of spatial variation from which the temporal processes giving rise to the variation are inferred. In fact, relatively little attention has been paid by evolutionary biologists to stages 1 and 2, and their connecting process of migration, and this is understandable since migration is already a well-documented ecological phenomenon (See Elton, 1958). Most evolutionary work has been directed at finding species in stage 3 of our simplified evolutionary cycle and inferring the prior action of directional selection; and at studying the incipient reproductive isolation of stage 4. (These two groups of investigations are covered in this chapter and the next respectively.) Occasionally, directional selection on a morphological character is actually observed happening temporally, within a natural population. An apparent example of this is the reduction in body size in moles, *Talpa europaea*, after a particularly severe winter, observed in a central European population by Stein (1951; see also Mayr, 1963). This reduction in size was interpreted as selection for lowered food intake during a period of food shortage. Although other interpretations of this particular case study are possible, and although studies of this general kind are relatively rare, such studies do provide a link between demonstrations of the effects of selection in laboratory populations (chapter 4) and the inference of prior selection acting in different directions in different natural populations as a cause of some of the patterns of interpopulation variation dealt with in subsequent sections of this chapter.

The study of spatial variation in genetically based characters is the classical province of ecological genetics. The main problem in this area, it might initially seem, is the distinguishing of selection from drift, as well as the identification of both the target character upon which selection is acting (Clarke, 1975) and the selective agent (see, for example, Arthur, 1982b). However, ecological geneticists have, as mentioned in the introduction, traditionally worked with characters whose inheritance is simple and total, as revealed by the appearance of Mendelian ratios in breeding experiments. Thus in studies of the spatial variation of such characters — often pigments, enzymes or chromosomal inversions — the problem of separating evolutionary mechanisms from direct environmental effects is easily overcome. In contrast, intraspecific variation in morphological characters is usually quantitative at the phenotypic level and polygenic in origin. Thus microevolutionary studies of these characters necessarily encounter the problem of separating genetic and environmental variation, and indeed this is often the major problem of such studies. On the other hand, distinguishing selection from drift, the major problem of microevolutionary studies on allozymes, is less problematic with morphological characters, since the value taken by such a character is unlikely to be completely neutral to selection.

Not all intraspecific variation in morphological characters is quantitative and polygenic. For example, sinistral coiling in the freshwater snail *Lymnaea peregra* is caused by a single maternally acting recessive mutation (Boycott *et al.*, 1930). Discussion of discrete variation in morphological characters will, however, be postponed until chapters 10 and 11, where its possible involvement in saltational evolution will be considered. Here, I discuss only the more usual quantitative variation which is the typical material of microevolutionary change. The following sections deal with four examples of interpopulation variation of this kind. As usual, many more studies have been conducted than are described, but it would be fruitless to attempt to give an exhaustive list. The examples chosen — two molluscs, a bird and a plant — illustrate both the problems and the successes of microevolutionary studies on morphological characters.

5.2 SHELL SHAPE IN *LYMNAEA*

Although many species of *Lymnaea* are extremely variable in shell shape, and some, notably *L. peregra* and *L. stagnalis*, have been extensively studied, the cause of interpopulation variation in this morphological characteristic is still not clear, and indeed it seems that similar variants of a species come to predominate, in different places, through fundamentally different mechanisms. The main problem in this case study is the determination of the degree of genetic *versus* environmental control of the character measured, and I will concentrate here on this problem.

Mozley (1935) studied variation in shell shape between several North American populations of the species *L. palustris* and *L. emarginata*. The measures of shape employed were various ratios of one morphological measurement to another, including H/A as described in chapter 2 but with the qualification that Mozley appears to have measured the aperture height (A) parallel to the measurement of overall height (H). Data on variation in H/A values are shown in Tables 5.1 (*L. palustris*) and 5.2 (*L. emarginata*). It can easily be seen that both species show a number of interpopulation differences which would be very highly significant, though Mozley did not in fact carry out any statistical analysis of his data. His conclusions, based on the results for H/A, and for other ratios, which behaved in a similar manner, were: (a) that *L. emarginata* formed 'local races' but *L. palustris* did not, and (b) that as regards the cause of the phenotypic differentiation observed in *L. emarginata*, "It seems reasonable to suppose that this diversity is a matter of multiple factor inheritance." The first of these conclusions is incorrect and the second is highly questionable. The difference between the species in their tendency to show local phenotypic variation is quantitative rather than qualitative, and it is in fact clear that, in *L. palustris*, snails from some sampling sites ("series numbers") are markedly different in mean shape from those of other sites — for example, compare sites 14 and 27.

The difficulty with Mozley's second conclusion is that there is a lack of evidence for his assertion that the phenotypic variation between populations is genetically based. In fact, species of *Lymnaea* have been known for some time (at least since 1914) to show a high degree of phenotypic plasticity in shell shape. Roszkowski (1914) showed through breeding experiments that shell shape variants of *L. peregra* found in deep-lake populations converged immediately (i.e. in the F_1 generation) in laboratory culture, and a similar result was obtained by Arthur (1982c) with *L. stagnalis* (Figure 5.1). Piaget (1929a,b) obtained varied results for *L. stagnalis*, with some variant forms

Table 5.1. Variation in shell shape (H/A) within and between some North American populations of Lymnaea palustris. The numbers in the main body of the table are numbers of shells. From Mozley (1935)

Series No.	1.55–1.59	1.60–1.64	1.65–1.69	1.70–1.74	1.75–1.79	1.80–1.84	1.85–1.89	1.90–1.94	1.95–1.99	2.00–2.04	2.05–2.09	2.10–2.14	2.15–2.19	2.20–2.24	2.25–2.29	2.30–2.34	2.35–2.39	2.40–2.44	2.45–2.49	2.50–2.54	2.55–2.59
1							1	1													
2														2	1			1			
3						1						1									
4				2	1		4	5	1				1								
5								3	5	7	6	2			1						
6											1										
7							1		1												
8									1		1				2						
9			1				1	8	7	10	6	9	4	1	1	1		1			
10						1															
11							1		1												
12						1	2	7	3	5	4	4	1								
13						1	2	2		2	1										
14	1	3	5	15	16	9	10	6	1		1										
15						2	4	4	3	3	1										
16		1	2			2	4	4	3	2	1		1								
17									1	1	2		1		1						
18					1	1	1	4		2											
19												1	1	1	1		1				
20					1	1	2	5	17	34	42	28	39	24	9	1	5	1			
21						1	1		3	1	1										
22								2	6	8	10	11	7	2	2	2					
23				1	1		2	3	2	9	8	1	1	1							
24									1		1										
25				1	2	9	19	44	45	54	38	33	13	4	4			2	1	1		1
26								2		2	1			1							
27							1		6	6	7	15	19	15	16	10	3	1	1		
28								2	2	8	9	9	4		1	1	1				
29									2	5	4	2	4	5		3					
	1	4	8	20	28	45	82	124	150	157	129	117	62	41	22	9	6	4	2		1

appearing to breed true, others not. Thus, in many cases, interpopulation variation in shell shape in *Lymnaea* has little if any genetic basis. The kinds of transplant experiments that have been carried out do not provide any information on the heritability of shell shape within populations, but this is not particularly important in the present context since the question is whether the observed population differentiation is the result of selection in the past, and not whether future evolutionary change is possible. The results of the early transplant experiments were not referred to by Mozley (1935) and he was presumably unaware of them since he proceeds to make the same unwarranted assumptions about the genetic basis of 'local races' in a study of variation in shell shape in *L. stagnalis* (Mozley, 1939).

Hubendick (1951), in an extensive morphological and taxonomic study of *Lymnaea*, reports on variation in shell shape in *L. peregra*, and his data show some parallels to that of Mozley (1935). Hubendick found (a) a fairly narrow range of variation in shape within populations, (b) considerable variation between populations from neighbouring localities, but (c) no overall geog-

Table 5.2. Variation in shell shape (H/A) within and between some North American populations of *Lymnaea emarginata*. From Mozley (1935)

Series No.	1.25–1.29	1.30–1.34	1.35–1.39	1.40–1.44	1.45–1.49	1.50–1.54	1.55–1.59	1.60–1.64	1.65–1.69	1.70–1.74	1.75–1.79	1.80–1.84	1.85–1.89	1.90–1.94	1.95–1.99	2.00–2.04	2.05–2.09	2.10–2.14	2.15–2.19
1								4	4	4	2	2							
2							2	4	1		1								
3								1	4	6	8	7	5						
4										2	1	2		1					
5											2	2	2	5	1				
6		2	4	12	16	4	1				1								
7			1	1			2		2										
8		3	3	4	1	1													
9							2												
10									1	2	4								
11						3		1	1				1						
12									1			1	1						
13					2	2	9	16	5	1									
14		1	2	1	1														
15					1		2		2	3	4	4	1	3					
16															2				1
17								4	7	16	19	28	13	5	8				
18				1		3	1												
19									1	1	1	2							
20										4			1	2	1	1	1		
21					1	6	10	19	17	21	10	4	1		1				
22									1			1	1		1			2	
23		1	2						1										
24					1	2	1	3	3	2	1								
25		2				1													
26	2		1		1														
	2	4	10	10	24	36	41	55	53	65	55	47	23	10	13		1	2	1

raphical pattern. Hubendick suggests that the low within-population variation is a result of populations being established through small founding numbers and self-fertilization. This may well be true, though the extent of self-fertilization in natural populations of *Lymnaea* is still uncertain. Hubendick appears to attribute the variation in shape between local populations to underlying genetic differences, though he does acknowledge, as have most malacologists, direct environmental influences on shell thickness. Again, no evidence is presented on the inheritance or otherwise of variation in shell shape, so the supposed genetic basis of the observed population differentiation is questionable. Transplant experiments carried out after Mozley's (1935) study but before Hubendick's (1951) discussion of variation in *L. peregra* showed that, as in Piaget's (1929a,b) investigations of *L. stagnalis*, variation in shape between populations of *L. peregra* sometimes disappeared on laboratory culturing, but sometimes did not (Boycott, 1938).

While it is desirable that the apparent genetic basis of some of Boycott's (1938) and Piaget's (1929a,b) variants be confirmed using improved experimental designs (Arthur, 1982c), we can tentatively conclude that these results are correct, and it appears, then, that the causation of shell-shape variation between populations is rather heterogeneous in *Lymnaea*. However,

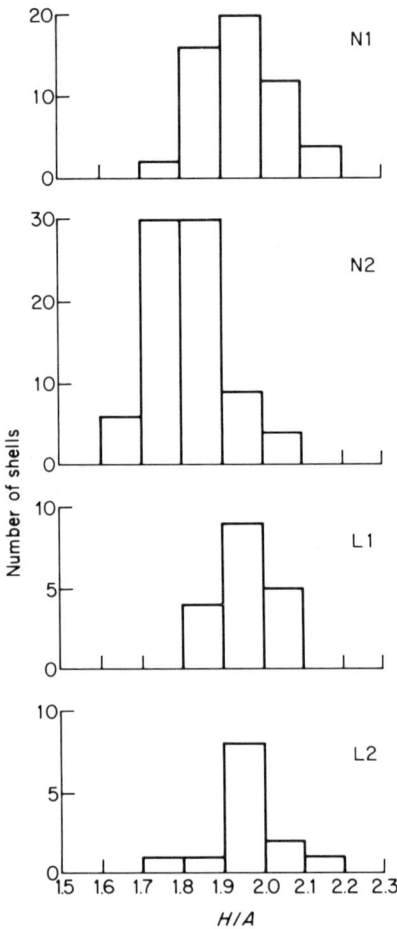

Figure 5.1. Shell shape distributions for two natural populations (N1, N2) of *Lymnaea stagnalis*, and for two laboratory cultures (L1, L2) established with juveniles from populations N1 and N2, respectively. The highly significant interpopulation difference in shape can be seen to disappear in laboratory culture. From Arthur (1982c).

Waddington (1975) has proposed a potentially unifying explanation for this variation; namely, that it was all originally environmental but in some populations has now become genetically assimilated. This proposal was made specifically in relation to Piaget's (1929a,b) experiments with *L. stagnalis* (though Piaget (1979) does not entirely accept it) but it can of course be extended to the other species as well. The hypothesis of genetic assimilation of shell shape has never been tested, even through laboratory experiments designed to see if it is possible to reproduce it artificially. However, the lack of such experiments is understandable since the generation time of *Lymnaea* (about a year) is probably too long to allow its successful use in this way. Thus we lack conclusive evidence of this potentially unifying theory of variation in shell shape in *Lymnaea*, and it must be said, in conclusion, that this group of microevolutionary studies has served better to illustrate the problems involved in attempting to demonstrate microevolutionary patterns in nature than to elucidate these patterns.

5.3 SHELL SIZE IN *CEPAEA*

Shell size cannot be easily used for evolutionary studies in *Lymnaea* because there is no definite stage when growth ceases, and no terminal lip is laid down. Thus the problem of separating the effects of different population age structures would be added to the difficulties discussed above. (This problem is not quite removed by considering shape: see Arthur, 1982c.) In *Cepaea*, however, a distinct adult lip is formed, after which no further shell growth occurs, except sometimes after the shell has been damaged. Thus adult shell size can be used in studies of variation between populations without the necessity of having to take into account variations in age structure. Many comparative studies of shell size have indeed been carried out using *Cepaea*. Most have dealt with *C. nemoralis*, to which species this discussion will be largely restricted, and these studies have involved a variety of populations in Ireland, Britain and mainland Europe. The spatial variation observed can be roughly divided into two categories, local and geographical, and I will deal with these in turn. The latter category will be used here to include only situations where a morphologically identifiable race inhabits a fairly extensive geographical region.

Local variation

Significant differences in mean adult shell size between populations are common, and indeed can be found, in *C. nemoralis*, between samples taken less than 50 m apart. Although we cannot exclude the possibility that some of these differences have a large genetic component, it is clear from the work of Williamson *et al.* (1976) that in many cases variation in shell size relates to population density in a way that is most easily explicable in non-genetic terms. The relationship between size and density is negative (Figure 5.2) and this parallels the results for a variety of other species in diverse taxa (see section 1.4). The usual interpretation of these negative relationships is that intraspecific competition in high-density populations leads, through resource limitation and/or some form of interference, to smaller individuals, though the offspring of these individuals will, if grown with plentiful resources, revert to larger sizes. It may be, as mentioned in section 1.4, that the ability to reach adulthood when resources are limiting by becoming reproductively mature at a smaller size is itself an adaptation, though the small size itself is not.

Some features of local variation in shell size are not readily explicable in terms of a response to population density. One of these is the parallel variation in size sometimes found between *C. nemoralis* and its congener *C. hortensis*, an

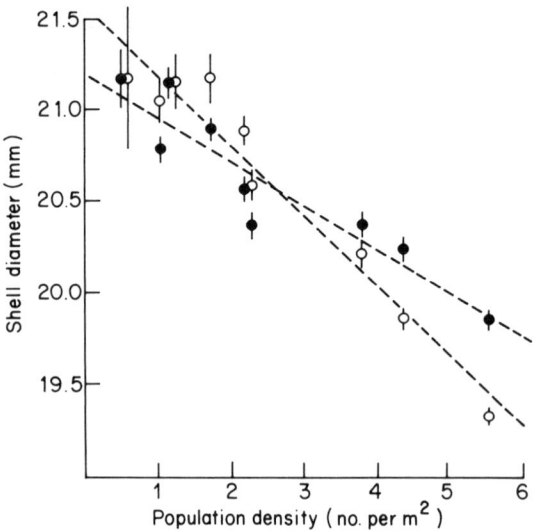

Figure 5.2. Relationship betwen mean adult shell size and population density in *Cepaea nemoralis*. (○), samples collected in May; (●), samples collected in August. Bars indicate ± 1 standard error. From Williamson *et al.* (1976).

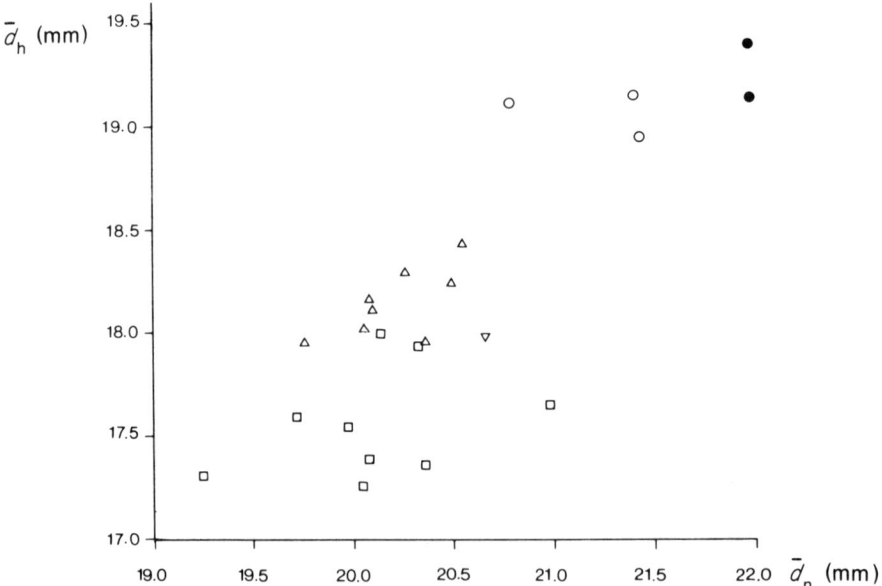

Figure 5.3. Parallel variation in mean adult shell diameter ($\bar{d}$) in *Cepaea nemoralis* (n) and *C. hortensis* (h). Each symbol represents a particular study area, and these were located in the following British counties or regions: (□), Northumberland; (△), Grampian; (▽), Somerset; (○), Derbyshire; (●), Gwynedd.

example of which is shown in Figure 5.3. This particular instance of co-variation includes both variation within survey areas of less than 2 km maximum span and also between a number of such areas; but all these areas are within Great Britain and so within the main shell-size race of *C. nemoralis*, so geographical variation (see next section) is not included. It is difficult to see how the sort of co-variation illustrated in Figure 5.3 could be caused by an effect of density on size if each species responds to its own population density, since the densities themselves appear not to co-vary. However, although many studies have been conducted on size–density relationships in a variety of species, little attention has been paid (with a few exceptions: see Atkinson, 1979) to the question of what component of density will be effective in mixed-species populations. This issue requires much further work for its resolution; in the meantime, it is only possible to say that co-variation in size of *C. nemoralis* and *C. hortensis* is potentially explicable as an effect of joint density, as a common phenotypic response to some other direct environmental influence, or as a form of coevolution.

Geographical variation

In most of the populations of *C. nemoralis* inhabiting eastern and central Ireland, Britain and northern and central Europe, the mean adult shell diameter varies between about 19 mm and 22 mm. However, in western Ireland, the Pyrenees and Cantabria, many localities are inhabited by a large-shelled race with mean adult diameter of around 24 mm. The existence of this race has long been known, and it has been studied by many workers including Comfort (1944), Stelfox (1945), and Cook and Peake (1960). Cook (1967) has examined the persistence of the difference in size between the large-and small-shelled races of *C. nemoralis* and has shown that there is negligible change in mean shell size in F_1 and F_2 generations in laboratory culture (Table 5.3).

Table 5.3. Constancy of shell size in large- and small-shelled races of *Cepaea nemoralis* over two generations of laboratory culture. The small shelled snails labelled S 45 originated from Co. Down on the eastern side of Ireland. Loughs Muskry and Diheen are in Co. Tipperary, on the western side of the country. From Cook (1967)

	Parents		F_1		F_2	
	N	Mean	N	Mean	N	Mean
Galtee, Lough Muskry	5	24.2 ± 0.5	10	24.4 ± 1.1	6	24.4 ± 1.2
Galtee, Lough Diheen	6	25.0 ± 1.4	1	25.3	–	–
S_{45}	2	21.9 ± 0.01	7	21.5 ± 0.7	9	21.2 ± 1.0

Given that the large-shelled race is an evolutionary rather than ecophenotypic phenomenon, is it an adaptation to the environments in which it occurs, and if so, what is the selective agent causing the interracial difference? The answers to these questions are not clear, partly because of the lack of geographical or ecological consistency in the distribution of the races. The main European and British range of *C. nemoralis* contains many areas as similar to the Pyrenees or western Ireland as each of these is to the other, yet large-shelled forms are not found. Admittedly, both areas inhabited by the large-shelled race are peripheral in relation to the overall range of *C. nemoralis*, but that fact in itself is not very informative.

Selection experiments aimed at elucidating possible selective agents acting on shell size have been performed, both (a) using large and small races (Cook and O'Donald, 1971) and (b) using snails of variable size but all of the same (small) race (Bantock and Bayley, 1973; Bantock et al., 1975). These two sets of experiments also differed in that they examined climatic and visual selection respectively. These are two of the most obvious forms of selection that might affect size both in *Cepaea* and in other land snails. Hotter climates may select for increased size since the lower surface-area/volume ratio of larger variants will reduce heat absorption and desiccation; and predators may select for decreased size by choosing the largest available items of food.

In Cook and O'Donald's (1971) experiment, samples were kept over winter in a series of boxes in an unheated room. Survival was examined in four samples of small-shelled snails from British localities and a fifth sample from Cledes, Spain, which consisted of large-shelled snails. The results showed, first, that the survival rate of the Cledes sample was within the range of values for the British samples, and secondly, that within some samples survivors were significantly larger than non-survivors. This second result is difficult to interpret, partly because only two of the five samples showed a significant difference, partly because of a lack of sampling information, as noted by Clarke et al. (1978), and partly because it is not clear why in fact the putative selection took place. Cook and O'Donald introduce their experiment by noting that the snails were kept in a "cool room" and end up suggesting that the selective factor may have been "the relatively high temperature". Thus it is difficult to draw any firm conclusion from this study.

The results of experiments on visual selection by predators (Bantock and Bayley, 1973; Bantock et al., 1975) are also inconclusive. These authors claim to find that predation by birds in artificial mixed-species colonies of *C. nemoralis* and *C. hortensis* results in selection for larger-shelled *C. nemoralis* and smaller-shelled *C. hortensis*. The latter species is smaller, on average, than *C. nemoralis* but the size distributions overlap considerably. Thus the purported result could be explained in terms of the avian predators predominantly eating snails from the middle of the combined size range. However, there are at least two problems in accepting the putative evolutionary changes as real: first, the size trends are not significant unless certain items of data are omitted; and second, the snails involved came from several colonies which may have differed in behaviour and this could produce spurious evidence of selection on size (Clarke et al., 1978).

Wolda (1963, 1967; see also Wolda and Kreulen, 1973) has shown that there is an association between size and several components of fertility in *C.*

nemoralis, an association which has also been found in many other species. This serves to stress that the main problem in understanding genetically based intraspecific variation in size lies in understanding the limitation of size rather than its evolutionary increase. One possibility is that further increase beyond a given size is inhibited because of increasing predation which eventually balances the gain in fecundity. If this is so, then populations in relatively predator-free environments should consist of larger individuals, on average, than others. Some evidence for this phenomenon has been presented by Bantock and Bayley (1973) who show that *C. nemoralis* snails on the island of Steep Holm in the Bristol Channel are larger than those on Flatholm, which correlates with a difference in predation of *Cepaea* by vertebrates on the two islands. Whether this sort of effect is important in the case of geographical differences in *Cepaea* is uncertain and to investigate this considerable work would need to be done on the level of predation in populations of the large-shelled race compared to the small-shelled. Also, it must be re-emphasized that local variation in shell size is often non-genetic, so even the causative inference made about the populations on islands in the Bristol Channel may not be correct.

As regards the genetic basis of geographical variation in shell size in *C. nemoralis*, Cook and Cain (1980) propose that the phenotypic difference between the two main races is caused by genetic differences at only a few loci. Several lines of indirect evidence support this proposition, including the fact that heritability estimates for shell size in *C. nemoralis* tend to be low when all parents are from within a single population, and high when they are not. This would be explicable if each population was fixed for a different allele at each of a few important loci. However, at this stage there is no detailed information on the genetic basis of the interracial difference in shell size.

5.4 BEAK SIZE IN *GEOSPIZA*

An enormous amount of evolutionary work has been done on this and other genera of Darwin's finches on the Galapagos Islands. A general account of their evolutionary biology and references to much of the earlier work can be found in Lack (1968). Here, I want to concentrate on variation in beak size in the ground finches *Geospiza magnirostris*, *G. fortis* and *G. fuliginosa*, and in particular *G. fortis*, since information has recently been obtained on the heritability of some morphological characters, including beak size, in this species (Boag and Grant, 1978).

The three species of ground finch are, in common with others in the genus *Geospiza*, found on a variety of islands and small islets within the Galapagos archipelago (Figure 5.4). Both mean beak size (Lack, 1968) and the variance of beak size (Grant *et al.*, 1976) vary between islands. Darwin's finches are unusual in that differences between species and between islands forms within species largely take the form of beak differences (Snodgrass, 1902). This contrasts with, for example, British finches which are less variable in beak

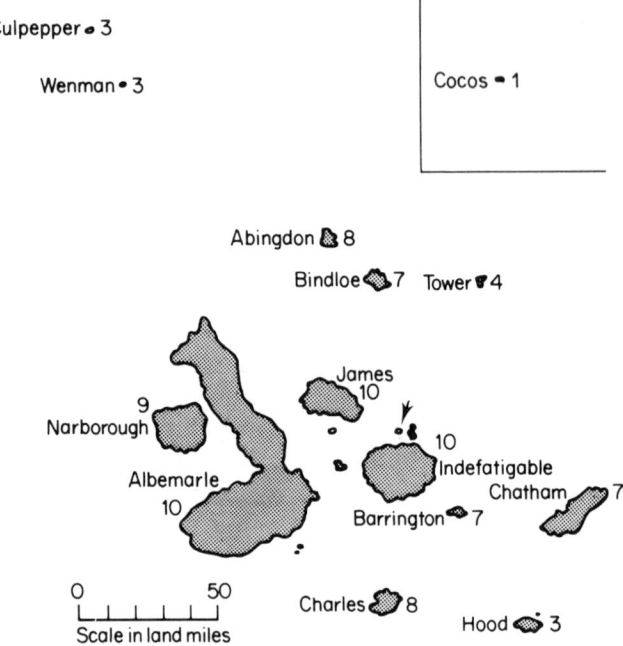

Figure 5.4. The Galapagos Islands, with the number of species of Darwin's finches shown for each of the main islands and some of the smaller ones. The arrow indicates Daphne. From Lack (1968).

between species but much more variable in male plumage. It is generally thought that the different species of *Geospiza* — all of which are restricted to the Galapagos — arose from migration of one original species between islands, followed by genetic differentiation of island forms, reproductive isolation and re-migration of newly formed species. Thus the pattern appears to conform closely to the 'evolutionary cycle' described in section 5.1, though the whole process has taken place, in this instance, in a particularly small geographical area. The main problems in confirming this view of the evolution of the finches, at least with respect to the pre-speciation component, are (a) establishing a genetic basis for the inter-island differences, (b) ascertaining whether the variation between islands is adaptive, and if it is, then (c) identifying the selective agent(s).

As regards the problem of inheritance, evidence was lacking until recently and is still very sparse. However, compared with biologists working on some other taxa, those working on morphological variation in Darwin's finches have been more aware of this problem. For example, Lack (1968) states: "It is now generally agreed that in animals the differences between geographical races of the same species are hereditary, though in Darwin's finches, as in most other birds, experimental proof of this statement is as yet lacking". Some such evidence has now been provided by Boag and Grant (1978) who give

Table 5.4. Heritabilities of several morphological characters in *Geospiza fortis* determined by four different methods. From Boag and Grant (1978)

	Midparent–offspring regression	Male–offspring regression	Female–offspring regression	Full-sibs correlation	Averages
Sample size	26	41	37	18	–
Weight	0.84 ± 0.23	0.74 ± 0.26	1.10 ± 0.42	0.80 ± 0.38	0.87
Wing length	0.53 ± 0.37	0.38 ± 0.36	0.97 ± 0.49	0.65 ± 0.41	0.63
Tarsus length	0.43 ± 0.22	0.46 ± 0.31	0.38 ± 0.30	0.27 ± 0.44	0.38
Bill length	0.62 ± 0.22	0.75 ± 0.30	1.17 ± 0.32	1.34 ± 0.25	0.97
Bill depth	0.82 ± 0.15	0.94 ± 0.23	0.95 ± 0.26	1.41 ± 0.23	1.03
Bill width	0.95 ± 0.19	1.06 ± 0.27	1.09 ± 0.42	0.99 ± 0.35	1.02
Bill length at depth of 4 mm	0.48 ± 0.32	0.61 ± 0.26	0.15 ± 0.32	0.78 ± 0.39	0.50
Averages	0.67	0.71	0.83	0.89	0.77

heritability estimates for a number of characters, including beak length, depth and width, for the *G. fortis* population on Daphne Major island (Table 5.4). As can be seen, these estimates are all high (averages over different methods for the three characters are all in excess of 0.95). However, both the appearance of some values above 1.0 and the rather high heritability estimates for body weight suggest that all the figures given may be over-estimates. It is worth noting, in this context, that correlations between relatives may be inflated due to non-genetic factors. For example, birds with larger than average beaks may be able to acquire more food and hence their offspring may be larger, and have larger beaks, on average, than those of parents with smaller beaks. Also, even assuming that the true values of the estimated heritabilities are indeed high, this information, relating to a single island, does not tell us much about the heritability of inter-island variation.

With this reservation in mind, it is of interest to examine just how large some of the differences between island variants are. Figure 5.5 shows variation in beak size in the ground finches, and illustrates in particular the differences between allopatric and sympatric populations of *G. fortis* and *G. fuliginosa*. Inter-island differences in other species of *Geospiza* are equally marked and in some cases, for example in *G. difficilis* and *G. conirostris*, the island variants have been given subspecific names (see Lack, 1968).

As regards the adaptive nature of the differences between island forms, early doubt on this has largely disappeared, as Lack comments, and most current discussion is centred on identifying the main selective agent. The central problem here, as stressed by Abbott (1980), is whether competition with other species is predominantly responsible for the evolutionary changes in beak sizes. Darwin himself placed considerable weight on interspecific competition, but recent views tend to stress food availability to a greater extent (Abbott *et al.*, 1977; Abbott, 1980). However, the distinction between these two views is not clear-cut since of course the kind and number of competitors will influence the availability of food. Thus the question of what is the main selective agent in the case of beak size is more restricted than in the case of many other characters because the ecological function of the character measured is obvious. There is

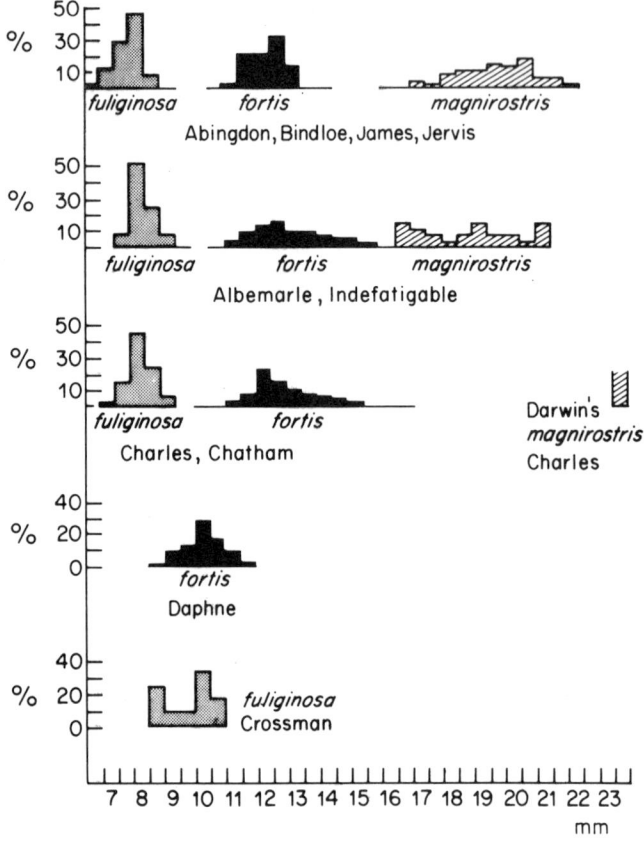

Figure 5.5. Inter-island variation in beak depth in three species of *Geospiza*, showing the difference between allopatric and sympatric populations in *G. fortis* and *G. fuliginosa*. From Lack (1968).

little doubt that the distribution of food items is important and the question then becomes what determines this — competitors or other factors.

If competition with congeneric species is indeed occurring, then it should be possible to demonstrate this. However, conclusive evidence for interspecific competition requires a replicated reciprocal explant experiment to establish mutual inhibition of population size. Such field experiments are difficult to perform successfully with any taxon, and are almost impossible with large, mobile vertebrates. Thus the evidence for competition among Darwin's finches is circumstantial, and largely takes the form of observations of dietary overlap. Some early data on this were provided by Snodgrass (1902), and more recently (Abbott *et al.*, 1977) the relationship between morphological similarity and dietary overlap among six geospizid species has been quantified (see Figure 5.6). This relationship supports the claim that increased difference in beak size in sympatric populations, as illustrated in Figure 5.5, represents character displacement, that is, selection against individuals in each of a pair of

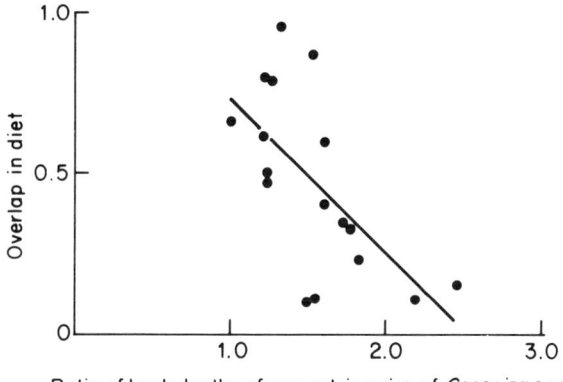

Figure 5.6. Negative relationship between difference in beak depth and overlap in diet in a number of pairs of sympatric species of *Geospiza*. From Abbott *et al.* (1977).

species most like the alternative species, though there are difficulties in reaching a firm conclusion on this (see Grant, 1972; Arthur, 1982a). Character displacement will be discussed in a more general context in chapter 7, where the reason for treating it separately from other microevolutionary studies will also be discussed. It may simply be concluded here that variation in beak size between islands is likely to be due to differences in the availability of various food items, and that whether these differences are determined largely by other species of *Geospiza* remains uncertain.

5.5 ECOTYPES IN *POTENTILLA*

The order of examples in this chapter reflects increasing success in establishing population differentiation as being the result of natural selection. In *Lymnaea*, variation in shell shape is often a direct result of differing environments; in *Cepaea*, geographical variation in shell size appears to be highly heritable but the selective agent is obscure; in *Geospiza*, there is evidence of a high heritability of beak size, albeit within a population, and the range of likely selective agents is fairly restricted. However, plants, and in particular a number of species of herbaceous and semiwoody angiosperms, provide some of the best evidence on the effectiveness of selection on quantitative morphological variation because the variants produced are often confined to particular habitats or climatic zones. This differs from the situation in *Cepaea*, where the large-shelled race is found in rather disparate geographical areas and is not confined to a particular habitat type; and also from *Geospiza*, where the variants found on any two islands are never identical, and unique explanations are often required.

Investigation of genetic variation between populations of plant species began with the classic work of Turesson (1922a,b, 1923, 1925, and many subsequent papers) who showed that, in a variety of species, genetically-based and habitat-related variants are common, despite the widespread ecophenotypic variation shown by many plants. Turesson regarded these variants as being distinct from each other, rather than intergrading, and they are frequently referred to as ecotypes. In some species, however, continuous variation (clines) is also common, and the series of variants forming a cline are collectively referred to as an ecocline if the morphological variation is associated with an ecological factor (Gregor, 1939). Despite early controversy on the relative importance of ecotypes and ecoclines (reviewed by Heslop-Harrison, 1964), it is now clear that both occur, and indeed that the distinction between them is not a rigid one. The species examined in this section exhibits variation which conforms more closely, but not precisely, to the discrete, ecotypic pattern.

In a series of publications, Clausen and co-workers have documented in considerable detail morphological variation between populations of the semiwoody perennial *Potentilla glandulosa*, a member of the rose family (Clausen et al., 1940; Clausen and Hiesey 1958, 1960). This species has a restricted range, being found only in western North America, where its populations are exposed to a wide span of altitudes from sea level to more than 3000 m. The morphology of the plants varies with climate, four major variants being recognized:

1. The coastal *P. g. typica*.
2. *P. g. reflexa*, a form growing in the warm foothills of the Sierra Nevada.
3. *P. g. hanseni*, occurring at mid and high latitudes.
4. The highest-living variant, *P. g. nevadensis*.

The growth forms on these four variants, together with their locations on an east–west transect through California, are shown in Figure 5.7. Part of the transect is duplicated in the figure because of overlap in the ranges of *reflexa* and *hanseni*. It should be noted, though, that within the overlap zone the two forms occupy edaphically distinct areas (Clausen and Hiesey, 1958). The four named variants are referred to both as subspecies and as ecotypes, though all are to some extent heterogeneous and are made up of a series of intergrading 'lesser' variants. Several additional subspecies of *P. glandulosa* have also been recognized, but the main work of Clausen and colleagues was conducted on the four forms described above.

The plants illustrated in Figure 5.7, which are representative of the four ecotypes, were grown in the same garden at Stanford, rather than being taken directly from their natural habitats. Thus the variation appears to be inherited. However, Clausen and Hiesey (1958) have gone much further than simply showing that the ecotypic differences have a genetic basis. They also performed crosses between some pairs of ecotypes to determine the kinds of genetic differences involved, and they examined all four ecotypes cytologically to establish their chromosome number. In fact, all individuals of all ecotypes were diploid with $2n = 14$, so it is clear that ecotypic differentiation in this species is not underlain by changes in ploidy. The crosses between ecotypes showed that

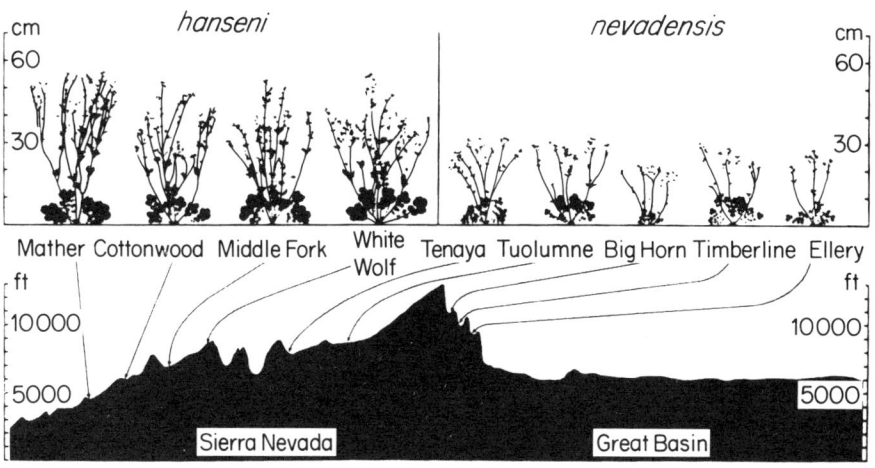

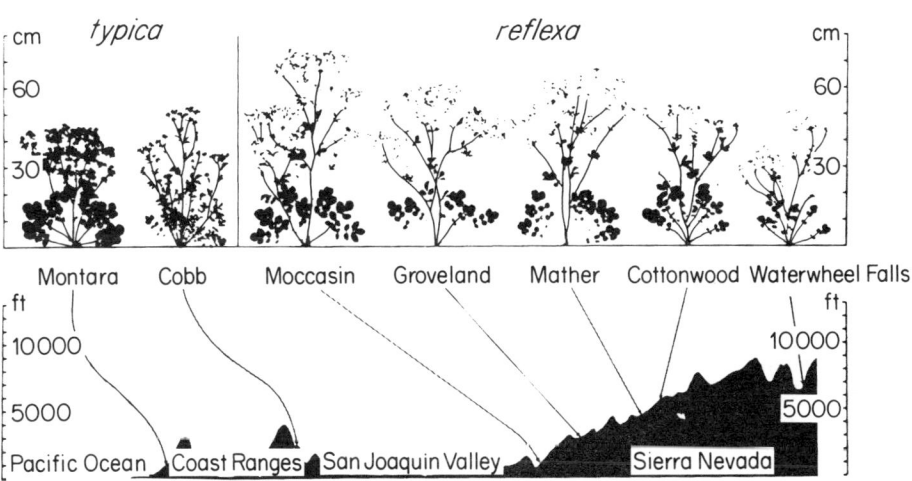

Figure 5.7. East–West transect through California showing the distribution of four ecotypes of *Potentilla glandulosa* and their approximate growth forms. From Clausen and Hiesey (1958).

most of the genes responsible for the phenotypic differentiation were of the "multiple kind which singly have relatively minor effects" (Clausen and Hiesey, 1958). These crosses also provided further confirmation of the inherited nature of the variation which, if only the transplant type of experiment had been conducted, could still conceivably have been caused by a non-genetic agent transmissible between generations, such as a virus.

Table 5.5. Percentage survival in three morphologically defined groups of *Potentilla glandulosa*. From Clausen and Hiesey (1960)

Station, and length of experiment	Classes of index values		
	20–29 Most subalpine-like	30–39 Intermediate	40–50 Most foothill-like
At Timberline 3000 m, 9 years	75.6	34.3	13.5
At Mather 1400 m, 9 years	49.5	71.5	76.5
At Stanford 25 m, 5 years	17.8	77.0	70.8
Number of clones:	90	330	89

Index values of parents: subalpine, 19; foothill, 54; F_1, 34.

Having established the genetic basis of the morphological differences between ecotypes, the next question to be answered is whether these differences are the result of directional selection for characteristics appropriate to each environment, or whether they are the result of stochastic processes operating, perhaps, on small founding populations. In fact, it is clear that the morphological differences are adaptive, as several selection experiments have been carried out with positive results. One of these, described by Clausen and Hiesey (1960), was more subtle than most selection experiments since, instead of examining survival of two types after reciprocal transplants, these authors proceeded as follows. First they measured a series of characters in parental plants of *reflexa* and *hanseni* (which are referred to in the later work as foothill and subalpine variants) and produced a compound index which took the values 19 (subalpine) and 54 (foothill). Next, they crossed foothill and subalpine plants to give F_1, then F_2, hybrids. The F_2 were scored for the same characters as the parental plants, and all fell within the range 22–50 of the compound index.

The survival of F_2 plants with index values of 22–29, 30–39 and 40–50 were examined at high, mid and low altitudes and these data are given in Table 5.5. One cautionary note is necessary in relation to the interpretation of the survival rates shown. A plant is considered to be a non-survivor if it "died once or twice" (Clausen and Hiesey, 1960). This illustrates that the death of the visible part of a rooted plant is not as clear-cut an evolutionary event as the death of an animal. Also, it should be noted that the figures for low-altitude survival are not directly comparable with the others since in this case survival over a five-year, as opposed to a nine-year, period was determined. However, the data clearly show that plants with index values close to those of the foothill parent did better than plants resembling the subalpine parent at low and mid altitudes, while at 3000 m the differential in survival was reversed. Thus selection experiments demonstrate that the morphological characteristics of each ecotype improve adaptation to its own climate. In species other than *P. glandulosa*, edaphic rather than climatic ecotypes have been described, one example of a species differentiating in this way being *Hieracium umbellatum* (Turesson, 1922b).

5.6 SUMMARY

Most species exhibit considerable variation between populations in a variety of morphological characters, and examples are given of four groups in which this is so. At the phenotypic level this variation is easy to measure and consequently patterns of local or geographical variation may be, and have been, described. However, despite the liberal use of subspecific names in early studies and the persistent tendency to refer to geographical races on the basis of phenotypic studies alone, the morphological differences between populations sometimes have a negligible genetic component. The problem is intensified by the fact that in many species, including *Lymnaea stagnalis* and *Cepaea nemoralis*, the same sorts of variant are in some places apparently produced by selection yet in others produced as a direct phenotypic response to environmental factors. This makes it imperative that, at least in certain taxonomic groups, authors claiming evolutionary differentiation must demonstrate the inheritance of their variants *from the particular localities with which they are dealing*. As regards the selective agent causing inherited interpopulation differences to evolve, there are some instances where the nature of the character studied and the ecology of the species concerned immediately suggest a particular agent. An example is provided by the evolution of beak size in Darwin's finches, which appears to take place largely in response to the availability of different types of food, though the role of competitors of other species in influencing that availability is still disputed. While quantitative ecological observations can strengthen the case for a particular selective agent, a more conclusive way of demonstrating selection is to conduct selection experiments in which differential survival is examined. The successful use of such experiments, together with evidence for inheritance of the phenotypic variation concerned, make studies on ecotypic variation in plants, for example in *Potentilla*, among the clearest demonstrations of inherited and adaptive morphological variation between natural populations. Finally, the intraspecific phenotypic variation observed in all four case studies discussed here, and in many others, is similar to, and sometimes no less marked than, the variation in the same characters that distinguishes closely related species in the taxa concerned.

Chapter 6

Speciation and Interspecific Differences

"This sort of differentiation draws mainly on the store of pre-existing variability in the population."

G. G. Simpson, 1944.

6.1 DIVERGENCE OF SPECIES AND DIVERGENCE OF OPINIONS

Although alternative interpretations are possible for some of the individual case studies so far described, the main theme up to now, that natural selection acting on polygenic variation is capable of causing morphological divergence of populations, is hardly controversial. Indeed, even the most severe critics of the evolutionary importance of selection on quantitative variation admit that such selection is effective in causing microevolutionary change at the intraspecific level (Goldschmidt, 1940). This chapter marks the beginning of the more controversial part of the book since it covers the level of evolutionary divergence — separation of genetically distinct groups into different species — at which opinions too begin to diverge. The conventional view is that reproductive isolation usually occurs simply as a by-product of gradual genetic divergence under selection in geographically separate populations (see Mayr, 1963), possibly reinforced by the Wallace effect (Murray, 1972) in which selection acts in neo-sympatry against individuals within each partially isolated group which tend to mate with members of the other. Alternative views on speciation abound, among them the views that reproductive isolation can occur in parapatry (Clarke *et al.*, 1978) or in sympatry (Thoday and Gibson, 1962), that speciation is commonly caused by the evolution of chromosomal differences (White, 1978), and that the pattern of morphological divergence and speciation observed in geological time is inconsistent with the idea of speciation as a by-product of the gradual accumulation of slight genetic differences (Eldredge and Gould, 1972).

The approach I take here is to discuss neontological work on speciation in the present chapter and to defer consideration of the recent challenge to conventional views from palaeontologists until chapter 8. The palaeontological treatment of speciation is necessarily a morphological one, so chapter 8 conforms to the general aim of the book in providing an integrated view of morphological evolution. The present chapter involves some digression into non-morphological topics because it is clearly impossible to treat the origin of biological species from an entirely morphological viewpoint. Thus some information on genic, chromosomal and behavioural aspects of speciation is included.

It may help to remark, at the start of the more controversial part of the book, that most evolutionary debates concern the relative importance of alternative mechanisms and processes rather than whether certain mechanisms are possible. This does not, of course, mean that the issues are unimportant. For instance, if sympatric speciation has occurred in one out of every million speciation events, this is a very different situation to its occurrence in one out of every two such events. But it is always worth scrutinizing, as alternative views are given, whether the argument seems to be about a process being possible, common, predominant or exclusive. The confusion of these issues has led to many an unnecessarily heated evolutionary debate.

6.2 THE MULTITUDE OF SPECIATION MODELS

Before proceeding to examine morphological differences between species and their connection with genic and chromosomal differences, it is worth briefly discussing the alternative models of speciation that are available. These models differ in three main respects: geography, genetics, and the mechanism by which reproductive isolation occurs.

As regards geography, the question is whether reproductive isolation always or very nearly always originates in allopatry — a view championed by Mayr (1963) and supported by most vertebrate zoologists; or whether isolation may frequently occur in parapatry and/or sympatry. Some recent authors clearly support Mayr's view. For example, Lewontin (1974) states that, "If there is any element of the theory of speciation that is likely to be generally true, it is that geographical isolation and the severe restriction of genetic exchange between populations is the first, necessary step in speciation". However, the predominance of allopatric speciation has been questioned by geneticists working on some groups of invertebrates and plants, and several recent studies in these groups have indicated that parapatric speciation may be more widespread than was previously thought. Clarke *et al.* (1978), reviewing genetic variation in the pulmonate molluscs, argue that "the study of pulmonates has generated a case for speciation without geographical barriers".

One of the groups of molluscs in which parapatric speciation may have occurred is the genus *Partula*, inhabiting the Pacific island of Moorea (Clarke and Murray, 1969). Here, two species complexes (which will be discussed in more detail later) each contain two closely related forms which hybridize in some localities but behave as distinct species in others, with the two types of colony separated by as little as a few hundred metres. Since *Partula* is extremely limited in mobility it is always possible to argue that a form of microallopatric speciation is taking place. That is, in the first stage of the evolution of reproductive isolation the diverging populations are separated by a zone which, though not a geographical barrier to an observer, is very large in terms of the distances the snails actually move.

There are other cases of incipient reproductive isolation of populations in close proximity to each other where an explanation in terms of microallopatry is impossible. One of these concerns populations of the grass *Agrostis tenuis* that are tolerant to heavy metals. Some tolerant populations of this species are now flowering earlier than the non-tolerant forms (McNeilly and Antonovics, 1968) and this constitutes a first step towards reproductive isolation. Yet the two forms occupy adjacent polluted and non-polluted sites with no separation between them and are easily within the range in which mutual pollination can occur. It appears that in this case the integrity of the diverging populations is maintained due to the very high selective coefficients involved. What remains to be seen, of course, is whether in this case of incipient isolation, the speciation process will go to completion.

The case for truly sympatric speciation is very weak, excepting speciation by polyploidy which is common in plants (Grant, 1981) and is exemplified by the marsh grass *Spartina* in which an allopolyploid hybrid species, *S. townsendii*, has recently arisen (Marchant, 1967, 1968). Evidence for evolution of sympatric reproductive isolation in *Drosophila melanogaster* as a result of disruptive selection was put forward by Thoday and Gibson (1962) but, as discussed in chapter 4, repeats of the relevant experiments by several other workers failed to confirm that isolation occurred in such circumstances. That sympatric speciation is theoretically possible, given strong enough selection, has been shown by Maynard Smith (1966) and Dickinson and Antonovics (1973). What is in doubt is that the process actually occurs in nature in the absence of polyploidy.

Turning to the genetics of speciation, one of the main issues has been whether reproductive isolation is usually achieved solely by allelic differences and without visible chromosomal changes such as inversions or translocations, or whether chromosomal alterations are often a primary factor, the latter view being strongly put by White (1978; but see criticism of Zouros, 1982). The existence of homosequential pairs or groups of species, 13 of which are known in *Drosophila* (Table 6.1), clearly shows that large chromosomal changes are not necessary for reproductive isolation. However, White argues that since the homosequentials constitute only a small fraction of the 2000+ species of *Drosophila* that are known, speciation generally involves chromosomal changes. Here we meet the problem dogging most inferences about speciation based on comparisons of closely related species. The existence of chromosomal differences between two species may indicate chromosomal involvement in the speciation event that led to their separation, but it may equally indicate that one or both species have undergone chromosomal restructuring after their

Table 6.1. Homosequential species complexes in the genus *Drosophila*. From *Modes of Speciation* by M. J. D. White. Copyright © (1978) by W. H. Freeman and Company. All rights reserved

Complex	Geographical distribution
1. *mulleri*, *aldrichi* and *wheeleri*	Western USA, Mexico
2. *meridiana* and *meridionalis*	Texas, Mexico, Brazil
3. *bostrycha*, *disjuncta*, *grimshawi orphnopeza* and *villosipedis*	
4. *glabriapex*, *pilimana* and *vesciseta*	
5. *limitata*, *sejuncta* and *ochracea*	
6. *balioptera* and *murphyi*	
7. *hawaiiensis*, *musaphilia*, *recticilia* and *silvarentis*	
8. *heteroneura*, *silvestris* and *planitibia*	Hawaiian archipelago
9. *neopicta* and *obscuripes*	
10. *hanaulae* and *oahuensis*	
11. *adiastola*, *cilifera* and *peniculipes*	
12. *ocellata*, *paucipuncta*, *punalua* and *uniseriata*	
123. *guaramunu* and *griseolineata*	South America

Table 6.2. Kinds of isolating mechanism

Pre-zygotic mechanisms	Post-zygotic mechanisms
1. Habitat differences	1. Hybrid inviability
2. Seasonal differences	2. Hybrid sterility
3. Behavioural differences	
4. Morphological differences	

separation, and it is usually impossible to distinguish between these two possibilities. This problem precludes an easy interpretation of the chromosomal differences between man and the great apes which are illustrated by Yunis and Prakash (1982). At present it is clear that speciation can occur without major chromosomal alteration, and it is equally clear that, at least in the special case of allopolyploidy, chromosomes may be heavily involved in speciation, but the relative importance of 'chromosomal' and 'genic' speciations remains uncertain.

As regards isolating mechanisms, it is conventional to group these as pre-and post-zygotic, with several divisions of each. One possible classification is given in Table 6.2. The various mechanisms are not, of course, mutually exclusive, and closely related species are often isolated by several mechanisms. For example, *Drosophila melanogaster* and *D. simulans* have different ecological preferences with respect to alcohol (Parsons, 1975) and biotin (Erk and Sang, 1966), different patterns of courtship behaviour (Manning, 1959a,b; Cowling and Burnet, 1981) and slightly different male genital structures; they also suffer varying degrees of hybrid inviability and sterility. Again, though, the question arises as to whether some of these isolating mechanisms evolved after isolation was effectively completed due to the action of only one of them.

There are many questions that remain unanswered in relation to isolating mechanisms. These include:

1. Which is most common?
2. Do different isolating mechanisms predominate in different taxa?
3. Which mechanisms are usually primary and which secondary?
4. Is every listed mechanism capable of producing complete reproductive isolation on its own?

Definitive answers cannot yet be given to these questions and this, combined with continuing uncertainty about the role of chromosomes and the importance of parapatric and sympatric speciation, means that there are many possible speciation 'models'. If we assume that, disregarding the relative commonness of the processes, speciation can occur in allopatry, parapatry or sympatry, can occur through genic or chromosomal changes or both, and can occur by any of the isolating mechanisms given in Table 6.2 or by any combination of them, then there are 567 ways in which speciation can theoretically take place. No doubt this number could be reduced by the exclusion of very unlikely schemes such as, perhaps, all forms of speciation where complete isolation occurs solely in response to seasonal differences, but we are still left with a very large number of possibilities. There is insufficient space in the single chapter devoted here to speciation to do justice to the various controversies surrounding the validity of certain speciation models, and only the main debates have been briefly outlined above. A thorough discussion of the matter is given by White (1978), though I do not share his view of the importance of chromosomal changes and do not agree with some of his other propositions, such as speciation via area effects of the type revealed in *Cepaea* by Cain and Currey (1963a,b) and recently shown to be less pervasive genetically (Jones *et al.*, 1980; Ochman *et al.*, 1983) than was previously thought (Johnson, 1976). My own view is that allopatric speciation through genic divergence and behavioural or post-zygotic isolation is commonest, with chromosomal changes and additional isolating mechanisms following later. The ensuing discussion may be coloured to some extent by this view. However, it should be stressed that, with up to 567 speciation mechanisms available, it is most unlikely that any one mechanism is universal, even within a particular phylum or class.

6.3 COMPARISONS OF CLOSELY RELATED SPECIES

Two problems are encountered when working with closely related species. The first is that, as mentioned eralier, we cannot be sure which of the differences now observed originated during speciation and which after. The second problem is that most genetic analyses of variation involve making crosses, but almost by definition crosses cannot be made between separate species. In fact, crosses producing viable and fertile offspring can sometimes be made between species which do not interbreed in nature due to pre-zygotic isolating mechanisms, but which have not yet evolved any post-zygotic barriers; one such case is discussed below. Despite these difficulties, comparisons of closely related species are useful in complementing direct studies of speciation in progress, which are more informative but less often possible. The emphasis in this comparative section is on morphology, and in particular on how morphological divergence relates to genic and chromosomal differences and to reproductive isolation.

Morphology and reproductive isolation

One result of species being defined 'biologically' is that the morphological differences separating closely related species are rather variable in magnitude. This contrasts with the situation prevailing when the morphological species concept was widely accepted because that acceptance almost guaranteed specific, or at least subspecific, status to any morphologically distinct group of populations. A good example of the reduction in the number of species in a genus due to the replacement of the morphological with the biological species concept is *Lymnaea*, for which Hubendick (1951) reports that several hundred 'species' were reduced to about 40. An even more drastic reduction in species number occurred in freshwater mussels of the genus *Anodonta*, where 251 'species' reduced eventually to one (White, 1978). In contrast, some groups where little morphological differentiation occurs have actually increased in their number of species due to the re-definition of species on reproductive grounds. A good example of this is the genus *Drosophila* where many new species have been and still are being named after detailed study revealed lack of interbreeding between groups with no, or negligible, differences in morphology. Species of this kind are referred to as sibling species (see Cain, 1971). Two examples of pairs of siblings within *Drosophila* are *D. melanogaster* and *D. simulans* in the *melanogaster* species group, and *D. pseudoobscura* and *D. persimilis* in the *obscura* species group. (See Patterson and Stone (1952) for a description of the various subgenera and species groups of *Drosophila*.)

Let us briefly consider the morphology of *D. melanogaster* and *D. simulans*, two of the most widely studied species. A fairly recent key to British *Drosophila* (Shorrocks, 1972) gives the following as morphological characters important in the separation of these two species:

1. The width of the gena (cheek) relative to the diameter of the compound eye.
2. The number of bristles at the ends of the maxillary palps.
3. The shape of a projection on the male genital arch.

Clearly, these are minor characteristics and, in addition, neither of the first two are diagnostic — they reflect differences in central tendency between the two species rather than reliable differences between pairs of individuals. Only the difference in genital arch is sufficiently reliable to be used as a criterion upon which to base identification, a fact that is reflected in the counting of males only (Moore, 1952a,b) or the use of genetic markers (Arthur, 1980a,b) in competition experiments involving mixed cultures of these two species.

Although *D. melanogaster* and *D. simulans* can be crossed to yield hybrids, they appear not to interbreed in nature. Also, I have observed in mixed populations where each species is marked with a different homozygous recessive mutation that wild-type hybrids only appeared when the frequency of one species was less than 1% of the mixed population, and even then the hybrids occurred in low numbers and in only one out of more than 50 population cages. It is clear, then, that we have here two good biological species with no morphological criteria by which females of one species can be separated from those of the other and with only one minor character distinguishing the males. Since convergent evolution of such close morphology after speciation is most unlikely, it can be concluded from this and other examples of sibling species that speciation may occur with almost no morphological divergence.

The converse situation, namely speciation accompanied by considerable morphological differentiation, is not demonstrable in a conclusive way through studies on closely related species because of the likelihood of some or all of the observed differences having accumulated after speciation. It is of course easy to find congeneric species differing considerably in morphology, and an example is provided by the land snail genus *Helix*. The garden snail *H. aspersa* and the edible snail *H. pomatia* are considerably different in size, shape and thickness of the shell, as well as in pigmentation and aspects of internal anatomy (see Ellis (1969) and Kerney and Cameron (1979) for descriptions). However, how many of these differences accompanied the speciation(s) separating the two species is unknown.

There are a few cases in which a major gene with a clear effect on development seems to have been involved in speciation. An example is the gene determining the direction of coiling in gastropods. This will be discussed in some detail in chapters 10 and 11; in the meantime it is sufficient to note that speciations involving such major genes may be rare but that they do indeed occur.

Morphology and chromosomes

Given that speciation can occur with or without chromosomal change and with or without substantial morphological change, is there any evidence for an association between sibling species and homosequential species that might indicate that chromosomal revolutions cause morphological ones? The general answer is almost certainly 'no', though there are occasional cases where chromosomal change does cause morphology to be altered. The answer to the above question and indeed the whole concept of homosequential species is clouded by the frequent existence of chromosomal polymorphism within species, though such polymorphism may itself be helpful in examining the link, if any, between chromosomal and morphological change. Although certain abnormal karyotypes in man have obvious morphological effects, for example Down's syndrome which involves trisomy of chromosome 21 (Lejeune *et al.*,

1959; Khush 1973), most cases of chromosomal polymorphism in natural populations of animals are not accompanied by morphological differences. This is true of inversion polymorphisms in *Drosophila*, as mentioned in chapter 1; also, B-chromosome polymorphisms in grasshoppers do not have marked morphological effects, though they may sometimes be associated with quantitative variation (Hewitt, 1973).

Taking the two pairs of sibling species of *Drosophila* discussed above, are these homosequential? The answer is 'no' in both cases. *D. melanogaster* and *D. simulans* have rather similar chromosomes, but they differ by a number of small inversions and translocations — up to 24 in all (Horton, 1939). *D. pseudoobscura* and *D. persimilis* are both highly polymorphic for inversions of the third chromosome and share only one out of at least 32 sequences (Figure 6.1). Thus in both of these cases, morphological similarity is not underlain by chromosomal similarity.

Turning to the groups of homosequentials, are the species within these groups always similar morphologically? Again the answer is 'no'. Most of the

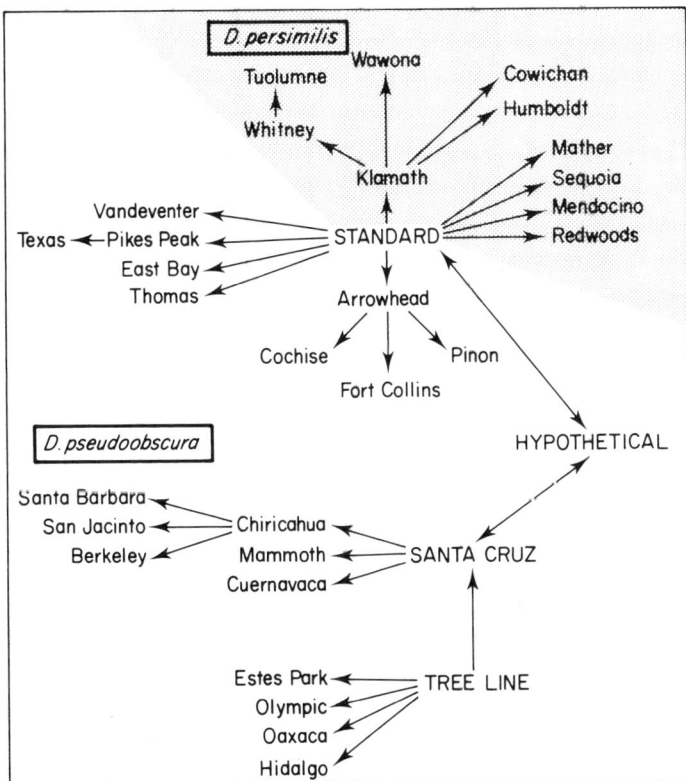

Figure 6.1. Diagram showing evolutionary relationships between third chromosome arrangements in the sibling species *Drosophila pseudoobscura* and *D. persimilis*. From White (1978).

groups shown in Table 6.1 belong to the Hawaiian Drosophilidae, a group characterized by explosive morphological radiation (see Carson *et al.*, 1970). These authors state that, "Not only is each species easily distinguished from the others morphologically, but in some cases the members of a homosequential group bear little superficial resemblance and would, in some cases, hardly be grouped together by ordinary taxonomic characters". The importance of this fact is, as Carson *et al.* (1970) point out, that it stresses the marked morphological divergence that can occur without large-scale chromosomal rearrangements.

Morphology and genic differentiation

So far, the message is one of non-correspondence between morphological change and other factors. Neither reproductive isolation nor chromosomal changes are necessary or sufficient conditions for morphological change to occur. The question that now arises is whether there is a link between genic and morphological divergence. It seems inevitable that in this case there is some connection, because morphological changes must clearly arise from genetic change, except in the case of environmentally induced variations, which are of no evolutionary consequence. I will examine the question of genic–morphological links in two ways. First, genic differences between morphologically similar and morphologically distinct pairs of species will be examined to see whether, in general, greater morphological and genic divergence go together. Second, the genetic basis of particular morphological differences between closely related species, as revaled by interspecific crosses, will be discussed. The question of how much genic divergence occurs during the speciation process will be considered in section 6.4.

One study directed at establishing a link between genic and morphological similarity was that of Hubby and Throckmorton (1968). These authors subjected a number of triads of *Drosophila* species, each consisting of two siblings and one more distantly related species, to electrophoresis of several enzymes. This allowed an investigation of whether, within a triad, the siblings were more genically similar than the non-siblings. Since nine triads were investigated, some assessment of the generality of the conclusions, at least within *Drosophila*, was possible. The results of this survey (Table 6.3) exhibit a remarkably consistent pattern. In all but one triad, siblings share more genes than non-siblings; however, a few cautionary comments are necessary here. First, identity of alleles has been assumed from identity of bands on a gel. This assumption may be incorrect, since it has been shown that in some cases, for example the xanthine dehydrogenase locus of *D. peudoobscura* (Singh *et al.*, 1976), these bands or 'electromorphs' are genetically heterogeneous. Also, even if bands are homogeneous within each species, the amino-acid sequences of the proteins represented by the same band in the two species may be different, since the same overall electrical charge can be obtained in a number of ways. The final problem, pointed out by Lewontin (1974), is that Hubby and

Table 6.3. Percentages of loci with identical alleles present in sibling species (I) and in comparisons involving one of the siblings and the third, non-sibling member of the triad (II). Modified from Lewontin (1974); data are from Hubby and Throckmorton (1968)

Triads	I	II
arizonensis mojavensis mulleri	42.1	6.3
mercatorum paranaensis peninsularis	55.0	11.8
hydei neohydei eohydei	43.8	3.2
fulvimaculata fulvimaculoides lemensis	50.0	13.2
melanica paramelanica nigromelanica	26.3	10.0
melanogasater simulans takahashii	52.9	7.9
saltans prosaltans emarginata	36.8	7.7
wilistoni paulistorum nebulosa	7.1	11.6
victoria lebanonensis pattersoni	64.3	0.0

Throckmorton (1968) used a single inbred strain of each species and so did not take into account the large amount of polymorphism within them.

Ashburner (1969) followed Hubby and Throckmorton's (1968) work by demonstrating a high degree of similarity in the puffing patterns of the salivary gland chromosomes of the siblings *D. melanogaster* and *D. simulans*. Thus not only are sibling species genically similar but, if Ashburner's results are representative, their patterns of gene activity are also similar. Unfortunately, Ashburner used a species pair rather than a triad so it is not certain that there is a difference between sibling and non-sibling pairs in the similarity of their puffing patterns. Perhaps future work will confirm that such a difference exists.

Studies involving genetic analysis of interspecific differences in morphology have an advantage over the kind of study just described because there is a clearer cause-and-effect relationship between the genes and the morphological differences. It can be argued in relation to Hubby and Throckmorton's work that sibling species have separated relatively recently, that they will therefore necessarily be less morphologically and genically divergent than non-siblings, and that the morphological and genetic differences are correlated only because both depend to some extent on time. We certainly cannot say that the morphological differences between non-sibling species are determined by all the genes involved in the excess genetic divergence observed between these species. The question therefore arises: what are the genetic bases of interspecific morphological differences?

One study designed specifically to answer this question was conducted by Val (1977) and Templeton (1977). These authors analysed morphological differences between two species of Hawaiian *Drosophila*, *D. heteroneura* and *D. silvestris*. These are homosequential species, both in relation to each other and to *D. planitibia* (see Table 6.1), a fact which no doubt contributes to their successful hybridization in the laboratory. Crosses yielding F_1, then back-cross or F_2, progeny may be conducted, and indeed Val (1977) gives information on progeny as far as the F_6 generation. The main morphological character examined in this study was head shape. This is illustrated, for males, in Figure 6.2, from which it can be seen that a rather marked difference has evolved. Females also differ, though to a lesser extent. Several aspects of external pigmentation were also dealt with by Val (1977) but these will not be considered here.

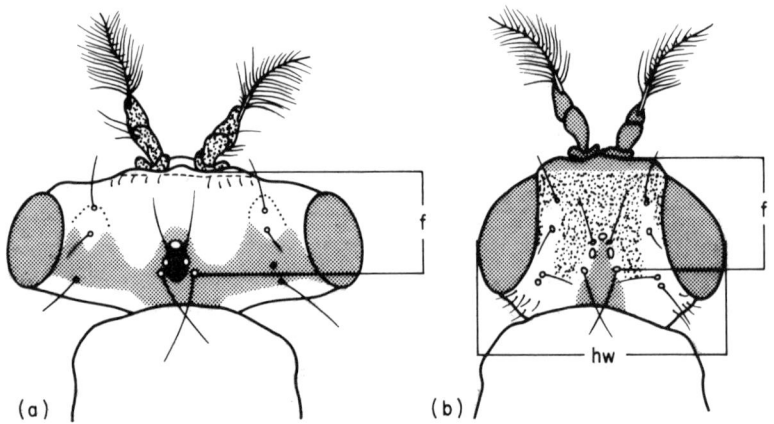

Figure 6.2. Illustration of interspecific difference in head shape and of morphological measurements relating to this difference. a, *D. heteroneura* male; b, *D. silvestris* male. f, frons; hw, head width. From Val (1977).

The pattern of variation in the pure species and in their F_1 and F_2 hybrids is compatible with a polygenic basis for the character since a range of values of head width (hw) and frons (f) was manifested rather than a series of discrete categories. Also, the F_2 generation showed considerably more variation than the F_1 (see Figure 6.3), an effect comparable to that seen in crosses between inbred lines with different mean values of a quantitative character. There appears to be a fairly major X-linked effect since the F_1 distribution is shifted

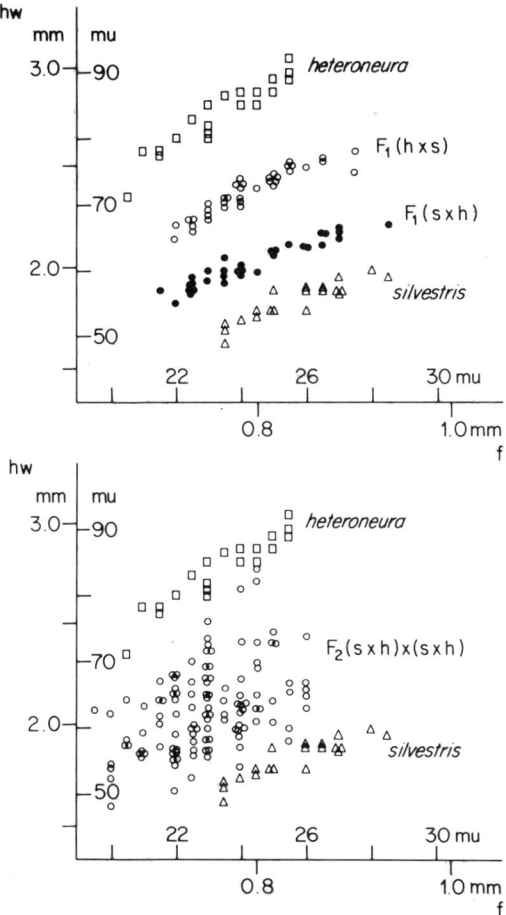

Figure 6.3. Measurements of frons (f) and head width (hw) in *D. heteroneura*, *D. silvestris* and their F_1 and F_2 hybrids. (Data are for males only.) The initial letter of a cross shows which species provided the female; for example, h × s means h♀ × s♂ (mu = micrometer units) From Val (1977).

towards the species that contributed the female parent. This effect is also sex-limited and does not occur in the F_1 females.

Templeton (1977) took the analysis of Val's (1977) results further, and estimated, using the methods of Tan and Chang (1972) and Cavalli-Sforza and Bodmer (1971), the number of loci contributing to the interspecific difference in head shape. The estimates obtained were nine and ten 'loci', indicating close agreement between the two methods of estimation. However, these methods involve certain assumptions, including free recombination between all loci which will clearly not occur. This means that the estimate obtained is a minimum; and Templeton (1977) concludes rather cautiously that his results tell us that "several" loci determine the interspecific difference in head shape.

Lande (1981), using a method referred to in chapter 3, re-analysed Val's (1977) data and obtained an estimate of 6–8 'genetic factors' contributing to the interspecific difference in head shape. Lande stresses that the estimates he obtains cannot exceed the number of independently segregating chromosomal elements, which is the haploid chromosome number plus the average number of recombination events per gamete. Since each chromosome probably undergoes only one or a few such events, the number of independent elements is unlikely to exceed a small multiple of the haploid chromosome number. Because *D. silvestris* and *D. heteroneura* have a haploid number of six and one of these six is a dot chromosome, Lande argues that every major chromosome is likely to carry one or more determinant of the interspecific difference in head shape. He concludes, from this and five other examples, that his results "strongly support the neo-Darwinian theory that large evolutionary changes in quantitative characters *usually* occur through the accumulation of multiple genetic factors with relatively small effects". I fully agree with this conclusion but have italicized 'usually' because I feel that unusual speciation events may be rather important. This point will be taken up again in chapter 11.

6.4 PARTIAL REPRODUCTIVE ISOLATION

All comparative studies of closely related species are open to the criticism that it is impossible to tell whether the observed differences — genic, chromosomal, morphological or behavioural — evolved during or after speciation. Indeed, some differences may have evolved before speciation. Thus there is clearly a need for direct study of speciation as it actually happens. Such studies are unfortunately few and far between because speciation comprises a very short section of the simplified evolutionary cycle outlined in the previous chapter. At any one time, most species consist of a series of genetically divergent populations, individuals from any of which can freely interbreed with those from any other. However, occasionally someone is lucky enough to study a population which turns out to be partially isolated reproductively from its parental species. Such a population may be regarded as having undergone the first stage of speciation, though whether it will subsequently undergo total separation cannot be conclusively established. The great advantage of studies on populations of this kind is that they tell us how much genic and morphological divergence is associated with speciation itself, though since speciation is incomplete, the estimates we obtain are clearly minimal ones. I will briefly consider two examples of partial reproductive isolation.

The first example stems from a study of *D. pseudoobscura* (Prakash, 1972) and is a clear case of initiation of reproductive isolation in allopatry. *D. pseudoobscura* is a North American species, but an outlying population is found at Bogota, Colombia (see Figure 6.4). Crosses between individuals of this population and individuals from a wide variety of populations from the main range of *D. pseudoobscura* revealed no behavioural barriers to reproduction, but showed that F_1 males were sterile if the parental female was

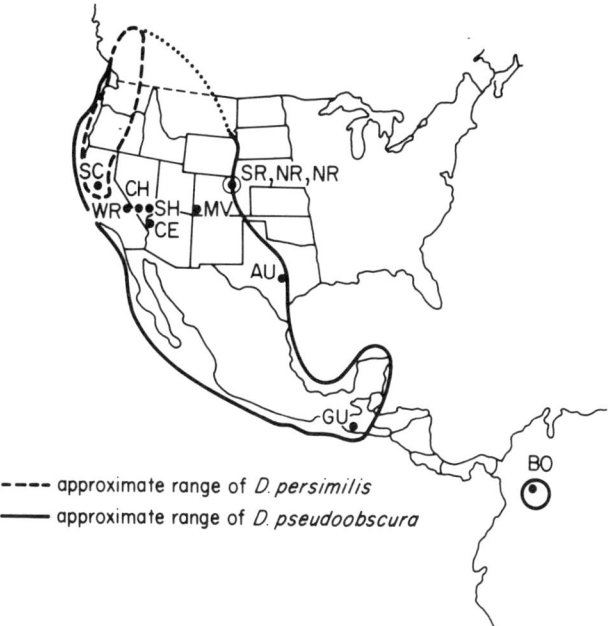

Figure 6.4. Approximate distribution of *Drosophila pseudoobscura*, showing outlying population at Bogota, Colombia (BO). The distribution of *D. persimilis* is also shown. From Lewontin (1974).

from Bogota. This partial reproductive isolation appears to occur in the absence of any new morphological or chromosomal features and with very little genic divergence. Of 24 loci examined by Prakash *et al.* (1969), 11 were identically monomorphic, none showed alleles unique to Bogota, and in many cases the population there was fixed for the commonest allele of a main-range polymorphism. These results must be interpreted with some caution since electrophoresis does not detect all genic variants, but nevertheless the impression is one of partial reproductive isolation with very little genic differentiation. Prakash (1972) estimates that as few as four loci may be involved in the barrier to the production of hybrid offspring.

The second example of partial reproductive isolation comes from the work of Murray and Clarke (1968b; see also Clarke and Murray, 1969) on land snails of the genus *Partula* on the Pacific island of Moorea. The taxonomic status of *Partula* is extremely complex, with 11 originally described species being tentatively allocated by Clarke and Murray (1969) to two species complexes, *P. suturalis* and *P. taeniata*. Within both of these complexes there is a pair of 'species' the members of which appear to interbreed in some places but not in others. One of these pairs comprises *P. suturalis* and *P. aurantia*, the other *P. taeniata* and *P. exigua*. I will restrict the discussion here to the latter pair since all snails within this pair from whatever locality have dextrally coiled shells. *P. suturalis*, on the other hand, is polymorphic for the direction of coiling and this issue will be discussed in chapter 10.

Murray and Clarke (1968b) measured *P. taeniata* and *P. exigua* for a number of functionally independent characters and examined whether, within each of a series of localities, there was a significant correlation between the values of the different characters using Kendall's coefficient of concordance (see Siegel, 1956). They took lack of correlation to indicate either the presence of a single entity or two hybridizing entities, and a positive correlation to show two separate, non-interbreeding, or at least only partially interbreeding, groups of individuals. In cases showing no correlation, hybrid populations could be distinguished from pure populations of *P. taeniata* or *P. exigua* by the mean score for all the variables examined. The results of this study showed separation into two distinct groups in most localities, but the degree of separation was variable and there was evidence of hybridization in some localities, though this was less strong than in the case of *P. suturalis* and *P. aurantia*.

While it is clearly desirable to back up these results with a breeding programme to confirm the variable degree of isolation between *P. taeniata* and *P. exigua*, it does indeed seem that we have here two species in the final stages of separation. As in the case of *Drosophila pseudoobscura*, little enzymic divergence has accompanied the evolution of reproductive isolation (Johnson *et al.*, 1977). The chromosomes of *Partula* are small and do not permit detailed comparisons, but all species in the genus are known to have the same chromosome number (B. Clarke, personal communication). As regards morphology, Murray and Clarke (1968b) list several characters in which *P. taeniata* and *P. exigua* differ. These include shell width and thickness, shell pigmentation and several characteristics of the genitalia. Thus it appears that, in this case, considerable morphological divergence has occurred during or before speciation.

6.5 THE WALLACE EFFECT AND REPRODUCTIVE CHARACTER DISPLACEMENT

If populations which diverge genetically in allopatry meet again and intermingle before any partial reproductive isolation occurs, their gene pools will clearly re-merge. If, on the other hand, they evolve total isolation in allopatry by whatever means, then the only effect they will have on each other's gene pools on re-contact is an indirect one: each may act, through interspecific competition, as a selective agent on the gene pool of the other (see next chapter). It is much less clear what will happen if two populations that have evolved partial reproductive isolation in allopatry, such as main-range and Colombian *D. pseudoobscura*, re-contact each other. One possibility that can be attributed to A. R. Wallace (1889), and is hence appropriately called the Wallace effect (Grant, 1966), is that natural selection will reinforce reproductive isolation. That is, individuals tending to mate with members of the alternative group will, because of hybrid inviability or sterility, decline in frequency. All that is necessary for this scheme to work is evolution of partial reproductive isolation in allopatry by post-zygotic mechanisms (as in *D. pseudoobscura*) and genetic variation for the relevant behavioural traits — something which seems very likely to occur and indeed has been demonstrated in one case by Eoff (1975).

The term 'Wallace effect' as normally used covers the evolution of any tendency to mate homotypically rather than heterotypically, regardless of whether it only involves modification of ecobehavioural characters or whether morphological change also occurs. Thus the Wallace effect must be distinguished from reproductive character displacement, a term first used by Brown and Wilson (1956), which refers exclusively to morphological characters. It is possible to have reinforcement of reproductive barriers via the Wallace effect with or without reproductive character displacement. We now need to ask whether there is any evidence for this form of character displacement.

The strongest kind of evidence for the Wallace effect in general, according to Murray (1972), is where the populations of two speciating entities are partially allopatric and partially sympatric, and characters whose similarity is required for interbreeding are seen to differ more strongly in the areas of sympatry. Most of the cases in which such a pattern is found concern ecobehavioural characters such as song patterns (for example, in anuran amphibians: see Blair, 1974), and so do not constitute reproductive character displacement. However, Levin and Kerster (1967) provide evidence, in the plant genus *Phlox*, of sympatric displacement in flower colour between *P. pilosa* and *P. glaberrina*. This divergence appears to reduce the mutual pollination by butterflies. A similar phenomenon, though this time involving flower size, has been reported by Whalen (1978) in the genus *Solanum*. The nuthatches *Sitta neumayer* and *S. tephronata* may exhibit reproductive character displacement in eye-stripe size and body size, though the evidence for this is not conclusive (Grant, 1975). Finally, a case of reversal in the direction of shell-coiling in *Partula* may constitute reproductive character displacement; this evolutionary change will be discussed in chapter 11. In conclusion, there are scattered cases where reproductive character displacement apparently occurs, but its general importance is not yet clear.

6.6 SUMMARY

The way in which speciation most commonly occurs in nature is still in dispute, both as regards the importance of geographical isolation and the frequency of chromosomal changes. In addition, the relative importance of different isolating mechanisms, both in general and in particular taxonomic groups, is in question. Regardless of these uncertainties, it is clear from the existence of sibling species that speciation events need not involve major morphological change. Other comparisons suggest that in some instances notable morphological change does accompany speciation. The extent of morphological divergence appears to be independent of the amount of chromosomal change, since homosequential species are often very different morphologically as evidenced by the Hawaiian Drosophilidae, and since sibling species often differ in karyotype. There is some evidence that morphological and genic change are correlated, and in cases that have been investigated there is a polygenic basis underlying the morphological divergence. Studies of populations actually undergoing speciation confirm that morphological divergence may or may not accompany reproductive isolation, since examples of morphologically identical and morphologically distinct entities with partial isolation are known. Where species separate initially in allopatry and then make contact again there is some scattered evidence of reproductive character displacement; that is, the morphological correlate of selection against heterospecific matings.

Chapter 7

Further Divergence After Speciation

"As species of the same genus have usually, though by no means invariably, some similarity in habits and constitution, and always in structure, the struggle will generally be more severe between species of the same genus, when they come into competition with each other, than between species of distinct genera."

Charles Darwin, 1859.

7.1 STAGES IN THE EVOLUTIONARY DIVERGENCE OF SPECIES

Speciation is sometimes regarded as some sort of end-point towards which evolutionary studies work for an explanation. Perhaps this stems from Darwin's choice of a title for his book and the gradual realization over the following century that the origin of species was, ironically, one of the topics least adequately dealt with in it. Perhaps it is because present-day neontological approaches to evolution are predominantly reductionist and microevolutionary, and are centred on a belief that hypotheses relating to evolutionary trends above the species level are vague and untestable; thus the evolutionary cycle outlined in section 5.1 provides the focus of attention, and speciation is an acceptable point at which to consider each in a radiating series of cycles as 'finishing'. Whatever the reason for viewing speciation as the highest level of evolutionary change worth serious investigation, it may be helpful to discard this outlook and to take a more holistic view. It is then possible to contrast a series of different stages in the evolutionary separation of two taxa and to develop a comprehensive scheme wherein divergence before, during and after speciation are all included. This has the advantage of allowing individual problems, such as whether chromosomal changes are intricately involved in speciation or alternatively occur after reproductive isolation, to be seen as part of the general question of the relative magnitude of evolutionary divergence at each of a series of different stages.

A start in this direction has been made by Lewontin (1974) who recognized three successive stages of evolutionary divergence. Lewontin's categories were (a) evolution of partial reproductive isolation in allopatry, (b) reproductive and competitive character displacement in sympatry, and (c) long-term phyletic evolution in which "the species continue to evolve but not in response to each other". This scheme is helpful but, for the present purposes, it may be useful to have a more comprehensive breakdown of the stages of evolutionary divergence, including the subspecific stage with which Lewontin was not concerned when he outlined his three-way classification. One possible scheme is as follows:

1. A phase of genetic and, sometimes, morphological divergence of geographically separated populations of a single species, without effect upon reproductive compatibility. Examples include the large- and small-shelled races of *Cepaea nemoralis* discussed in chapter 5.
2. A phase of further divergence, still in allopatry (or parapatry) where partial reproductive isolation evolves as a by-product of genetic differentiation occurring for ecological reasons. Colombian and North American populations of *Drosophila pseudoobscura*, discussed in the previous chapter, exemplify this phase of divergence.
3. A phase of neo-sympatry during which selection operates against any characteristics, whether behavioural or morphological, that result in heterospecific matings. Divergence in floral size in *Solanum* may constitute an example of this phase.
4. A more prolonged phase of sympatry during which the now isolated species continue to influence each other's genetic structure, but without exchange of genes. Both act as a selective agent on variation in the other, the selection being mediated through interspecific competition. A possible example of this phase, *Hydrobia*, will be discussed in section 7.3.
5. Lewontin's long-term post-speciational and post-competitive phase where further evolutionary changes occur in both species in response to a variety of factors, No doubt the bulk of many interspecific differences, including those between man, gorilla, chimpanzee and orang-utang, is due to changes accumulated during this phase, which is terminated only by extinction. it should be noted that the 'punctuational model' discussed in the next chapter denies this phase any importance, except for the rather special case of further speciation events.

The splitting of Lewontin's (1974) second stage of divergence into phases (3) and (4) of the above scheme may or may not be justified depending on the results of future work on the two aspects of character displacement. In the present state of uncertainty in this area, it does seem useful to make this split, because there are reasonable grounds for suspecting that reproductive character displacement will precede competitive, in cases where both occur. Reproductive displacement will only occur while there is continuing hybridization and will cease when 'good' species have been produced. Studies on competitive character displacement, on the other hand, have assumed (in the case of models) or been based upon (in the case of experiments and observations) species which do not interbreed. Although it is conceivable that some competitive character displacement could occur before total reproductive isolation, gene exchange would hinder this process. Certainly, the theory of limiting similarity (MacArthur and Levins, 1967), which provides a theoretical basis for character displacement, clearly relates to separate species, rather than to partially interbreeding groups of dubious taxonomic status. Thus the two aspects of character displacement will be treated here as distinct, though it is worth noting, as Lewontin (1974) has done, that some characters may be affected both by selection against heterospecific mating and by selection against similarity of diet.

It should be stressed that phases 3 and 4 of the above scheme are 'optional'. It is quite possible that in many cases these two phases are both by-passed by genetic, morphological and behavioural divergence proceeding far enough in allopatry to allow the species pair to go straight from phase 2 to phase 5. Some evidence was provided at the end of the last chapter that at least some species pairs go through phase 3; in the present chapter I will examine the question of whether at least some pairs also go through phase 4.

Before moving on to look in some detail at competitive character displacement, two general points need to be made. The first of these is in response to the question that may have arisen in some readers' minds of the wisdom or otherwise of interposing a chapter not dealing with speciation between two which do deal with just that subject. The reason for this intrusion is that when palaeontologists refer to speciation, they are in fact referring to morphological observations on a series of fossils from which they infer speciation; but the exact stage at which interbreeding ceased is always uncertain. This means that morphological changes occurring immediately after reproductive isolation may well be included in the palaeontologist's 'speciation event'. It is thus worth examining what these changes may be before considering (in chapter 8) palaeontological observations at the species level.

The second point is that uncertainty still exists over whether competitive character displacement is common and indeed whether it occurs at all. There is a tendency in science towards uncritical acceptance of concepts from related fields (except where these pose some threat to one's own system of ideas), contrasting with the high degree of criticism to which most scientists will subject a new concept in their own area. This comment should not be taken as implying disapproval of scientific attitudes, but simply as a statement of fact; it would be very difficult for scientists to proceed at all on any other basis. But a spin-off from this arrangement is that concepts in interdisciplinary areas sometimes receive too little criticism, and too ready acceptance, from those working in the flanking disciplines. I would contend that character displacement falls into this category. It is widely accepted (even if sometimes mis-stated: see Hassell, 1976) by ecologists (Odum, 1971) as well as evolutionists (for example, White, 1978); yet both major reviews of the concept (Grant, 1972; Arthur, 1982a) have been rather dubious of much of the purported evidence for it. Thus the present chapter is not simply documentation of a phenomenon of recognized importance; rather it is a discussion of a phenomenon whose very existence is in doubt.

7.2 COMPETITIVE CHARACTER DISPLACEMENT: AN INTRODUCTION

The basic ideas underlying the concept of competitive character displacement are as follows. First, competing species often overlap partially but not totally in their utilization of a common limiting resource when we consider some 'dimension' upon which selection of that resource will depend; this is depicted in Figure 7.1a. An often-cited example of the quantitative variable by which food selection occurs is seed size in granivorous birds. Second, there is a morphological analogue to the ecological overlap shown in Figure 7.1a, with the two species partially overlapping in their distributions of a morphological character connected with the acquisition of food, as shown in Figure 7.1b. For example, corresponding to the preferred seed size in birds we may find a difference in mean beak size; and there will also be corresponding interspecific overlap in seed size and beak size distributions. Third, part of the variation in resource utilization within a species (e.g. from x to y in Figure 7.1a) is due to differences between the various phenotypes that comprise the distribution in Figure 7.1b, rather than all of the variation in the range x–y being attributable to within-phenotype behavioural flexibility. This distinction was particularly clearly stressed by Roughgarden (1972). Finally, selection is thought to occur

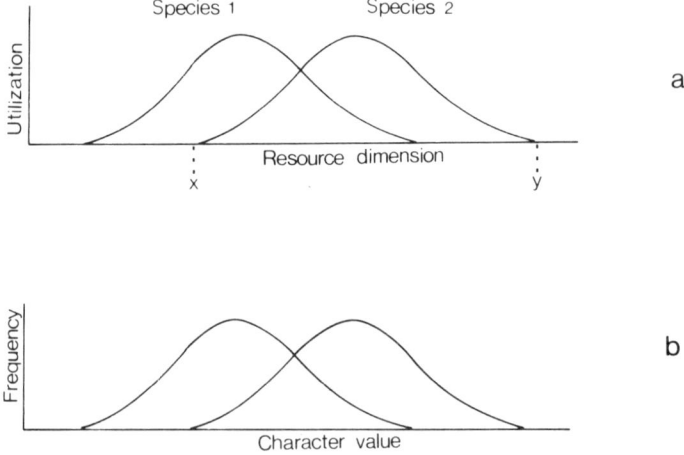

Figure 7.1. Idealized view of partial overlap in (a) dietary and (b) morphological variables in a pair of closely related species. x and y indicate the limits of the resource-utilization curve for species 2.

against individuals within each species that are morphologically similar to the alternative species, in particular those in the overlap zone of Figure 7.1a, and, if there is a straightforward correspondence between phenotypes and resource utilization, those in the overlap zone in Figure 7.1b. The reasoning behind this proposal is that these individuals suffer both intra- and interspecific competition and are hence at a disadvantage to the other individuals which are involved only in competition with their own species. This final proposal is perhaps less readily acceptable than the other three, since it requires particular shapes of resource-utilization curves. I have followed May and MacArthur (1972) and others in depicting these curves as normally distributed, but in practice this may often not be the case. It is easy to imagine very concave curves where the amount of intraspecific competition among individuals with modal values will exceed the combined intra- and interspecific competition experienced by 'overlapped' individuals. It should also perhaps be noted that the equating of the strength of competition with the degree of overlap in resource utilization has been questioned (Pianka, 1981).

The above scheme, like many generally stated ideas, seems fairly plausible. However, in order to proceed further we need to ask two questions about competitive character displacement: (1) When a mathematical model of the relevant biological situation is constructed, does it confirm that displacement will indeed occur, without the necessity of making altogether unreasonable assumptions? (2) Is there any evidence, from studies of natural populations, that displacement does actually occur in regions of sympatry between closely related species? Because I have recently attempted to answer these questions elsewhere (Arthur, 1982a), I will restrict myself here to making some brief comments on the former and, in section 7.3, discussing what appears to be the best example to date of character displacement occurring in nature.

Models of competitive situations in which evolution of the competitors is permitted have been formulated by a large number of authors including MacArthur and Levins (1967), Levin (1969, 1971), Léon (1974), Bulmer (1974), Crozier (1974), Lawlor and Maynard Smith (1976), Roughgarden (1976), Fenchel and Christiansen (1977) and Slatkin (1980). Most of these authors have reached the conclusion that displacement will occur, though Bulmer (1974) has questioned whether it will produce as much divergence as is actually observed in some purported examples of character displacement in natural populations. All the models make assumptions about the correspondence between phenotypes and resource-utilization functions which, if violated, could result in different predicted coevolutionary patterns. One notable feature of the example given in the following section is that it includes evidence for a relationship between morphological phenotypes and resource utilization, which many other studies have failed to investigate.

7.3 THE 'NEO-CLASSICAL' EXAMPLE

Anyone who has located this section because they think that I am about to attempt some sort of connection between character displacement and the neutralist (or neoclassical) theory of molecular evolution can return to the contents list! In fact, I use the term 'neo-classical example' to describe Fenchel's (1975a,b) study of character displacement in mud snails of the genus *Hydrobia* because, as Brown and Wilson (1956) anticipated, their analysis of Vaurie's (1950, 1951) data on nuthatches initially became the 'classical case' of character displacement (see also Grant, 1975), but this has to a large extent been superseded by Fenchel's study as the most frequently referred to, and perhaps the most conclusive, or least inconclusive, example.

The populations studied by Fenchel were mostly in the Northern Jutland region of Denmark, and consisted of individuals belonging to four species, *H. ulvae*, *H. ventrosa*, *H. neglecta* and *Potamopyrgus jenkinsi*. The major part of the study related to *H. ulvae* and the slightly smaller *H. ventrosa*, and I will largely confine the present discussion to these two species. These snails are small, dextrally coiled gastropods which live in estuarine mud flats and feed by ingesting particles of substrate and assimilating attached micro-organisms. Allopatric populations of *H. ulvae* and *H. ventrosa*, as well as sympatric populations, are all found within a relatively small geographical area. In addition to the main study area (Limfjord), two others were also surveyed — a second Danish area and a Finnish one.

The morphological character monitored in Fenchel's (1975b) study was shell length, which showed consistent displacement in areas of sympatry. Figure 7.2 shows data from 15 sympatric and 17 allopatric populations, and it can be seen that in sympatry *H. ulvae* is noticeably larger than *H. ventrosa* in every case, while in allopatry the interspecific difference is much smaller. The relationship between shell size and food-particle selection was experimentally demonstrated by monitoring the size of particles in the faeces of snails of known shell length. A positive correlation between shell size and particle diameter was observed (Figure 7.3), with *P. jenkinsi* behaving, as might be expected, rather differently to the three species of *Hydrobia*. This ecomorphological relationship provides a basis for understanding why selection might favour large *H. ulvae* and small *H. ventrosa* in areas of sympatry.

There is one important gap in Fenchel's (1975b) case for competitive character displacement in *Hydrobia*. I have previously (Arthur, 1982a) drawn attention to this but it is probably worth repeating the criticism as it may turn out to be important. The problem is that Fenchel assumes, rather than demonstrates, that the difference in mean shell size between allopatric and sympatric populations is genetically based. There is some cause for concern on this count because of the short time-span involved in the apparent evolution of character displacement. The Limfjord system is, according to Fenchel, only 150 years old; that is, a maximum of 150 *Hydrobia* generations. Moreover, some particular localities included in Fenchel's study are considerably younger. For example, locality 23A, which shows 'displaced' populations (see Figure 7.2), had only been in the ecological form in which Fenchel found it for about ten years due to the building of a causeway. Of course, we do not know

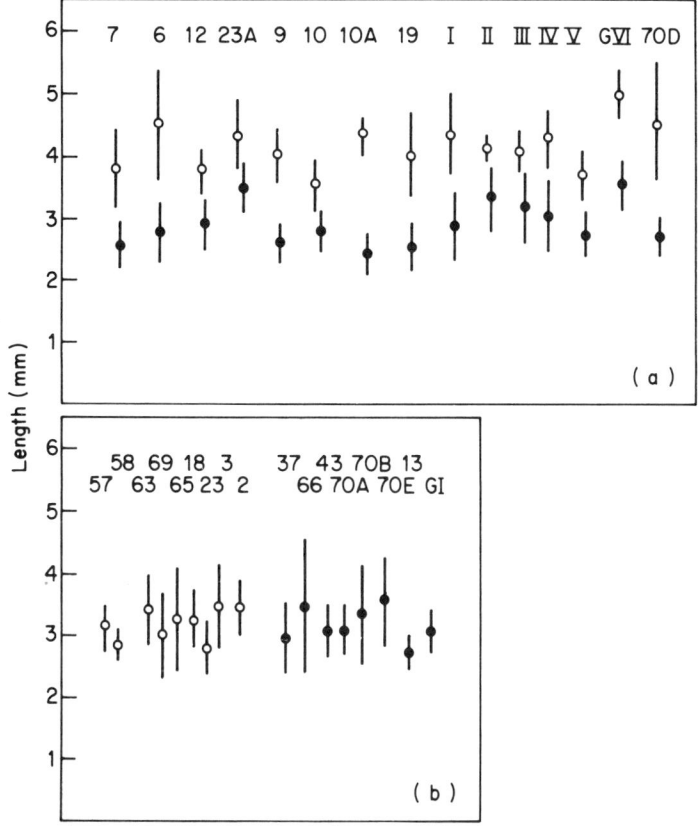

Figure 7.2. Mean shell length in sympatric (a) and allopatric (b) populations of *Hydrobia ulvae* (○) and *H. ventrosa* (●). Bars indicate ±1 standard deviation. From Fenchel (1975b).

where the snails colonizing that site come from, and character displacement does not have to evolve separately at each sympatric site. Nevertheless, even 150 years is a very short period of evolutionary time for an organism with an approximately annual life-cycle. Fenchel (1975b) argues, from the rapid responses to artificial selection observed in experiments with *Drosophila*, that a period of 150 generations is quite sufficient for microevolutionary change of the type he proposes to take place; but the quantitative equating of the responses to artificial and natural selection may be misleading because of the stronger connection between reproduction and variation in the former (see chapter 4). In addition to the question of whether genetically based character displacement could occur in the time available in this case, it is also possible (Arthur, 1982a) that non-genetic processes wherein body size responds to density may 'mimic' character displacement. For these reasons, Fenchel's study cannot be regarded as conclusive. To my knowledge there is not a single example of competitive character displacement where the genetic basis of the

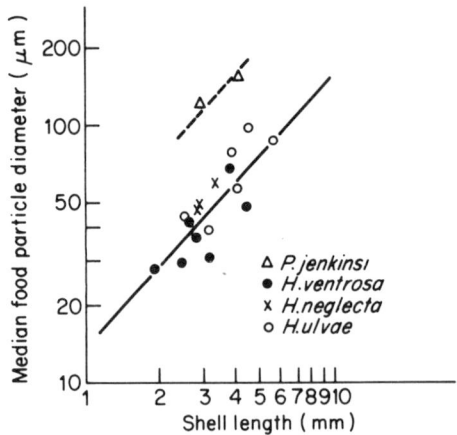

Figure 7.3. Positive relationship between shell size and food-particle size in *Hydrobia* and in *Potamopyrgus jenkinsi*. From Fenchel (1975b).

phenotypic differences between allopatric and sympatric populations has been clearly demonstrated. Attempts to get around this problem by comparing polymorphic characters in allopatry and sympatry (Murphy, 1976; Arthur, 1978, 1980c, 1982b) have not been able to show convincing evidence for a single-locus 'analogue' of character displacement despite theoretical predictions of this (for example, Crozier, 1974).

7.4 SPECIES RANGES AND SPECIES RICHNESS

Although there is not yet any totally conclusive evidence for competitive character displacement, this may be due to the difficulties involved in obtaining such evidence rather than to a lack of occurrence of this proposed evolutionary process. In anticipation of future work which does clearly demonstrate displacement, I will clarify very briefly its likely importance in species pairs and species groups in which it occurs.

Newly separated species pairs

If a partially reproductively isolated population re-invades the range of its parental species, then, assuming the two gene pools do not re-merge, we might expect that, following initial character displacement due to selection against interspecific hybridization, there would follow, if the newly separated species compete with each other, a further phase of displacement due to selection against dietary similarity. If so, then the importance of displacement in the evolution of our newly separated species pair depends on the degree to which the two species' ranges eventually overlap. If the ranges coincide exactly then of course both species will be uniformly 'pushed apart' throughout all their populations, and we may picture the situation diagrammatically as in Figure

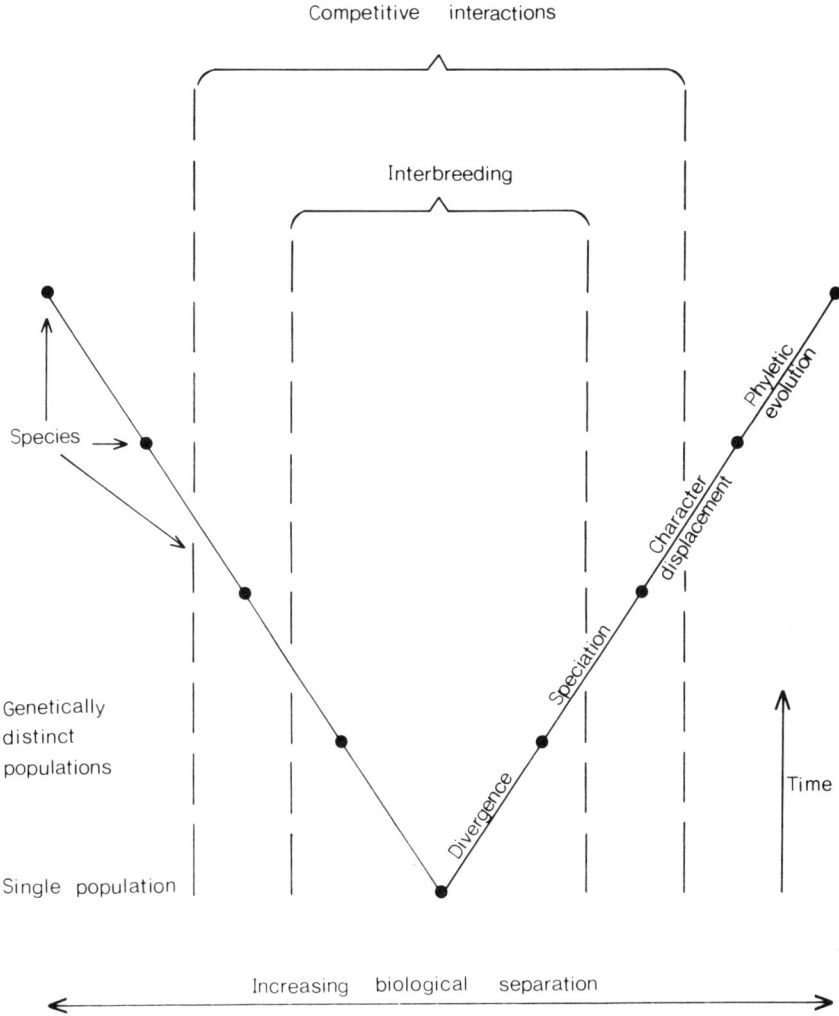

Figure 7.4. Diagram showing how competitive character displacement may be seen as part of a general scheme of evolutionary divergence. (Reproductive character displacement is not shown here as a distinct process but is considered to be included in 'speciation'.)

7.4. However, more common events are inclusion of the range of one species within that of another, for example *Drosophila persimilis* within *D. pseudoobscura* (see Figure 6.4), and partial but not total overlap of ranges, for example *Cepaea nemoralis* and *C. hortensis* (see Taylor, 1914). Although displacement will, if it occurs, affect all of an included species such as *D. persimilis*, inclusive and partially overlapping species will experience a degree of genetic and morphological heterogeneity between their populations as a result of character displacement in some of them. Looked at in this way, competition can be seen in the same light as any other selective agent causing differentiation of populations, and indeed it is often difficult to distinguish its effects from the effects of other selective agents, as was noted in chapter 5 in relation to inter-island variation in beak size in *Geospiza fortis* and *G. fuliginosa*. Thus while displacement may, if it occurs, be important both in cases of approximately coincident and non-coincident ranges, the kind of evolutionary effect it has will depend strongly on which of these two range relationships pertains. It may also depend on the extent to which the two species concerned coexist at the ecological level within the area of geographical overlap. Statements of the latter alone may be misleading; for example, Diver (1940) noted that only about 16% of colonies of *Cepaea nemoralis* and *C. hortensis* were mixed, while geographically the species ranges overlap by about 80%.

Groups of closely related species

So far, I have discussed competitive character displacement as if it only occurs between two species which have recently speciated from each other. While this is indeed the situation in which the strongest competition (and hence perhaps the strongest selection for displacement) might be expected, as recognized by Darwin (1859) and many others since, competitive interactions may be widespread among whole groups of closely related species, especially in species-rich genera such as *Drosophila*. It may be that as one species of such a group migrates between communities and its populations come to experience different assemblages of closely related species in different places, a process of coevolutionary fine-tuning occurs, with character shifts occurring in response to the unique combination of competitors found in each particular community. This is a fascinating thought, but until conclusive cases of character displacement between particular pairs of species have been established, it seems premature to attempt the analysis of complex, multispecies systems of this kind.

7.5 SUMMARY

The various phases of evolutionary divergence which two groups of individuals may go through are distinguished, and it is noted that a pair of species may undergo a phase of competitive character displacement in between speciation and long-term phyletic evolution. In this phase the species do not hybridize but they do affect each other's gene pools by acting as mutual selective agents, the selection being mediated through interspecific competition. After outlining the general concept of competitive character displacement, one of the most conclusive examples is discussed in some detail and a serious gap in the evidence for evolutionary displacement is noted, namely a lack of evidence for a genetic basis to the observed phenotypic differences. No other examples have conclusively plugged this gap and it thus remains possible, though perhaps unlikely, that competitive character displacement does not occur at all.

However, in anticipation of future demonstrations that this evolutionary process does indeed occur, at least in some taxonomic groups, the relation between the degree of overlap in species ranges and the evolutionary effect of character displacement is briefly discussed. Finally, the possibility of character shifts in relation to competitive interactions among a larger group of closely related species is mentioned, but it is stressed that studies undertaken in the near future would be better to concentrate on the conclusive demonstration of competitive character displacement in the relatively simple case of particular species pairs.

Chapter 8

Long-term Evolution at the Species Level

"Local varieties will not spread into other and distant regions until they are considerably modified and improved; and when they do spread, if discovered in a geological formation, they will appear as if suddenly created there, and will simply be classed as new species."

Charles Darwin, 1859.

8.1 INTRODUCTION: PUNCTUATED EQUILIBRIUM VERSUS PHYLETIC GRADUALISM

In the last three chapters I have tried to piece together a neontologist's view of population divergence and speciation, with particular reference to morphological changes occurring before, at and shortly after the period of reproductive isolation. Although there is not total agreement on all the issues, the picture is a coherent one in that speciation is seen as a result of natural selection acting in opposite directions in different populations. The available evidence suggests that changes in size and shape usually occur through selection on polygenic variation rather than on systems of discrete morphological variants (Lande, 1981). The rate of genetic and morphological divergence is expected to vary according to both the strength of selection on any particular population and the variability of selective pressures between populations. The speciation process itself, however, has no fixed relation to the rate of morphological divergence except that strong selection may sometimes be generated against interspecific hybridization in neo-sympatry, in which case we might expect rapid divergence in one or more characters to accompany the completion of reproductive isolation.

This general picture, in which speciation and morphological divergence are only weakly associated, has been questioned by Eldredge and Gould (1972; see also Gould and Eldredge, 1977) and by a number of other palaeontologists (Stanley, 1979; Williamson, 1981a). These authors argue for a dominant role for speciation in generating new morphological variants and a very minor role for selective forces acting within the lifetime of a species; that is, between successive speciation events or cladogeneses. The pattern of morphological evolution through geological time that would result from the restriction of major morphological changes to speciation events would be long periods of morphological stasis interrupted, or punctuated, by much shorter periods during which rapid and significant morphological change occurs. This pattern, known as punctuated equilibrium, is contrasted by its proponents with phyletic gradualism, an alternative evolutionary pattern in which morphological change occurs throughout geological time with a variable rate but without the periods of rapid change and of cladogenesis necessarily corresponding.

The controversy which has developed over punctuated equilibrium, which has become rather heated of late (see Williamson, 1981a,b, and Jones, 1981), is an unusual one in that the two sides are not directly opposing. Eldredge and Gould (1972) argue for a particular distribution of evolutionary *rates* from which they suggest that if rates have been adequately sampled inferences about *mechanism* can be made. However, most neontological evolutionists are preoccupied with establishing mechanisms and are only indirectly concerned with long-term evolutionary rates measurable in geological, rather than ecological, time. This 'tangential opposition' is, however, accompanied by a more direct confrontation of opposing views between the punctuationalists and those palaeontologists holding the extreme gradualistic view that if the fossil record was complete enough in any particular case, a series of intermediates between two species would commonly be found.

At an early stage in the discussion, the punctuational pattern was seen as a direct result, in a series of fossils from a particular locality, of allopatric speciation by divergence of a peripheral isolate (Eldredge, 1971). However, in the decade between Eldredge's proposal of the punctuated equilibrium model based on palaeozoic invertebrates (and the trilobite *Phacops* in particular) and Williamson's (1981a) support for the model based on samples of cenozoic molluscs, the relationship between punctuated equilibrium and speciation appears to have undergone a notable change. While Eldredge (1971) proposed punctuated equilibrium as a result of a conventional mechanism of speciation, Williamson (1981a) argues that "speciation is a qualitatively different phenomenon from gradual, intraspecific microevolutionary change". Thus a model which apparently started off as an attempt to incorporate conventional neontological speciation theory into palaeontological studies seems to have ended up, at least according to some of its supporters, requiring some novel (though largely unspecified) mechanism of speciation.

Because of the complex nature of the controversy over punctuated equilbrium, and because I view this model as helpful or misleading depending on the interpretation placed on the observed patterns of morphological change, I will devote some space to clarifying certain aspects of the controversy before proceeding, in the next section, to look at individual case studies. The two 'models' are often depicted as in Figure 8.1, which shows two alternative evolutionary trees. However, the contrasting patterns of Figure 8.1a and 8.1b

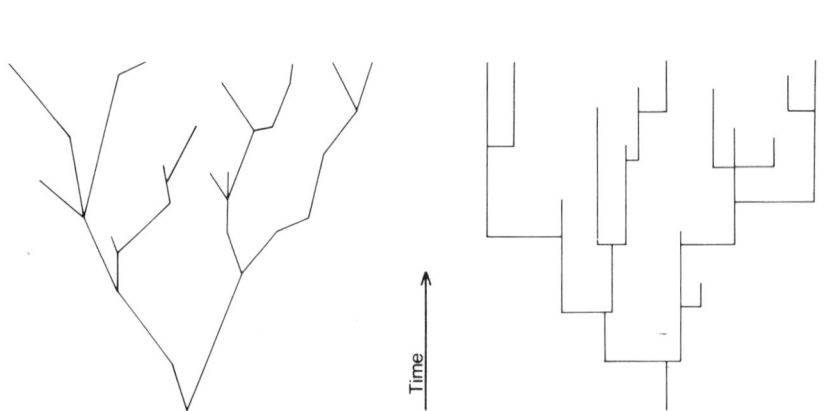

Figure 8.1. Alternative kinds of evolutionary tree representing extreme gradualist (a) and punctuationalist (b) views of the pattern of morphological change through geological time.

give an inadequate and oversimplified view of the controversy, and several points need to be made to forestall some misconceptions that may otherwise arise.

1. Darwin was not entirely a gradualist

The similarity of Figure 8.1a and the diagram facing p. 117 of *On the Origin of Species*, together with Darwin's insistence that 'natura non facit saltum' (see, for example, Darwin 1859, p. 194), has led some authors to the conclusion that Darwin was a proponent of an extremely gradualistic view of evolution. Gould (1982) reinforces this conclusion by quoting from *On the Origin of Species* the following passage: "If it could be demonstrated that any complex organ existed, which could not possibly have been formed by numerous, successive, slight modifications, my theory would absolutely break down". However, the interpretation of these various points as evidence that Darwin viewed evolution as a gradual process at the level of the species, proceeding according to the pattern of Figure 8.1a rather than of Figure 8.1b, may be seriously in error for two reasons. First, because Darwin regarded his diagram as a simplified and general scheme and not a definitive comment on the distribution of evolutionary rates. Indeed, A. Huxley (1982) quotes a passage from the fourth edition of *On the Origin of Species* which reveals that Darwin's views were to some extent punctuational: "the periods during which species have been undergoing modification, though very long as measured by years, have probably been short in comparison with the periods during which these same species remained without undergoing any change". Secondly, the large changes that Darwin rejected with his 'Natura non facit saltum' were, as noted by A. Huxley (1982), mutations of large effect, or sports as Darwin called them. These are a different matter to the much smaller sort of phenotypic jumps implied in the punctuated equilibrium model. In fact, T. H. Huxley (1859; see F. Darwin, 1887) expressed concern that Darwin might be incorrect to exclude large-scale changes and I share this concern, as does A. Huxley (1982) who expresses the view that with respect to mutations of large effect: "It may yet turn out that these are important". I will expand on this possibility in chapter 11 but in the meantime will simply conclude that as regards the typical degree of variation in evolutionary rates *at the species level*, Darwin's views appear to have been a good deal more punctuational than has sometimes been thought.

2. *Different characters have different evolutionary trees*

Eldredge (1971), Eldredge and Gould (1972) and Stanley (1979) all provide outline evolutionary trees broadly similar to Figure 8.1b and all label their x-axes 'morphology' or 'morphologic change'. The question arises of what aspect of morphology is represented and in what form it is measured — a single linear measurement, a ratio or some multivariate complex. Whichever of these variables is used, different morphological characters in the same species will have different evolutionary trees. An equivalent phenomenon is well known in the field of protein evolution, where some proteins, notably histones H3 and H4, evolve very slowly, and others, such as the globins, much more rapidly (seee Table 11.3 in Lewin, 1980). However, this point has been insufficiently stressed with respect to morphological characters in discussions of punctuated equilibrium. In particular, it must be noted that characters directly connected with reproduction would often be expected to show a very close correspondence to the idealized punctuational pattern, the stasis between punctuations being maintained by strong normalizing selection. Because the evolution of reproductive characters will be highly constrained in this way, most of the remaining discussion in this section will be based on the assumption that the main function of the characters under consideration is not a reproductive one. Within this group of characters it may be that some evolve in a punctuated way, others more gradually; indeed Eldredge (1971) notes that this appears to be the case in Kauffmann's (1933) study of the trilobite *Olenus*.

3. *Spatial aspects must be more fully considered*

In his diagrams of possible evolutionary patterns, including punctuated equilibrium, Eldredge (1971) clearly specifies that he is referring to the evolutionary sequence occurring "in one local area"; however, many case studies of punctuated equilibrium have paid insufficient attention to which area is being dealt with. If allopatric speciation occurs via a small peripheral isolate, then a fossil sequence taken from anywhere within the main range of the species concerned may show a punctuated pattern, but whether it does so will depend on ecological rather than evolutionary factors. In particular, the outcome of competition between the recently separated species after the establishment of neo-sympatry will be crucial. Competitive exclusion of the old species from large areas will lead to what I will call an *ecological punctuation* appearing in the fossil record. The alternative exclusion would probably lead to no observed change in the fossil sequence as the period of transient coexistence before the daughter species was eliminated would probably be too short to allow recovery of fossil material. Stable coexistence or patchy competitive exclusion would result in the sudden appearance of a second entity alongside the first, continuing, one and this would constitute another kind of ecological punctuation.

From an evolutionary viewpoint, the more important question is what pattern a sequence of fossils taken from the area occupied by the speciating isolate would show. Eldredge (1971) considers this sequence as exhibiting a punctuation because of rapid morphological change accompanying speciation and only slow, erratic change occurring at other times. This assertion is debatable in the light of frequent lack of morphological change at speciation, as discussed in the context of sibling species in chapter 6, and the frequent occurrence of notable morphological differentiation within species which we saw in chapter 5. However, the debate cannot even be started, for any particular case study, until it is clear that the fossil sequence under investigation comes from the area inhabited by the speciating population.

4. *Care is needed in equating punctuations with speciations*

Since it is impossible to define palaeospecies in terms of hybridization, these species being delineated solely on morphological grounds, there is something tautological about a theory which states that most morphological change in palaeospecies accompanies speciation. That morphological changes *sometimes* accompany speciation is hardly in doubt; indeed, lists of isolating mechanisms always include morphological (or 'structural' or 'mechanical') barriers as one possibility (Dobzhansky, 1951; Mayr, 1963; White, 1978). Nor is it doubted that natural selection can produce ecologically rapid and geologically instantaneous change. Thus observation of morphological punctuation in a series of fossils taken from the area of a diverging peripheral isolate is expected to occur in some cases. What is in doubt is (a) that this is common and (b) that at other times there is a high degree of stasis. The apparent commonness of evolutionary (as opposed to ecological) punctuations relative to speciation events not involving punctuations will be exaggerated by the invisibility of speciations giving rise to sibling species, and it is difficult to see how the frequency of such events can be estimated. Also, it will often be difficult to distinguish large morphological changes occurring at speciation from those occurring shortly after. The latter might be expected to occur as a result of rapid adaptive morphological differentiation, including character shifts of the kind outlined in section 7.4, as a new species expands its geographical range.

5. *Care is needed in equating large morphological change with large genetic change*

So far, our discussion of punctuated equilibrium has been centred on morphological issues. But the question arises of what genetic mechanism is supposed to underlie the putative punctuations. Gould and Eldredge (1977) claim that, "The data of molecular genetics support our assumption that large genetic changes often accompany speciation". It is difficult to know what is meant by this statement, and at least three alternative interpretations of 'large genetic changes' may be distinguished: gene-frequency changes at *many* loci; fixation of one or more mutation of *large individual effect*: and *chromosomal* changes. The non-correspondence beteen chromosomal and morphological divergence was noted in chapter 6. Huxley's (1982) distinction between punctuations and mutations of large effect seems entirely reasonable since these mutations (for example the arthropodan homoeotics: see Ouweneel (1976) and chapter 10) produce changes far in excess of typical intrageneric variation. If large means extensive and punctuations result from changes at many polygenic loci then morphological punctuation is hardly a revolutionary concept. Gould and Eldredge's (1977) claim that regulatory genes are involved is unconvincing since they do not say what is meant by a regulatory gene. There are in fact several distinct kinds of regulatory gene; this problem will be addressed in chapter 10.

It is clear from the above series of points that different aspects of the theory of punctuated equilibrium are acceptable to different degrees. For example, the view that ecological punctuations will occur over much of the range of a species which has recently given rise to a non-sibling daughter species by divergence of a peripheral isolate is readily acceptable. It is also an advance over some early palaeontologists' expectations of finding a finely graded series of intermediates at most localities if the fossil record was complete enough. However, the view that over their entire history between cladogeneses most species remain morphologically static is totally unacceptable and can be dismissed by a host of observations of considerable morphological differentiation within many living species.

At this point I hope that enough clarification of the general controversy surrounding the theory of punctuated equilibrium has been given for a discussion of specific case studies to be meaningful. I have selected two for consideration: human evolution, due to its intrinsic interest and because it has recently been examined in relation to the theory of punctuated equilibrium by Cronin *et al.* (1981); and the molluscs of the Turkana basin (Williamson, 1981a), since this study was based on a more complete fossil record, and has generated more heated controversy, than any other comparable investigation.

8.2 CASE STUDIES OF MOLLUSCS AND MEN

Molluscs of the Turkana basin

Williamson (1981a) undertook a multivariate morphometric study of 13 species representing 12 genera of molluscs (a mixture of bivalves and gastropods)

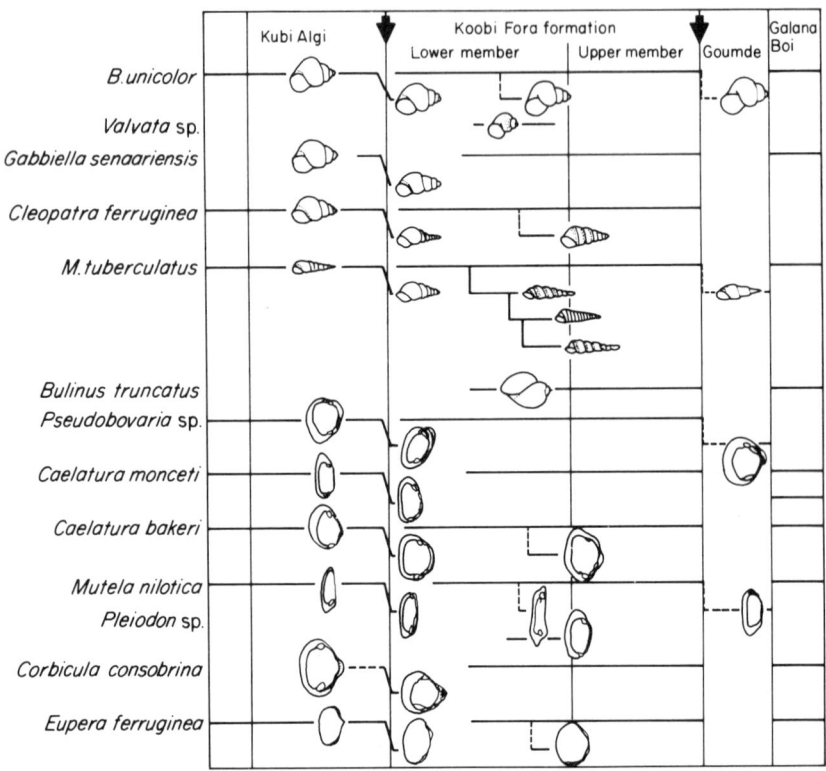

Figure 8.2. Patterns of morphological stasis and change in 13 species of molluscs. From Williamson (1981a).

inhabiting Lake Turkana, Kenya, in the Cenozoic era. The pattern of change observed in these lineages was, except in the few cases which exhibited stasis throughout, highly punctuated. No gradual morphological trends were apparent. A simplified, pictorial account of the punctuations is given in Figure 8.2. One of the most noteworthy points about the overall pattern of change is that punctuations occur simultaneously, geologically speaking, in all or nearly all lineages extant at the appropriate times. The distinctness of the variant forms and the lack of intermediates connecting these with the norm of their parental species is illustrated for *Bellamya unicolor* in Figure 8.3, which shows the results of a canonical variate analysis with each population being represented by its centroid on the first and second canonical variates.

The first question that arises in relation to these data is: Are the observed punctuations real? The main problem in establishing the reality or otherwise of apparently sudden morphological changes in fossil sequences is that these sequences often contain gaps during which no fossilization occurred. While Eldredge and Gould (1972) have correctly pointed out that we cannot write off all sudden changes as artefacts produced by gaps in the record, it is important not to go to the other extreme and accept abrupt punctuations as real

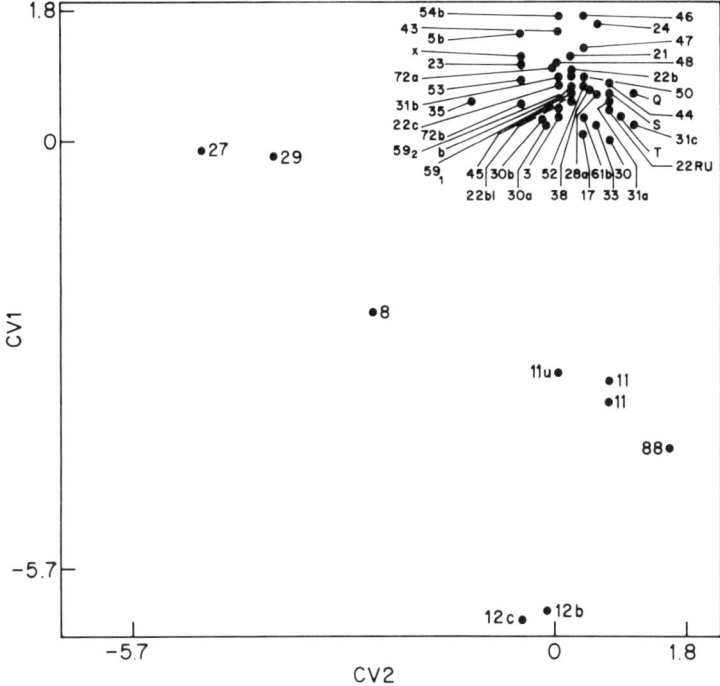

Figure 8.3. Results of canonical variate analysis of data on *Bellamya unicolor*, showing the cluster of samples belonging to the main morphological 'type' and the distinct variants produced by punctuational change (8, 11, 12, 27, 29, 88). From Williamson (1981a).

too readily. However, Williamson's (1981a) study is based upon one of the most complete fossil sequences known. Schindel (1982) has distinguished between the *temporal scope* of a palaeontological sequence, which is the total time-span covered, the *microstratigraphical acuity*, which is the time-span represented by each individual sample within the sequence, and *stratigraphical completeness*, which is the proportion of the total time-span represented by fossil-bearing strata. Clearly, for unambiguous demonstration of punctuational change, a high degree of stratigraphical completeness is of the utmost importance. Schindel notes that Williamson's (1981a) study was based on sequences with up to 73% completeness, which compares favourably with other examples, some being as low as 2 or 3%. Thus it does seem that the pattern of alternating stasis and rapid change is real, and this has been accepted even by critics of Williamson's interpretation, for example Jones (1981).

Having established the reality of the morphological punctuations, the next thing that needs to be ascertained is whether these are underlain by genetic changes or whether they merely represent ecophenotypic variation; that is, variation caused by direct environmental effects. The latter alternative is suggested by the parallel variation in different lineages and if it cannot be refuted then the case for evolutionary punctuations is seriously weakened. This problem of heritability has been raised in connection with Williamson's (1981a) study by Mayr (1982), Boucot (1982) and Arthur (1982c). Of course, no experimental work can be brought to bear directly on the problem but studies of the inheritance or otherwise of morphological variation in living members of the same species would provide some relevant information.

In the absence of any direct evidence for an inherited basis of his variant forms, Williamson (1981a, 1982) has followed three main lines of argument to back up his claim that the punctuations are genetically based:

1. The changes took more than 5000 years to complete.
2. The variants produced are outside the range of variation of present-day populations.
3. Some distinct variants coexist with each other.

None of these points taken separately, nor even all of them together, can refute the hypothesis that the morphological changes were ecophenotypic. As regards the fact that they took thousands of years, whether this would be expected of ecophenotypic variation depends on how the environment changes and to which environmental variables the shells respond. Williamson (1981a) notes that his 'speciation events' correspond with major lacustrine regressions, and if this is so then there is little doubt that large and prolonged changes in many environmental variables were occurring, though the precise patterns of change of different variables will have been different and cannot now be documented. Turning to the fact that the variant forms produced by some punctuations are outside the range of present-day morphological variation in the same species, we again meet the problem of the unknown values of environmental variables at the time of the punctuations. If the environments encountered then were more extreme than any experienced now, then ecophenotypic variants would be expected to be correspondingly extreme. Coexistence of two ecophenotypic variants would indeed be remarkable if the two were found in the same small pond or at the same sampling site within a large lake. However, significant ecophenotypic variation in the shell shape of *Lymnaea stagnalis* has been shown to occur within a single lacustrine system, that is a lake and its surrounding ponds (Arthur, 1982c), and doubtless this may occur in other species also. Palaeontological studies lack the spatial resolution necessary for observations of 'coexistence' to be interpreted unambiguously.

Williamson (1981a) also contrasts the gastropods of his own study, which are prosobranchs, with the basommatophoran group of pulmonates, which includes *Lymnaea* and other genera showing marked ecophenotypic variation, and states that shell form is highly heritable in the lineages with which he worked. No data are given to back up this statement and it is perhaps worth re-emphasizing the point made in chapter 5 that high heritabilities and pronounced ecophenotypic variation may occur in the same character in the same species (e.g. shell size in the stylommatophoran pulmonate *Cepaea nemoralis*) and that, related to this, the heritability of variation within and between populations need not be the same. A final point worth noting is that it is odd that the obligate asexual species *Melanoides tuberculata* shows the most numerous punctuations if the changes involved are indeed largely genetic.

It is clear, then, that no definite answer can be given to the question of whether and to what extent the observed morphological punctuations are underlain by genetic changes. This is, of course, very different from being able to state that the changes were definitely ecophenotypic, and it is worth briefly considering what evolutionary inferences we might draw from the Lake Turkana molluscs, supposing for the moment that the morphological changes were largely genetically based. In particular, Williamson (1981a) argues that his results show that "speciation is a qualitatively different phenomenon from gradual intraspecific morphological change" and later (1981b) that the pattern of punctuated equilibrium observed presents "real problems for neo-Darwinism". We need to enquire whether these conclusions are justified.

The fact that punctuations occur at all is of course no problem for neo-Darwinism since pronounced evolutionary changes can be readily produced by selection on polygenic variation as shown in chapter 4 and pointed out in the context of Williamson's study by Jones (1981). Nor is a pattern of alternating change and stability, with long periods of the latter under the influence of normalizing selection, inexplicable from a neo-Darwinian viewpoint. The main problem is the restriction of most morphological change to speciation events. Here we may note as before that speciation is inferred from morphological data in palaeontological studies, so this statement is rather circular; but if we assume that speciations really did occur then their intricate link with morphological change is unexpected. However, the lack of correspondence between morphological change and speciation noted in chapter 6 was a general rather than exclusive conclusion, and there are bound to be some ecological situations that favour both morphological change and reproductive isolation. One such situation would be environments with a high degree of habitat heterogeneity and much geographical fragmentation of populations. The first of these is likely in a changing environment such as a lacustrine regression; also, Williamson (1981a) talks of his speciating populations as peripheral isolates, indicating fragmentation. Thus it may be that the Lake Turkana environment was such that both morphological change and speciation would be expected.

One problem with studies of punctuated equilibrium is that they tend to eschew neo-Darwinian explanations without putting forward any coherent alternative. It was noted in relation to Gould and Eldredge's (1977) paper that 'regulatory' genes were implicated in a rather vague way. Williamson (1981a,b, 1982) appears to be thinking along similar lines since one of his main themes is the 'developmental instability' of transitional forms. The pattern of punctuated equilibrium is attributed to long periods of developmental constraints followed by the rapid breakdown of these and their reconstruction around some new 'mean phenotype'. Apart from the fact that it is not at all clear why such a pattern of change in developmental constraints should occur, this explanation is weak on the grounds that the increased phenotypic variance from which the developmental instability is concluded may be an artefact of sampling during a period of rapid change (Charlesworth and Lande, 1982).

In conclusion, the molluscs of Lake Turkana show a clear pattern of morphological stasis punctuated by short periods of rapid change. Whether these changes are even partially genetic is not known. If they are, the evolutionary pattern observed shows a closer correspondence of morphological change and speciation than neontological studies indicate is usual, but the punctuational events concerned may have taken place in environments which favoured both morphological change and reproductive isolation. There is no need to conclude that speciation is qualitatively different from intraspecific variation except in the obvious sense that it involves the establishment of barriers to interbreeding; nor that the observed pattern of morphological change is inconsistent with neo-Darwinism.

Hominid evolution

Hominid evolution over the past 3–4 million years has recently been analysed, with particular emphasis on rate variability, by Cronin et al. (1981). These authors have concluded that, contrary to Gould and Eldredge's (1977) comments on human evolution, the relevant data are best interpreted in terms of gradual change with a variable rate rather than as a series of sharp punctuational changes interrupting long periods of stasis. The analysis was based upon data on the evolution of the genera *Autralopithecus* and *Homo*, during the period after the divergence of the hominid and pongid lines. This divergence has been estimated, from molecular data, at between 4 and 6 MYBP (Sarich and Cronin, 1977).

In contrast to the molluscs of Lake Turkana, the ancestor–descendant relationships between the various species of hominid are not entirely clear. Cronin et al. (1981) give four possible evolutionary trees representing the main alternative views currently held; these are shown in Figure 8.4. Apart from phylogeny d, all are agreed that one or both of the gracile Australopithecines, *A. africanus* and *A. afarensis*, are ancestral to *Homo*. Also, all schemes provide a side-branch to the robust Australopithecines, *A. robustus* and *A. boisei*, which are not regarded as human ancestors. Again, there is reasonable agreement that *H. habilis*, *H. erectus* and *H. sapiens* form an ancestral–descendant series, though this is not universally accepted (see, for example, Delson et al., 1977). One of the main distinctions between the alternative trees is the position of *A. africanus*; as we will see shortly, the uncertainty over the position of this species affects the interpretation of the analysis.

Cronin et al. (1981) analysed data on two morphological variables, cranial capacity and estimated body weight, in a series of fossils spanning nearly 3 MY; the results are shown in Figure 8.5. The authors state: "The trends seem to show no jumps or discontinuities." This is expecting a bit much of the data, and such a conclusion, based on a sequence of four points, can hardly be sustained.

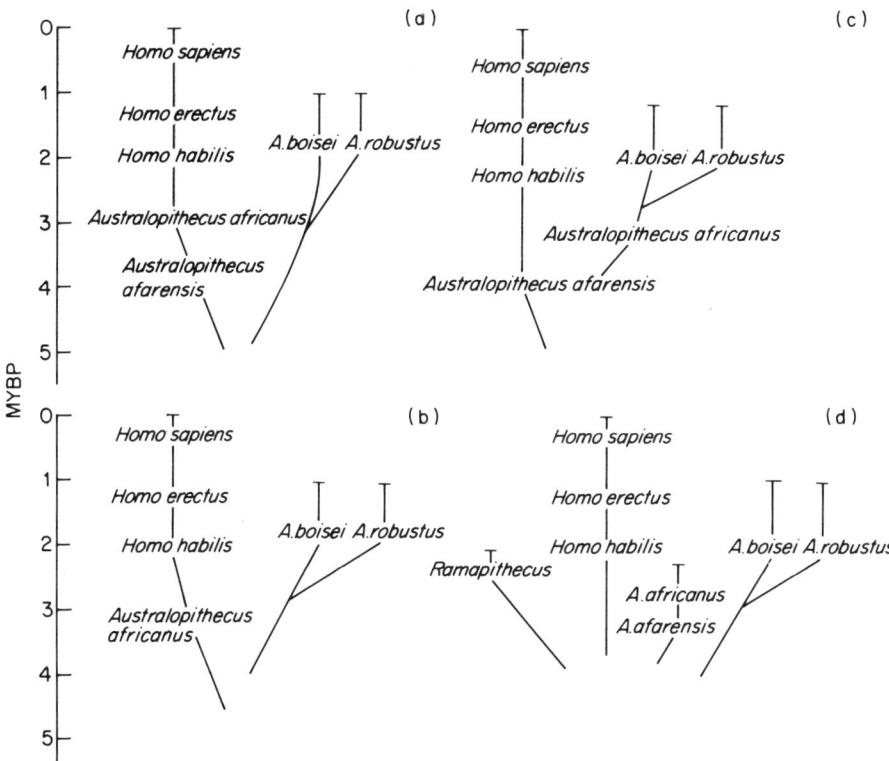

Figure 8.4. Four alternative evolutionary trees showing possible relationships between species of *Australopithecus* and *Homo*. From Cronin et al. (1981).

Cronin et al. state that the most parsimonious approach is to interpret the evolutionary change that has occurred in terms of a *line* of best fit. However, it has frequently been pointed out in evolutionary studies (for example, in a different context, by Jones et al., 1977) that the most parsimonious explanation is not always the correct one. Thus Figure 8.5 can hardly be taken as strong evidence of gradualism. Furthermore, the four points on the graphs shown represent *A. africanus*, *H. habilis*, *H. erectus* and *H. sapiens*; yet the position of *A. africanus* is uncertain, and one of the authors concerned (Y. Rak) favours phylogeny c in Figure 8.4, in which *A. africanus* is not ancestral to *Homo*. If this phylogeny is correct, then the inclusion of the lowest point on both graphs of Figure 8.5 is unwarranted and the number of points thus reduces to three. Since a single punctuation separating two periods of stasis requires a minimum of four points to demonstrate it, any use of the data to support or reject the existence of punctuations in hominid evolution is strongly dependent on the phylogenetic position of *A. africanus*. Cronin et al. (1981) state that they wish only to examine the evidence for punctuated equilibrium and do "not try to argue for or against the various possible phylogenetic trees presented"; but these two issues cannot really be separated.

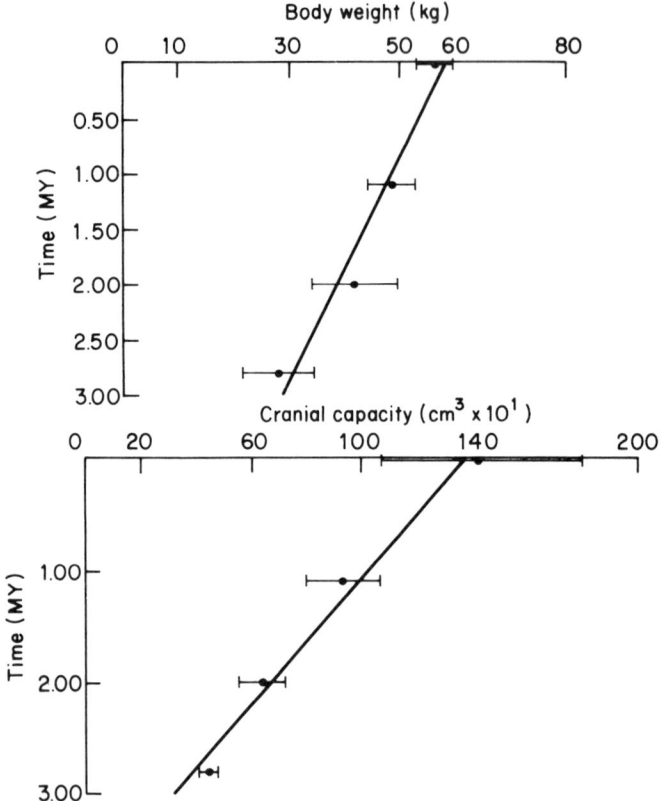

Figure 8.5. Trends in body weight and cranial capacity in hominid evolution. In each graph the points represent (bottom to top) *A. africanus*, *H. habilis*, *H. erectus*, *H. sapiens*. Bars indicate observed range for *H. sapiens* and ±1 standard deviation for the other species. From Cronin et al. (1981).

In general, information on human evolution is plagued by gaps in the record, insufficient and damaged fossil material from which reconstruction of overall morphology is attempted, and continuing uncertainty over the correct phylogenetic tree. Although future discoveries of fossils will no doubt improve the position, it seems unreasonable at present to use the limited available data to argue strongly for or against punctuated equilibrium, despite the intrinsic interest in knowing the precise evolutionary pattern of our own genus and its nearest relatives.

8.3 A CRITIQUE OF PUNCTUATED EQUILIBRIUM

The idea that if the fossil record were complete enough all localities would show a finely graded sequence of intermediates between ancestral and descendant species seems untenable, given that speciation frequently occurs through the divergence of spatially restricted populations. Indeed, where speciation involves morphological change in such a population, most localities would be expected to show an ecological punctuation, as described in section 8.1. The theory of punctuated equilibrium as propounded by Eldredge (1971), Eldredge and Gould (1972) and Gould and Eldredge (1977) has been largely instrumental in exposing the expectation of common intergradation as unwarranted in the light of modern views on cladogenetic speciation. However, this theory has unfortunately been put forward as more of a revolution than it is (for example, the claims that it stands against neo-Darwinism) and with too little regard for some serious problems. It has consequently been reacted against rather strongly by many population geneticists, who see it as an unwarranted attack on establised evolutionary theory.

Several points of criticism have been made in the previous two sections, some general and some in relation to particular examples of punctuated equilibrium. In the present section I will summarize what seem to be the three main problem areas that need to be dealt with before the typical pattern of long-term evolution at the species level can be established and its significance assessed. I have kept this section very brief, since the points have already been alluded to, albeit in a restricted context.

The first problem is that spatial details need to be given more attention if an evolutionary rather than an ecological punctuation is claimed. If speciation commonly occurs by the splitting off of populations occupying small geographical areas then the odds against a particular study site turning out to be the area inhabited by the speciating population are presumably high, and clear evidence needs to be given to show that these odds have been overcome. In studies conducted so far, such evidence is lacking.

The second problem stems from the recent shift in palaeontological work from higher taxa to the species level. Although it is clearly desirable to have palaeontological data on the one 'real' taxon, this downwards taxonomic shift has carried with it the danger that the small phenotypic changes observed may in some cases be due largely to direct environmental effects. Although this problem may be particularly severe in molluscan shells, some of which rival plants in their plasticity, a genetic basis should rarely be taken for granted in the case of small variations. Also, the fact that shells are easily modified by the environment is in itself rather worrying because they provide some of the best fossil material and are likely to be frequently used in testing for morphological punctuations.

The third problem is that the mechanisms underlying genetically based morphological punctuations need to be more clearly thought out before it is apparent whether these punctuations require us to question neo-Darwinism. In particular, the tendency to equate large morphological changes with large (in an unspecified sense) genetic changes should be avoided. Also, there seems to be a tendency to contrast developmental and polygenic change despite the fact that these are not alternatives to each other; polygenic modification of adult morphology, like any other modification of it, must act through the alteration of development.

In conclusion, the theory of punctuated equilibrium is an advance on some prior palaeontological thinking in that it has begun to introduce spatial considerations into studies of long-term temporal change. However, it needs to be further refined and tested; and also needs to be detached from the generalized attacks on neo-Darwinism which often accompany it. Charlesworth (1981), in discussing such attacks, remarks "I hope this is only a passing phase". I can only add that if it does not pass then the attacks should become more specific and the supposed alternative evolutionary mechanisms should be clearly stated so that their validity can be assessed.

8.4 SUMMARY

The theories of punctuated equilibrium and phyletic gradualism are briefly outlined, and several points are made in order to clarify the controversy over which of these theories provides a more adequate picture of long-term evolution at the species level. Two case studies are then discussed, one of which concludes that punctuated equilibrium is the prevalent pattern of morphological evolution in the group concerned, the other that gradual change predominates. The problems involved in the two studies are discussed and it is seen that neither study allows an unambiguous interpretation of the relevant data. The three main general problems, namely location of the area inhabited by a speciating population, demonstration of the heritable nature of the morphological changes involved in punctuations, and establishment of the developmental and genetic mechanisms underlying these changes, are re-emphasized out of the context of any particular case study in the hope that future investigations may tackle these problems more determinedly. Finally, the desirability of detaching support for punctuated equilibrium from generalized attacks on neo-Darwinism is noted.

Chapter 9

Development and Evolution

"What evolutionary theory requires is not so much a hypothesis of the ultimate physico-chemical mechanisms of development, but rather a picture of the possible interactions between developmental processes."

C. H. Waddington, 1975.

9.1 DEVELOPMENT, EVOLUTIONARY THEORY AND SALTATIONS

Frequent attempts have been made to incorporate some aspects of development into evolutionary theory. These have ranged from early proposals that ontogeny 'repeats' phylogeny, which have been exhaustively discussed by Gould (1977), to recent attempts to extend population genetics theory into the realm of regulatory genes (Hedrick and McDonald, 1980). Claims by Williamson (1981a) that patterns of punctuated equilibrium are a reflection of alternating phases of developmental constraint and developmental instability provide another example. These diverse studies have little in common except for a striving towards the incorporation of developmental processes or the regulatory genes that underlie them (see chapter 10) into a more synthetic theory of evolutionary mechanisms than neo-Darwinian theory, which sees these mechanisms largely in terms of genetics and ecology.

Why is it necessary to take a partially developmental approach to evolution, and why have workers in such disparate evolutionary disciplines felt the need to attempt such an approach? There are at least two answers to this question, one of which is independent of, and the other firmly dependent upon, the extent to which evolution occurs by phenotypic saltations, and I will deal with these respectively.

The first answer has been clearly stated by Lewontin (1974), who noted that although the connection between Mendelian and biometrical approaches is now apparent, nevertheless we still have two bodies of evolutionary theory, one framed in terms of the frequencies of genes and genotypes, the other in terms of phenotypic measurements and their distributions. Neither is sufficient in itself as an overall theory of evolution because, while it is genes rather than phenotypes that pass from one generation to the next, it is phenotypes rather than genes that are subjected to natural selection. Thus a complete theory of evolution needs to make reference to both genetic and phenotypic variables and, among other things, needs to incorporate the laws which govern the production of a phenotype from a genotype; that is, the laws of development. Maynard Smith (1972) was focussing on the same issue when he commented: "Lacking a theory of epigenesis, we cannot say how many gene substitutions are required to convert a fin into a leg, or a monkey's brain into a human one."

What we need, at least in the first instance, are the general principles underlying development rather than a complete molecular breakdown of developmental processes. This was stressed by Waddington (1975; see quotation at the start of this chapter) and it is worth noting that the same applied to the incorporation of the laws of genetics into evolutionary theory. This incorporation was begun after the re-discovery of Mendel's work and did not have to await the revelation of the structure of DNA and the molecular details of the genetic code. The trouble is, of course, that sometimes elucidation of principles may follow from, rather than precede, detailed molecular investigations, and at this early stage in our understanding of development it is quite possible that the former pattern may hold. However, we are not totally ignorant about development, and the remainder of this book is largely devoted to attempting to incorporate what little is known or can reasonably be postulated about development into an expanded version of neo-Darwinism.

Despite the readily apparent incompleteness of an evolutionary theory lacking a developmental component, many population geneticists have shown a remarkable ability to pursue microevolutionary projects without any references to developmental aspects of the specific problem with which they are dealing. One example of such a study, in which I have unfortunately contributed to this non-developmental approach (Arthur, 1978, 1980c, 1982b), is the case of colour and banding polymorphism in the land snail *Cepaea nemoralis*. The genetic basis of variation in colour and banding in this species is well known (see review by Murray, 1975) and the microevolution of these polymorphisms has received an enormous amount of attention (see Jones *et al.* (1977), Clarke *et al.* (1978), and references therein). However, nowhere in the vast literature on the evolutionary genetics of *Cepaea* is there any significant discussion of developmental aspects of the patterns of pigmentation. In the case of the colour polymorphism this omission may not be too serious. The major locus determining shell colour is not a 'regulatory' gene in that it appears to be turned on in all of the shell-forming cells of the mantle, and in consequence the base colour of the shell is uniform. However, the loci controlling the banding pattern of a *Cepaea* shell are pattern-forming genes in that they cause a repeatable heterogeneity of pigment distribution across each whorl, which differs between different genotypes. Their action may be viewed in terms of a distribution of morphogen, and one possible model is given in Figure 9.1. This is entirely speculative but it serves to emphasize that the

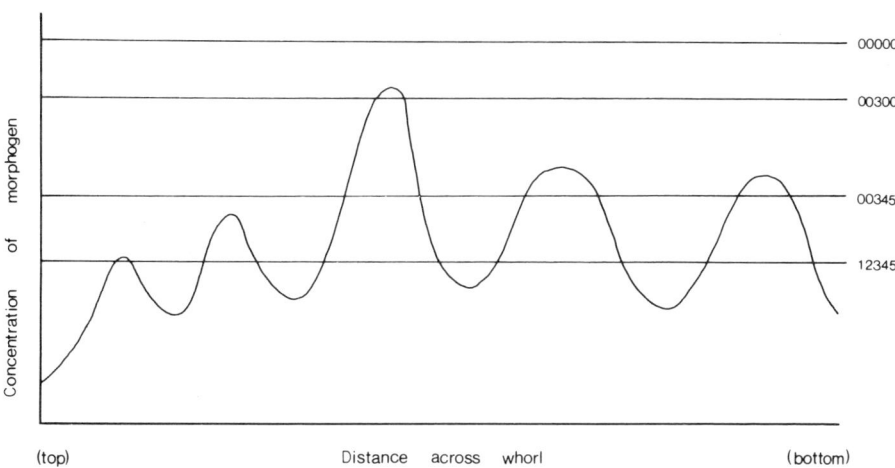

Figure 9.1. Scheme showing how several banding morphs of *Cepaea nemoralis* could be explained in terms of a morphogen/threshold system with genetic variation in the position of the threshold. The phenotypic codes indicate (by a zero) which of bands 1 (top) to 5 (bottom) are missing.

development of banding patterns in *Cepaea* is one aspect of this microevolutionary system that requires further investigation. The scheme proposed here is a direct transposition of Waddington's (1969) attempts to explain the more complex pattern of shell pigmentation of *Conus* in developmental terms.

Returning to the central issues of size and shape, it is apparent that the microevolution of these characters also has been dealt with in a non-developmental, but this time a quantitative and biometrical, way. Here, not only the connection betwen genotype and phenotype, but even the genotype itself, is poorly understood. The connection between changing gene frequencies and changing phenotypic values in artificial selection experiments illustrated in Figure 4.11 was discussed in a rather abstract way and, as mentioned in chapter 4, progress in our understanding of what polygenes actually do has been slow.

Despite this problem, there is clearly considerable understanding of the microevolution of size and shape, particularly in response to artificial selection but also, to a lesser extent, in response to selection in natural populations. This has been possible because of the negligible effects of individual polygenes. The fact that these effects are so small means that the 'developmental gap' in our understanding of the evolutionary processes concerned is, or at least appears to be, slight. It is sufficient to know that different alleles cause slight differences in morphogenesis without knowing exactly *how* they do so, though clearly our understanding of morphological microevolution would be more complete if this information was available.

This is an appropriate point at which to turn to the second reason for taking a developmental approach to evolution; that is, the reason that applies only in the case of phenotypic saltations. In conventional neo-Darwinian theory saltational evolution of morphology is rare or non-existent, stemming from Darwin's 'Natura non facit saltum' mentioned earlier; thus the developmental problem is minimized. However, anyone proposing a saltational theory of evolution cannot escape so lightly because the onus is on them to demonstrate how the saltations that they propose actually take place at both the *individual* level, which is relevant here, and the *population* level (see chapter 11). So far, no saltational theory, ranging from that of de Vries (1905) to punctuated equilibrium, has been given an easy ride and some have foundered at least partly because their putative developmental changes or constraints have not been accompanied by a satisfactory and general explanatory mechanism — a point which I have already stressed in the context of punctuated equilibrium in the previous chapter.

My own view on the degree to which evolution occurs in a saltational manner is somewhat heretical. I do not share the punctuationalists' view that the typical speciation event is accompanied by a morphological punctuation caused by developmental changes qualitatively different from those involved in microevolutionary differentiation of populations. Some speciation events no doubt are accompanied by rapid morphological change but, as has been pointed out by Jones (1981) and Charlesworth and Lande (1982), this is easily explicable within a neo-Darwinian framework of polygenic modification. However, I also do not share the extreme neo-Darwinian view that nature 'does not make jumps'. Rather, I take the view that evolution makes jumps very rarely indeed but that the importance of these occasional jumps is out of all proportion to their frequency, and that they are, at least in some instances, largely responsible for the origin of a novel major taxon such as an order, class or phylum. It seems unlikely that all major taxa arise in this way, and it is certainly not true that all fixations of mutations with large morphogenetic effects lead to new higher taxa (see chapter 11), but the lack of a one-to-one correspondence between these events should not lead us to rule out an important evolutionary role for mutations of large phenotypic effect.

Although the excesses of de Vries (1905) and Eldredge and Gould (1972) have met with considerable opposition, many eminent biologists of less extreme viewpoints have also stated, or hinted at, a belief that some form of saltational evolution has taken place. Probably the first to do so in the context of natural selection was T. H. Huxley (see F. Darwin, 1887). However, the case has been most eloquently put by D'Arcy Thompson (1942), who contrasts the discontinuous nature of differences between phyla with the continuous variation among genera to which his own method of transformations applies, and comments that: "Our geometrical analogies weigh heavily against Darwin's conception of endless small continuous variations; they help to show that discontinuous variations are a natural thing, that 'mutations' — or sudden changes, greater or less — are bound to have taken place, and new 'types' to have arisen, *now and then*." The italics are mine and are employed here to stress that Thompson did not apparently think — and nor do I — that phenotypic saltation was the common mode of evolutionary change. As suggested earlier, this form of evolution may be rare but of considerable importance. Since this is the view taken here, it is clearly necessary to state precisely how saltational changes are thought to occur at both individual and population levels. In chapter 10, which deals largely with individuals, some mutations giving rise to large phenotypic changes are described. The evolution of mutations of this kind, and in particular the mechanism by which they spread at the population level, is discussed in chapter 11. In the remainder of the present chapter a broad outline of some aspects of development will be given and the question of what developmental phenomena or principles are likely to be important in an evolutionary context will be examined.

9.2 CELL DIFFERENTIATION, PATTERN FORMATION AND MORPHOGENESIS

It is neither possible nor desirable to build all aspects of developmental biology into evolutionary theory. Ideally, a few key principles should be sufficient, as noted by Waddington (1975), but unfortunately these principles largely remain to be established. However, developmental phenomena, as opposed to the mechanistic principles that underlie them, have been well documented in a large number of organisms, and are often categorized as in the above heading. I will be concentrating here on pattern formation and morphogenesis rather than on cell differentiation; a few words are thus necessary on (a) the distinctions between these different aspects of development, and (b) the reasons for concentrating on some of them.

Most organisms begin life as a single cell, usually a fertilized egg cell containing male and female pronuclei which fuse to form the first zygotic nucleus and restore the diploid condition. The situation is more complex in some groups such as ferns which exhibit sporophyte and gametophyte generations, but I will restrict the discussion here to organisms having a simpler life-cycle with only one unicellular stage, such as vertebrates, insects and molluscs. The adult forms of such organisms are composed of a very large number of cells. A *Drosophila* imago, for example, is reckoned to contain approximately one million cells (Struhl, 1982). In addition to producing an increase in the number of cells and hence in organismic size, development also produces a very precise and repeatable spatial pattern of the thousands or millions of cells involved. This is readily apparent from a comparison of the adult morphology of different individuals of the same species which, notwithstanding the slight quantitative variation discussed in earlier chapters, resemble each other very closely.

At the cellular level, this increase in size and complexity that constitutes development is brought about entirely by three processes: cell division, cell migration and cell differentiation. Cell growth could conceivably be thought of as a separate process, but it seems reasonable to consider externally visible changes in any particular cell, including changes in size and shape, as part of cell differentiation along with internal changes such as the switching on or off of particular genes in particular cell types, the proliferation or destruction of organelles, and so on.

In books on developmental biology (for example, Garrod, 1973) there are sometimes separate chapters on pattern formation and morphogenesis, though the distinction between these two processes is usually left rather obscure. Garrod seems to see morphogenesis as involving a considerable amount of cell migration, and pattern formation as a process taking place largely through cell division, but this distinction cannot be easily made. Gastrulation is given as an example of morphogenesis, yet this process involves both division and migration of cells. (See Sonnenblick (1950) for a description of cell movements and proliferation in gastrulation in *Drosophila*.) Also, Garrod (1973) gives amphibian limb regeneration as an example of pattern formation, yet this process involves cell migration as well as cell division and differentiation (see review by Tank and Holder, 1981). Attempting to interpret the words literally does not help to achieve a distinction either. Morphogenesis means the formation of shape; and shape, of course, is a pattern in two- or three-dimensional space. Thus morphogenesis essentially means pattern formation, albeit it refers to a restricted type of pattern. Perhaps 'pattern formation' could be used to refer to situations where not only is an overall shape being moulded but also a spatial pattern of cell differentiation is being laid down, while 'morphogenesis' could be used to refer to situations in which only the former process is occurring (such as cleavage). However, since the two processes usually occur together this distinction seems unhelpful.

Because of the difficulty in achieving a clear distinction between the two processes I will treat pattern formation/morphogenesis together and will consider either of these terms to refer to the among-cell rather than within-cell component of development; that is, the attainment of a particular three-dimensional arrangement of various tissues through controlled cell migration and division combined with spatially ordered differentiation. The mechanisms underlying morphogenesis are presumably different to those causing a particular cell to differentiate in a particular way, though the former may be considered to subsume the latter.

The reason for concentrating here on morphogenesis rather than cell differentiation is that morphological evolutionary changes usually do not involve the origin of novel cell types whereas they do, by definition, involve changes in morphogenesis. This has been stresed by Wolpert (1978), who points out that histologists recognize roughly 200 different cell types in man and the same 200 in the chimpanzee. The notable differences between ourselves and *Pan troglodytes* are differences in the spatial arrangement of tissues and organs as well as in overall shape and size (and, of course, behaviour) rather than differences in the array of available cell types. (It has also been shown that the structure of many individual proteins differs negligibly between man and the chimpanzee: see King and Wilson, 1975.)

A recent attempt at a developmental approach to evolution, apparently inspired in part by the contrasting enzymic similarity and morphological differences between ourselves and the great apes, was made by Hedrick and McDonald (1980). However, these authors based their work on regulatory genes similar to those of Britten and Davidson's (1969) model of eukaryotic gene control despite the fact that Britten and Davidson's model relates more to cell differentiation than to the among-cell component of development, namely morphogenesis. The discussion herein will not follow this course and will be centred instead on morphogenesis and the genes which affect it. These genes may, as will be seen in the following chapter, be regulatory *or* structural genes in the sense that geneticists normally use these words; for example, in the context of the *lac* operon in *Escherichia coli*.

That an explanation of morphogenesis requires principles over and above those necessary for an understanding of cell differentiation is easily seen from a consideration of the shapes of different muscles in a single organism. For example, in humans both the biceps and the infraspinatus muscle (which extends across the dorsal surface of the scapula) are composed of striated muscle cells; also, both have tendons for attachment to bone and both have a nerve and blood supply intimately associated with the cells of the muscle itself. These 'accessories', like the muscles, are composed of the same cell types and extracellular materials in biceps and infraspinatus, as well as in the many other muscles of the human body. Despite this cellular similarity, the familiar shape of the biceps and the 'flattened triangular' shape of infraspinatus are very different. So although an individual tissue type always involves differentiation into the same array of cell types, the overall spatial pattern of differentiation can be very varied among different pieces of the same tissue.

There are at least two fundamentally different ways in which a spatial pattern to cell division and differentiation could arise. First, the cell could have access to 'historical' information. For example, if the cell registered each mitotic event in some way and kept a record of the number and orientation of the mitoses it had undergone, this information could perhaps be used to terminate cell division at a certain stage. The main alternative to such a process is for cells to use positional information. Cells possessing this latter kind of information 'know' where they are in a developing organism and behave accordingly. It is difficult to see how historical information alone could account for several important phenomena including regeneration, and also development of the regulative, rather than mosaic, kind where experimental removal of part of an embryo does not affect the production of a normally proportioned adult. Positional information, on the other hand, can potentially explain both of these phenomena. Thus it seems reasonable to conclude that development involves the determination of cell activities either by positional information alone or by a combination of positional and historical information. In either case, positional information is of considerable importance.

While Wolpert (1968, 1969, 1978) has done much to stress the involvement of positional information in development, several other authors have made suggestions as to how this information might be conveyed. The suggested mechanisms include gradients of diffusible substances (Child, 1951; Crick, 1970) and alternatives such as the phase-shift model of Goodwin and Cohen (1969) wherein cells measure their position by monitoring two distinct pulses of biochemical or electrical activity originating from a so-called pacemaker region. As Wolpert (1978) has pointed out, this latter system is roughly equivalent to an observer working out his position relative to that of a thunderstorm by measuring the time interval between thunder and lightning. Whether any one mechanism of specification of positional information is universal and, if it is, then whether it is always achieved in the same way biochemically, remains to be seen. In the next two sections I hope to show why it is possible to make some headway in a developmental approach to evolution even though these uncertainties remain.

Much of the work done on positional information has not been explicit about the involvement of genes. However, a separate and largely genetic approach to morphogenesis has been adopted, with considerable success, by Garcia-Bellido, Lawrence, Morata and others working on development in *Drosophila* (see, for an overview, Garcia-Bellido *et al.*, 1979). Here, the switching on and off of certain key genes appears to be involved in providing information about which of a series of developmental compartments a cell is in. This work will be discussed in chapter 10. At present it is only necessary to note that, while the genetic bases of proposed chemical gradients are largely unknown, so are the biochemical aspects of morphogenetic gene action in *Drosophila*. Because of this problem, these two approaches continue to be rather separate, though as our knowledge of development progresses, the gap between them should gradually close.

9.3 THE CENTRAL PROBLEM OF MORPHOGENESIS

Having decided to concentrate on morphogenesis rather than cell differentiation as the main aspect of development that needs to be incorporated into a comprehensive theory of morphological evolution, it is necessary to ask which facts or hypotheses in the morphogenetic area are likely to be most important. The two things which most readily suggest themselves are the *genes* that control morphogenesis and the *general pattern* of their interactions in so far as this can be ascertained.

The identification and categorization of genes which affect morphogenesis is clearly important since, when this has been done, it should be possible to investigate the evolutionary role of each category. Classical evolutionary theory in the form of neo-Darwinism (see section 1.5) has tended to suppress fruitful discussion in this area because of its assertion that only genes with very minor effects on morphogenesis are usually involved in evolution. These genes are, precisely because of their small phenotypic effects, difficult to identify, so a belief in neo-Darwinism leads to a rather pessimistic prognosis for a genetic and developmental, rather than just a biometrical, approach to morphological evolution. However, since the view of evolution proposed here (chapters 11–15) is not as restricted as conventional neo-Darwinism and does not rule out an important evolutionary role for genes of large phenotypic effect, a classification of different types of gene with special reference to their effect on morphogenesis will be included (chapter 10).

Although identification and classification of morphogenetic genes is useful, this falls a long way short of providing a comprehensive conceptual framework for morphogenesis. Clearly it is desirable to have some such overall framework so that evolutionists can build developmental generalities rather than just developmental details into an expanded version of neo-Darwinism. The first step in establishing the general principles of development, or of any other process for that matter, is asking the right questions. This point is exemplified by the early history of attempts to explain inheritance. The basic Mendelian theory is both remarkably simple and remarkably general but, despite many eminent nineteenth-century biologists including Darwin devoting much time to considering possible mechanisms of inheritance, only one of them (Mendel, for course) came up with the right answer to the problem and even then its importance went unrecognized for 34 years. Fisher (1930) has commented: "It is a remarkable fact that had any thinker in the middle of the nineteenth century undertaken, as a piece of abstract and theoretical analysis, the task of constructing a particulate theory of inheritance, he would have been led, on the basis of a few very simple assumptions, to produce a system identical with the modern scheme of Mendelian or factorial inheritance."

The main problem, presumably, lay in seeing on which axis the old theory and the then unknown correct one represented opposing poles. Early ideas on inheritance could be described by a variety of terms. It was only after the discovery of particulate inheritance that these early ideas became strongly identified as 'blending inheritance'. Had they always been so labelled, no doubt someone could have attempted to devise an alternative theory of particulate inheritance from first principles long before Mendel discovered such inheritance through his experiments on garden peas.

To an outsider, the field of theoretical developmental biology, if such a field can be said to exist, appears to be in a state resembling that of early nineteenth-century 'genetics'. This view seems to be shared by Medawar (1967), who states that in developmental biology "something is wrong; or has been wrong. There is no *theory* of development, in the sense in which Mendelism is a theory that accounts for the results of breeding experiments." Although there has been a good deal of theoretical work on development since Medawar made the above comment, it still applies. The theories such as they are do not provide an overall conceptual framework, and they certainly are not predictive in the sense that Mendelism is — though perhaps this latter goal cannot be fulfilled due to the different nature of the phenomenon to be explained.

The continuing lack of a satisfactory theory of development may be at least partially due to developmental biologists asking the wrong questions. For example, if we decide, as Watson (1976) has, that "The heart of embryology is the problem of cell differentiation", then we will be led to ask about the general principles governing how a cell differentiates. However, while an elucidation of these principles would be of enormous value in other contexts, it would not lead to an understanding of morphogenesis because it would relate to what goes on within individual cells rather than within large populations of cells.

Another question that is often asked of development is: How does homogeneity and simplicity change into heterogeneity and complexity? This at first seems a fairly sensible question to ask since most organisms show little evidence of cell differentiation up to the blastula stage and are hence regarded as more-or-less homogeneous entities during this earliest phase of the life-cycle whereas later on they are clearly extremely heterogeneous at the among-cell level. However, concentrating on how heterogeneity is initiated may not be particularly wise. There are already four potentially sound answers to this question, each of which may apply at least in certain groups or organisms, but none of which leads to a large conceptual advance in our understanding of morphogenesis.

One way in which heterogeneity may be initiated is through contact, at the egg stage, with interior maternal structures which are themselves already highly differentiated. Body axes (anteroposterior, *etc.*) are in some way, presumably molecular, passed on from one generation to the next. This means of transmission of heterogeneity appears to operate in *Drosophila* (Sonnenblick, 1950) where the alignment of the egg while still inside the mother determines its axes. An alternative to using maternal heterogeneity to initiate heterogeneity of the offspring is to make use of the ubiquitous heterogeneity of the environment. A zygote beginning development in the external environment rather than in a maternal reproductive tract may, for example, make use of gravity to determine primary organismic axes. Even in the case of a blastula consisting of a number of identical cells and inhabiting that non-existent place, the homogeneous environment, initiation of cellular heterogeneity (gastrulation) could be achieved purely through stochastic molecular events (Turing, 1952). A fourth means of initiating organismic axes is to use the point of entry of the sperm into the egg to determine them. There are several examples of species which appear to use this mechanism, including the mollusc *Chaetopterus* (Morgan and Tyler, 1938). Thus, in conclusion, there is no shortage of explanations for the production of cellular heterogeneity, but none of these explanations are satisfactory in themselves as explanations of the broader process of morphogenesis.

In fact, it is apparent from the above comments on the eggs of *Drosophila* and from the action of some maternal genes in the ensuing generation as a result of agents in the egg cytoplasm (see section 10.3), that eggs are not uniform biological entities, even if the cytoplasm is considered in isolation from the nucleus. What actually happens in development is *not* the production of a complex, heterogeneous phenotype from a simple, homogeneous one. Rather, development involves the production of one highly complex, heterogeneous entity from another, through a series of intermediate and intergrading stages. The fact that the heterogeneity of egg cytoplasm is much less obvious to a human observer than that of an adult snail, horse or fruitfly should not lead us to conclude that eggs are internally uniform. Of course, most biologists would agree that, despite the lack of an acceptable scale of measurement, phenotypic heterogeneity and complexity increase during development (and evolution: see chapter 13) but this is a rather different matter. Recognition of this point leads to a suspicion that the key questions to ask of development should relate not to the non-existent initiation of phenotypic heterogeneity but rather to its *transmission* and/or *spread*. Thus we may ask:

1. How does one kind of biological heterogeneity within a developing organism lead to another? and
2. What causes the developmental increase in the complexity of the phenotype?

Much theorizing in developmental biology can be seen as attempts to answer the first of these questions rather than the second, though the authors concerned do not usually make this clear. A case in point is the so-called French flag problem introduced by Wolpert (1968, 1969) to illustrate how a gradient of a diffusible substance could give rise to a spatial pattern of different cell types (symbolized by patches of red, white and blue cells). (Wolpert also used the French flag model to investigate the problem of size invariance; that is, the production of invariant relative proportions even in the face of surgical reduction of the size of the developing system. However, I will not be concerned with the problem of size invariance here.)

Although most discussion of the French flag model seem to stress only the heterogeneity of the end-result, i.e. the flag, this model is actually one in which there are three forms of heterogeneity in a definite cause-and-effect sequence. Using Crick's (1970) version of the flag system which illustrates this point most clearly, we see (Figure 9.2) that a primary type of heterogeneity, namely the system of 'source', 'sink' and 'neither' cells, leads to a second kind of heterogeneity — the gradient of the diffusible substance — which leads in turn to a third and final form of heterogeneity, the three sheets of differently coloured cells. Thus the French flag analogy shows how the transmission of heterogeneity could occur, not how heterogeneity is spread (or initiated). So it is an attempt to answer, or at least to refine our ideas on, question 1 above rather than question 2.

It is my view that this concentration on the transmission of heterogeneity rather than its spread is one of the main obstacles to the formulation of an overall picture of development that will be useful to evolutionists. The question of the causes of the developmental *spread* of heterogeneity and associated increase in phenotypic complexity, on the other hand, seems to be the central problem of morphogenesis from an evolutionary viewpoint and perhaps even

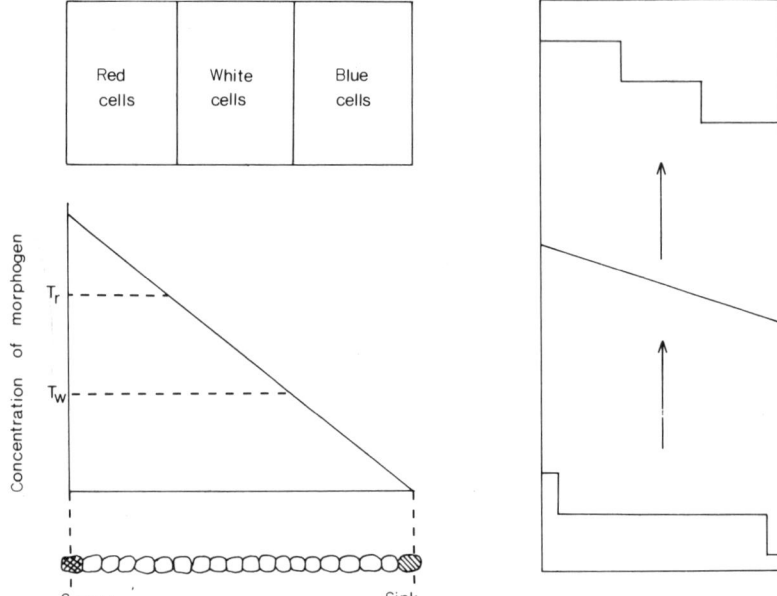

Figure 9.2. The French flag model. The extreme left-hand cell in the bottom line produces a diffusible morphogen that is destroyed by the cell at the right-hand end of the line. This produces a linear gradient in the concentration of the morphogen which in turn produces sheets of red, white and blue cells, the positions of which depend on where the gradient crosses the thresholds T_r and T_w. The pattern of heterogeneity characterizing each of the three components of this system is indicated on the right. Based on Crick (1970).

for an understanding of development itself. How this increase might be brought about by interacting systems of morphogens is considered in the following section.

9.4 THE MORPHOGENETIC TREE: A HYPOTHESIS

Throughout this section I will be referring to entities known as morphogens and, since their activities are central to the hypothesis to be formulated, it is necessary to state at the outset exactly what is meant by 'morphogen', though the word is to some extent self-explanatory. Basically, a morphogen is anything whose presence, state or concentration causes cells to move, divide or differentiate or to stop doing any of those three things. The physicochemical nature of morphogens may be quite varied, and although they are often thought of as molecules, there is no reason why other entities such as a pulse of electrical activity could not also constitute a morphogen. Crick (1970) has suggested that morphogens are unlikely to be either large molecules like proteins, because of their poor diffusibility, or small ions such as K^+, because of their non-specificity, but rather something intermediate in size, perhaps cyclic AMP or a steroid. Whether or not this turns out to be the case is not particularly relevant here since it is the causal connections between different morphogens that provide the focus of attention, and such connections are, in some respects at least, independent of the physicochemical nature of the morphogens concerned.

Two kinds of effect that a morphogen may have, which can be labelled as *terminal* and *serial* effects, now need to be distinguished. A terminal effect has already been illustrated in the French flag analogy. If we consider the diffusible substance of Figure 9.2 as the morphogen, its effect is to produce cells of three distinct kinds in a particular spatial pattern. That effect is an end-result in that the three types of differentiated cells represented by the flag do not go on to cause further morphogenetic processes. For the purposes of illustrating a serial effect, we can consider a slight extension to the French flag model. Suppose that some additional morphogen is involved in the differentiation of the bottom line of cells in Figure 9.2 into source, sink and 'neither' cells. The effect of this morphogen may be described as serial because it does not stop after one phase of pattern formation, but leads directly to a second phase of morphogenetic activity. The difference between serial and terminal effects, then, is that in the former the morphogen causes differentiation into a number of cell types, some or all of which produce another morphogen. In the case of terminal effects, the differentiated structure produced as a result of the morphogen's presence is itself morphogenetically inert.

It is difficult to see what can be gained from theorizing about terminally acting morphogens. Indeed, the most interesting thing to find out about these would seem to be their physicochemical nature. However, serially acting morphogens must be widespread in developing systems and it is of interest to consider what kinds of interconnection may exist between such morphogens; that is, what sort of network they form. At least three elementary kinds of network (and many compound ones) are conceivable, and these three may be referred to as *line, tree,* and *house-of-cards* networks; they are illustrated in Figure 9.3. It is important to note that this figure is designed to show causal connections and, although it thus also illustrates temporal sequences (in that a cause must precede its effect), some temporal questions are left unanswered, such as whether one morphogen ceases to be produced once it has stimulated the production of others.

Although the morphogenetic networks underlying actual developing systems cannot be as simple as any one of the three elementary networks illustrated in Figure 9.3, it is worth enquiring whether they conform more closely to any one elementary network than to the others. It seems likely that the usual type of network underlying a system like vertebrate development, where a unicellular stage gives rise to an adult through a series of gradually larger and more complex intermediates, is basically a tree, though some small parts of it may depart from this pattern. This proposal, which I will call the morphogenetic tree hypothesis, is supported by the following argument.

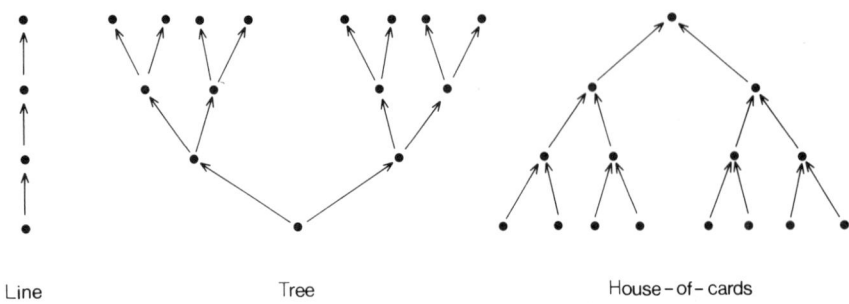

Figure 9.3. Three possible networks of serially acting morphogens. Each dot represents a morphogen; arrows indicate causative links.

First, the mere fact of increasing size suggests that the number of morphogenetic fields should increase during development if each field has a limit to its size. Such limits are widely accepted, and maximum sizes of around 50–100 cells have been suggested (Wolpert, 1959; Crick, 1970). Second, increasing complexity regardless of size suggests an increase in the number of active morphogens through development. Until the molecular nature of morphogens is established we will not know whether a binary, on–off type of action or a more complicated switching action such as the three-way switching of the French flag's diffusible morphogen is the rule. But assuming that the average complexity of switching action does not increase during development, it seems likely that the number of morphogens increases instead. In fact, size and complexity usually increase together in development and as they do so the number of different parts of the organism with different morphogenetic activities characterizing them increases. This argues strongly in favour of different morphogens acting in different places, during late development, rather than the same morphogen, produced in several source regions, controlling all morphogenetic fields. Finally, although the above argument was formulated with diffusible morphogens in mind, it should be noted that both a diffusible substance and a gene whose state (on/off) alters the effect of that substance in receptor cells are morphogens, as defined above. Thus a morphogenetic tree could take the form of proliferating 'signal morphogens' or proliferating 'receptor morphogens' or both. It seems likely that morphogenetic trees are produced by a synergistic spread of both these two kinds of morphogen; however, this proposal, like the tree itself, is hypothetical.

So far, nothing has been said about the role of genes in producing the diffusible kind of morphogen. Whatever sorts of molecule these turn out to be (and however widespread they are in developing systems) it seems inevitable that they are produced enzymatically from some morphogenetically inert precursor. A simple scheme of this kind is illustrated in Figure 9.4. Undoubtedly the real situation will turn out to be more complex than this, but at least the picture presented can serve as a basis for discussion of the role of genes in morphogen production and I will return to this picture in chapter 10,

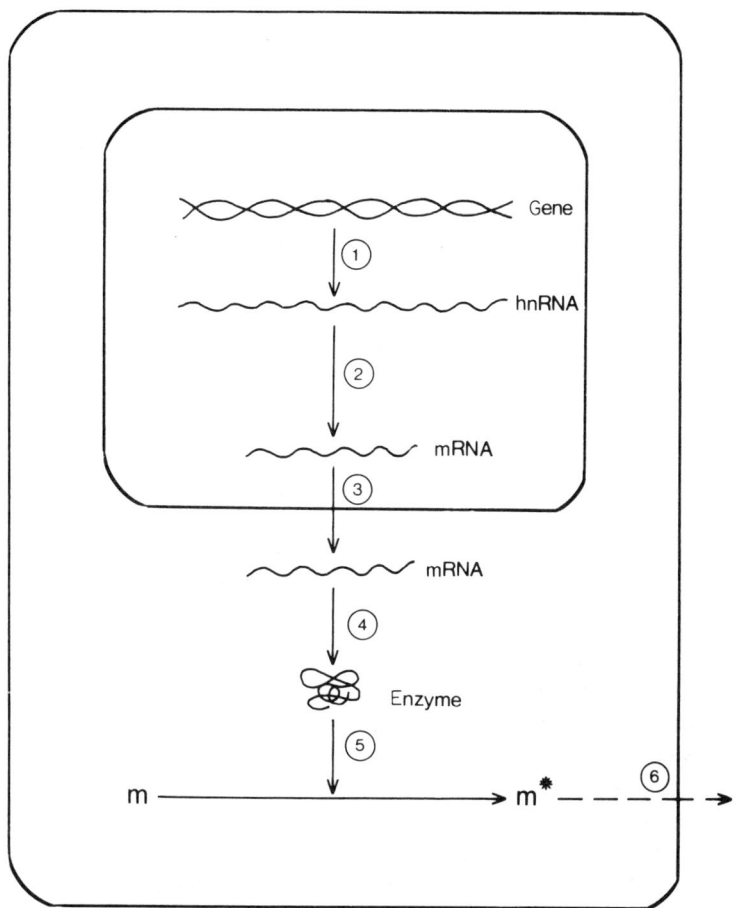

Figure 9.4. Diagram showing how genes could be involved in the production of diffusible morphogens. Steps 1–6 are, respectively: transcription, intron removal, passage through nuclear membrane, translation, production of morphogen (m*) from inert precursor, (m), diffusion out of cell.

where some specific examples of genes which affect morphogenesis will be examined.

Organisms with complex life-cycles are unlikely to be explicable, developmentally, in terms of a single tree arrangement of serially acting morphogens. However, it may well be that the developmental system involved still has a tree-like basis rather than a linear, or house-of-cards, one. Possible dual-tree networks for the alternating generations of ferns and the imaginal disc/metamorphic system of insects are presented in Figure 9.5.

9.5 SUMMARY

Reasons for attempting to incorporate concepts from developmental biology

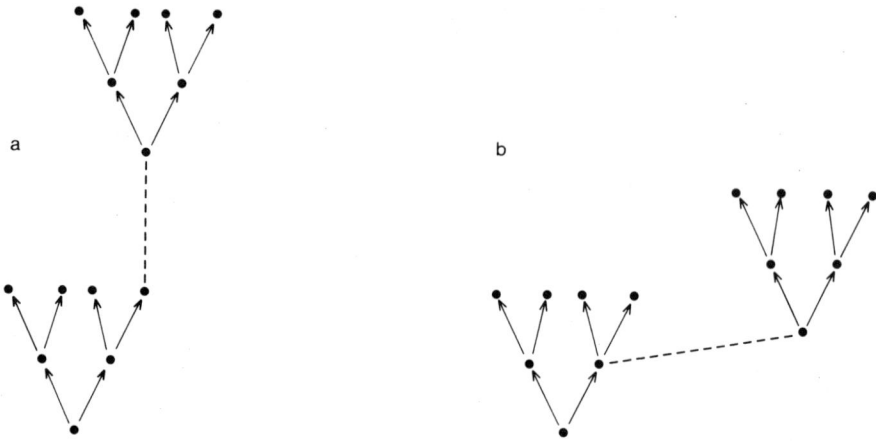

Figure 9.5. Dual-tree networks of morphogens potentially applicable to (a) ferns and (b) insects. Dashed lines show the interconnection between different stages of the life-cycle (sporophyte/gametophyte and larva/imago, respectively). See also legend of Figure 9.3.

into evolutionary theory are given and some previous attempts of this kind are briefly discussed. It is noted that developmental principles rather than details are required and that among-cell spatial ordering patterns are more pertinent for the evolutionary morphologist than cell differentiation *per se*. Possible questions that can be asked about this among-cell component of development are considered and it is suggested that too much current work is directed at the problem of how cellular heterogeneity is transmitted in developing systems and too little at how it is increased. This increase then provides the focus of attention and a tree-like network of interacting morphogens is proposed as a possible general scheme underlying morphogenesis in organisms with relatively straightforward life-cycles. Finally, a few comments are made about the role of genes in morphogenesis; and some refinements to the morphogenetic tree hypothesis are proposed which may help to extend it to organisms whose life-cycle is relatively complicated.

Chapter 10

Developmental Genetics

"I have tried for a long time to convince evolutionists that evolution is not only a statistical genetical problem but also one of the developmental potentialities of the organism."

R. Goldschmidt, 1940.

10.1 INTRODUCTION

The ideal genetic follow-up to the material discussed in chapter 9 would be a description of the genes which are involved in the production of certain morphogens, including details of the link between gene and morphogen and of that between morphogen and phenotype. Unfortunately, a complete treatment of this kind cannot be given because, as noted earlier, the products of key morphogenetic genes are largely unknown, as are the molecular identities of morphogens. The general theory of morphogen action in terms of gradients of diffusible substances has thus remained rather distinct from developmental genetics. However, our present inability to provide a complete genetic and molecular account of morphogenesis should not be allowed to act as a deterrent to attempts to consider the evolutionary implications of genes which have an identifiable morphogenetic effect. In many cases, much is known about the genes themselves and about the phenotypic effects of mutations in them. For example, the chromosomal location of a gene, the dominance relationships between its alleles, and whether or not it is polymorphic in natural populations, have been established in some of the cases to be discussed in sections 10.3 and 10.4. Also, much is known about the effects of some of these genes not only on the adult phenotype but also on the developmental processes leading to that phenotype. Thus for some genes a substantial fraction of the overall picture of their role in morphogenesis can be seen.

The number of genes with identifiable morphogenetic effects is so large that it would be impossible to describe even a small proportion of them in any detail. In *Drosophila melanogaster* alone, several hundred mutations which affect morphogenesis have been documented (see Lindsley and Grell, 1968), including many well-known and routinely used ones such as bar eye (*B*) and vestigial wing (*vg*). However, many such mutations (extreme neo-Darwinists might say all) are associated with marked fitness-depression and have probably not been involved in evolution except to be removed by natural selection as soon as they occur. What is thus needed is a discussion of a *few* genes affecting morphogenesis, so that a fairly intensive approach can be adopted, the few actually dealt with being chosen because they have at least a *possible* evolutionary role.

I have chosen two groups of morphogenetic genes for discussion: those affecting the direction of shell coiling in gastropods (section 10.3) and the insect homoeotics — particularly well studied in *Drosophila* — each of which causes all or part of a segment to develop in a manner inappropriate to itself but in approximate conformity to the normal pattern of development of some other segment (section 10.4). Genes that switch the direction of shell coiling have almost certainly been involved in evolution on several occasions, and they are also of interest because their effect is delayed for a generation — the genotype of the mother determines the phenotype of her offspring. The homoeotics are better understood genetically; for example, their precise chromosomal locations are known. They have also been involved in evolutionary debate though their role in evolution is much less certain than that of the gastropod shell-coiling genes. Homoeotics have been included by Goldschmidt (1940, 1952) in his category of 'hopeful monsters', dismissed as evolutionarily irrelevant by critics of Goldschmidt's work, and, more recently, suggested as having an evolutionary role by those currently working on them from a developmental angle (for example, Garcia-Bellido, 1975; Lawrence and Morata, 1976). Because major advances in our understanding of the compartmentalization of development in *Drosophila* have taken place over the last ten years, a re-examination of the possible evolutionary role of the homoeotics — which control this process — may be timely. In the present chapter these two groups of morphogenetic genes, the homoeotics and the genes controlling shell coiling, will be described; evolutionary discussion will follow in chapter 11.

10.2 THREE CATEGORIES OF GENES

Before discussing particular examples of genes affecting morphogenesis in certain species, it is useful to consider into what general category these genes fit. It is especially important to dispel the notion that 'regulatory genes' and genes affecting development are the same thing. This point is illustrated in Figure 10.1 (see also Figure 9.4) in which a gene producing an enzyme that converts a molecule a to a related form a' is controlled in some way (either positively or negatively) by a regulatory gene. This system may incorporate negative feedback of a' on the regulator gene or its product, with the result that the concentration of a' fluctuates within a relatively narrow range of values. The system may also be influenced by a or other molecules entering the cell so that production of a' can be boosted or switched off when certain conditions prevail.

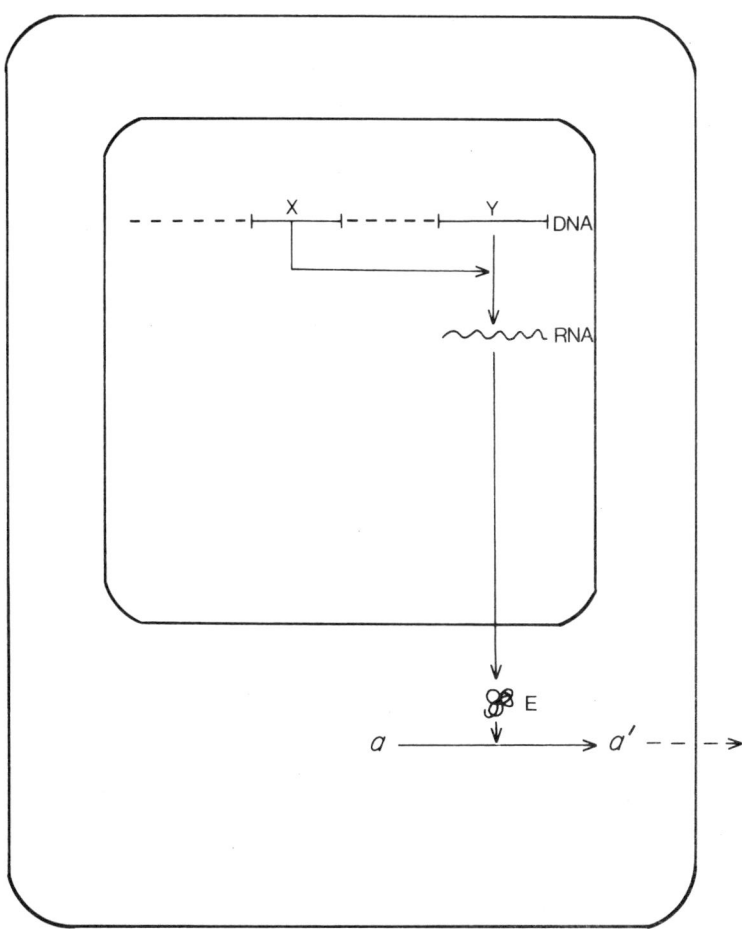

Figure 10.1. A gene control system in which a structural gene, Y, which produces an enzyme, E, is regulated in an unspecified way by a controlling gene, X. Whether X and Y regulate development depends on whether the enzyme's product, a', is a morphogen

An important distinction between the hypothetical eukaryote system described above and the bacterial operon is that the former is located in a cell which is itself embedded in a mass of other cells, and any substances produced in one cell as a result of gene activity therein may be transported, actively or passively, to neighbouring cells and may influence developmental processes occurring in those cells. If this happens in the system outlined in Figure 10.1, then a' is a morphogen and both the 'structural' gene which produces the enzyme E and the 'regulatory' gene which influences that structural gene, are regulatory in a developmental sense, and mutations in either gene may alter the pattern of morphogenesis. I have called this combined category *D*-genes (developmental genes: Arthur, 1982d) to distinguish them from those genes, such as the regulator of Figure 10.1 if a' is not a morphogen, which are regulatory in the genetic sense of the word but not in a developmental sense. This latter category may be referred to as R-genes. The third category of genes that it is useful to distinguish contains those structural genes whose products are not directly involved in morphogen synthesis; these will be labelled S-genes. This scheme of classification provides three *mutually exclusive* categories of genes and hence is slightly different from my earlier (Arthur, 1982d) scheme.

It is clear that if genes are classified as suggested above, we need, for an understanding of morphological evolution, to concentrate on D-genes. Studies of the genetic control of the production of routine metabolic enzymes such as alcohol dehydrogenase (ADH) in *Drosophila* (see Thompson *et al.*, 1977, and Clarke *et al.*, 1979) are, unless the enzyme turns out to have a cryptic morphogenetic function, studies of R-genes and as such are not central to our purpose here, though they may be of considerable importance in furthering our understanding of eukaryote gene control *per se*.

Whatever their gene products and the precise role of those products in morphogenesis, the insect homoeotics and gastropod shell coiling genes are clearly D-genes. Also included in this category, though their effects are relatively minor, are the polygenes of quantitative genetics. Basically, morphological evolution is based on evolution of D-genes, and one of the key questions is the relative importance of different types of D-genes. It may, however, be misleading to think of 'types' within the D-gene category; homoeotics and polygenes are more likely to represent extremes of a continuum representing the magnitude of effect of individual loci on morphogenesis. Indeed, within both of these 'types' the magnitude of effect is rather variable though this variation is small in relation to the difference between them. Looking at D-genes, then, in terms of this continuum, the question is how evolutionary importance and magnitude of effect are related. Here we must distinguish between the frequency of fixations and the biological effects of the mutations that are fixed. As regards frequency, it is generally agreed that the larger the magnitude of effect of a mutation, the lower its probability of fixation. This conclusion was reached by Fisher (1930) through an abstract argument which will be discussed in some detail in chapter 12. The exact pattern of this relationship is of interest, though, and it is in terms of this pattern that a distinction between strict neo-Darwinism, as defined in section 1.5, and the view taken in this book can be seen (Figure 10.2). Although the precise shapes of the two curves are rather arbitrary (indeed it is even difficult to imagine a satisfactory scale for the x-axis) the important aspect of the figure is the relative shapes of the two curves and in particular the tail on the right-hand curve, indicating a low as opposed to zero frequency of fixations of mutations with very large phenotypic effects.

If all that was at issue here was the slight difference in the shape of the two curves in Figure 10.2, it would hardly be worth further discussion. However, the important point about the evolutionary significance of D-genes with large

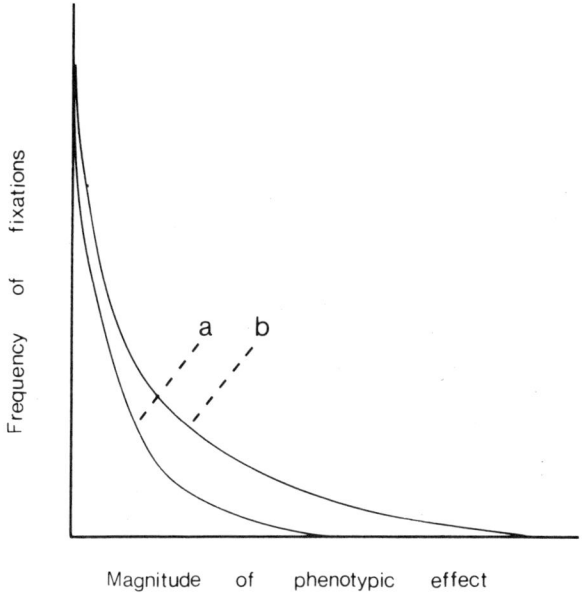

Figure 10.2. Illustration of the difference between a strict neo-Darwinian view (curve a) and the view taken here (curve b) of the relationship between a mutation's magnitude of effect and its probability of fixation.

effect is not that these genes do occasionally participate in evolutionary change; this they almost certainly do. It is rather whether on some of those rare occasions when a large-effect mutation is fixed, this event takes the population concerned to a valley in some uninhabited piece of the adaptive landscape (to use Wright's (1931) forceful analogy) from which point polygenic modification can produce a new adaptive radiation to surrounding peaks. The answer to this question, which will be returned to in later chapters, is of considerable importance to our understanding of the origins of new higher taxa.

10.3 MATERNAL D-GENES. SHELL COILING IN *LYMNAEA* AND *PARTULA*

In most gastropod species with helicospiral shells, all individuals are coiled dextrally when viewed from above, with the result that the aperture of the shell is on an observer's right if he holds the shell upright and with the aperture towards him. The shells of Figure 2.2, for example, are all coiled in this manner. Relatively few species are known in which the direction of coiling is consistently sinistral. Neither of these groups of species is informative as regards the genetic determination of the direction of coiling because, of course, there are no variant forms that can be crossed in order to observe the kind of phenotypic ratios that appear in hybrid progeny. Luckily, however, there are a few species including *Lymnaea peregra* and *Partula suturalis* in which both sinistral and dextral individuals are found in natural populations. Both of these species have been used to examine the genetics of coiling and, as will be seen below, the results are remarkably similar despite the fact that *Lymnaea* and *Partula* belong to different suborders within the Pulmonata.

Lymnaea peregra

This is a widespread freshwater species, populations of which may be found in suitable habitats anywhere from North-western Europe, including Iceland, to North Africa and Afghanistan (Ellis, 1969). In most of these populations, all individual snails have dextrally coiled shells; a few populations, however, have a small proportion (often 1–5%) of sinistrals. Localities with heterogeneous populations of this kind include ponds in Surrey, Durham and Yorkshire (see Diver *et al.* (1925) for details).

Investigation of the genetic basis of sinistrality in *L. peregra* began with the work of Boycott and Diver (1923) who performed a number of crosses using dextral and sinistral snails. Although these authors recognized certain key features of the mode of inheritance of sinistrality, they did not provide a completely adequate genetic explanation. However, on the basis of their work, Sturtevant (1923) proposed a simple system involving a single Mendelian locus with two alleles, that for dextrality being completely dominant. The novel feature, compared to other Mendelian systems known at the time (and to most discovered since), was that the phenotype of any given individual is determined not by its own genotype but by that of its mother.

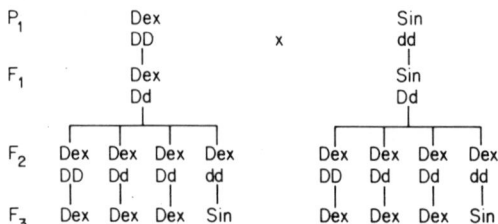

Figure 10.3. Genotypes and phenotypes of parents and progeny (up to the F_3 generation) in a cross between dextral and sinistral *Lymnaea peregra*. From Boycott *et al.* (1930).

Sturtevant's hypothesis was confirmed by Boycott *et al.* (1930) in an exhaustive breeding programme involving about a million snails. Figure 10.3 shows the genotypes and phenotypes of the F_1, F_2 and F_3 progeny resulting from a cross between homozygous dextral and sinistral snails followed by two generations of self-fertilization. It can be seen from the figure that both dominance and segregation are apparent one generation later (F_2 and F_3, respectively) than in ordinary Mendelian systems. Also, it should be noted that all offspring from any particular mother are identical as regards their

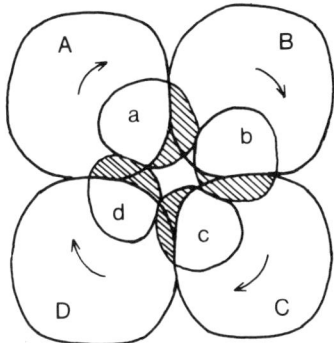

Figure 10.4. The results of the third wave of cell division in a system with spiral cleavage. Gastropods in which the micromeres (a–d) are shifted clockwise in relation to the macromeres (A–D) as shown end up with dextral shells. Those in which the shift is anti-clockwise have sinistral shells.

direction of coiling. The 3:1 phenotypic ratio observed in the F_3 is based upon numbers of *broods*, not numbers of snails within broods.

The fact that the direction of coiling of a snail's shell is determined by its mother's genotype is not entirely surprising since that direction is actually fixed at an extremely early stage in development. All molluscs, except cephalopods, exhibit spiral cleavage, in which oblique orientation of spindle axes causes a somewhat irregular (but highly repeatable) positioning of blastomeres in relation to each other. This is readily apparent after the third cleavage division, the result of which is shown in Figure 10.4. According to Sturtevant (1923), Crampton showed in 1894 that snails originating from a system of clockwise rotation of third-division micromeres, as in the figure, have dextrally coiled shells, whereas sinistral shells result from a mirror image of this cleavage pattern. In fact, it was even possible for Crampton to distinguish dextral *Lymnaea* from sinistral *Physa* from the nature of the first and second cleavage divisions. Within-species variation in the direction of shell coiling in *L. peregra* has the same link with the reversal of the cleavage pattern as has the between-genus pattern studied by Crampton (see Raven, 1964). Thus presumably some agent in the egg cytoplasm, itself determined by the diploid maternal cells surrounding the unfertilized egg and not the egg itself, which is haploid, acts to determine the planes of the initial cleavage division following fertilization, which in turn determines the ultimate direction of coiling. Thus we have here a system where, for a particular D-gene, we know:

1. its genetics, i.e. single locus, two alleles, complete dominance, maternal effect;
2. the time at which it begins to influence development;
3. its initial and final phenotypic effects; and
4. the frequency of its variant form in at least some natural populations.

This is a remarkably well-known system by present standards (at least outside of *Drosophila* work), but there is of course one major gap in our knowledge of the determination of the direction of coiling in *L. peregra*, namely the nature of the gene's action. Another remaining problem is the cause of some anomalies found only infrequently by Boycott *et al.* (1930; see also Diver and Anderson-Köttö, 1938) but too often to be the results of mutations. These were: occasional mixed broods, the aberrant snails in which sometimes bred true and sometimes did not; and the appearance, in a few sinistral lines, of a complete brood of dextrals which did breed true. Mixed broods are also found occasionally in *Partula* (see below) and an 'ontogenetic origin' has been suggested in this case (Murray and Clarke, 1966). However, Diver and Anderson-Köttö (1938) pointed to the possible action of modifiers, since they were able to isolate lines showing an above-average frequency of production of mixed broods. It seems likely that there is a polygenic, partially heritable variation superimposed upon the action of the major locus, but a full understanding of occasional departures from the simple delayed-Mendelian pattern of inheritance may have to await discovery of the mechanism of gene action.

Partula suturalis

It would be difficult to find a species of pulmonate differing in more respects from *L. peregra* than does *Partula suturalis*. The latter species is terrestrial rather than freshwater; it is a stylommatophoran as opposed to a basommatophoran; its distribution is restricted (to the tiny Pacific island of Moorea) rather than widespread; it gives birth to young snails and does not lay egg masses; and it is highly polymorphic for colour which *L. peregra* is not. As regards the natural occurrence of the sinistral and dextral forms, the two species are again strikingly different. Whereas in *L. peregra* there are no totally sinistral populations or any heterogeneous ones with more than 10% sinistrals, *P. suturalis* exhibits wholly sinistral and wholly dextral populations as well as polymorphic ones in which a wide range of relative frequencies of the two variants is found (see map in Clarke and Murray, 1969).

Similar genetic bases of the coiling polymorphisms in two such distinct species are perhaps less likely than in, for example, a pair of congeners, but if found would provide much stronger evidence for a *general* genetic basis of coiling in snails, including the large number of invariant species which are not amenable to genetic analysis. It is of considerable interest, then, that Murray and Clarke (1966, 1976) have demonstrated a pattern of inheritance of coiling in *P. suturalis* that is essentially the same as that in *L. peregra*, the only notable distinction being a reversed direction of dominance, i.e. sinistral dominant to dextral.

The single-locus, delayed-Mendelian pattern of inheritance (which confirmed Crampton's (1932) suspicions on finding only homogeneous broods in his dissections) is illustrated by the results of one of Murray and Clarke's crosses, shown in Figure 10.5. The letters refer to phenotypes; genotypes are not given. However, since the snails used as parents in the initial cross came from

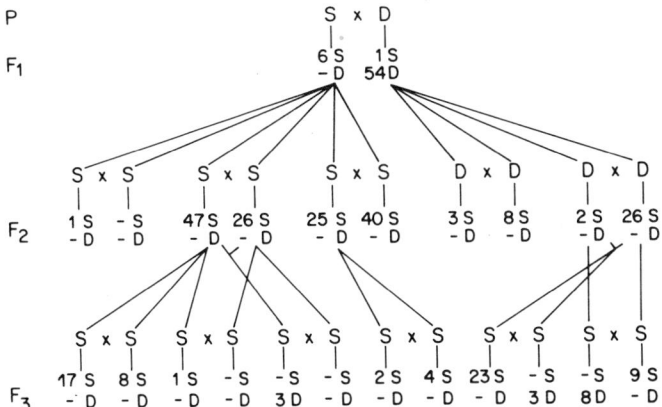

Figure 10.5. Phenotypes of parents and progeny (up to the F_3 generation) in a cross between dextral and sinistral *Partula suturalis*. A fork at the beginning of one of the lines indicates uncertainty over parentage. From Murray and Clarke (1976).

phenotypically homogeneous populations, they are presumably homozygotes. Murray and Clarke (1976) refer to the locus determining the direction of coiling as the H-locus, with alleles H^S (dominant) and H^D (recessive). Using this form of symbolism, the genotypes of the parents in Figure 10.5 are (left to right) $H^S H^S$ and $H^D H^D$. The coil of F_1 heterozygotes depends on which parent acted as the mother (though there is one aberrant snail) while the dominance of H^S is shown by the lack of any dextral phenotypes in the F_2. Broods in the F_3 generation were all homogeneous, with an excess of sinistral ones, though statistical testing against a 3:1 ratio would clearly require many additional broods. Thus in its essential features the pattern of inheritance of dextrality in *P. suturalis* is the same as that of sinistrality in *L. peregra*.

The occasional departures from this simple pattern of inheritance exhibited by *P. suturalis* are, like their counterparts in *L. peregra*, too common to be explicable as mutations. Murray and Clarke (1976) found 11 snails with the 'wrong' coil out of a total of 286. These were not restricted to a particular family, and an unspecified ontogenetic origin is suggested by the authors. As with *Lymnaea*, it may not be until the gene product is identified and its mode of determination of the planes of cleavage divisions ascertained, that we will fully understand the reasons for occasional departures from the otherwise simple pattern of inheritance.

Other species

Although there is at least one additional case in which a single-locus, delayed-Mendelian inheritance of coiling is known (*Laciniaria biplicata*: Degner, 1952), lending additional weight to the case for a general genetic mechanism underlying the direction of coiling, there are many other species (more than 200 of them: see Pelseneer, 1920) in which very occasional inverted-coil individuals appear, and at least in some cases these are developmental accidents rather than genetically determined effects. These non-genetic phenomena, however, do not necessitate any revision of our views of the genetic basis of coiling, just as the existence of ether-induced bithorax phenocopies of *Drosophila* do not in any way invalidate the conclusions that have been drawn as to the genetic basis of this phenotype (see next section). The fact that coiling variants with a genetic basis have not been observed in most species of gastropods is probably due to a lack of study combined with a very infrequent occurrence of such forms by mutation and their rapid removal, by selection, in most ecological situations.

10.4 ZYGOTIC *D*-GENES: INSECT HOMOEOTICS

The studies on shell coiling in gastropods have shown that very early stages of morphogenesis (and hence any later characteristics determined at an early stage) are controlled by the maternal, rather than the zygotic, genome. Other investigations of various kinds support this conclusion and extend it to other taxonomic groups. In both echinoderms and amphibians, morphogenesis can proceed until the blastula stage even if all zygotic chromosomal material is eliminated (Ede, 1978); thus no zygotic *D*-genes are effective at this early stage in these groups. In *Drosophila* it appears that synthesis of proteins coded for by zygotic genes does not begin until after blastoderm formation (Roberts and Graziosi, 1977). Thus the general picture that emerges is one of initiation of morphogenesis by largely unidentified agents in the egg cytoplasm, followed by a gradual or abrupt takeover of control by zygotic *D*-genes, the exact pattern of which no doubt varies between different taxa.

Although some characteristics of the adult phenotype are determined at a very early stage in development and hence are controlled by maternal rather than zygotic *D*-genes, the development of many phenotypic characters does not begin until after the switch of developmental control to the zygotic genome. Variant forms of such phenotypic characters should show an ordinary Mendelian pattern of inheritance, rather than the delayed pattern exhibited by the maternally acting *D*-genes of *Lymnaea* and *Partula* discussed in the previous section.

One important class of *D*-gene, most of whose members are of this kind (i.e. zygotic as opposed to maternal) is revealed by homoeotic mutations in insects (see review by Ouweneel, 1976). These mutations, and especially certain examples of them, are the subject of the present section; but before discussing them I will briefly outline some features of insect development which need to be understood before the effects of homoeotic mutations can be fully appreciated. The account that follows relates to *Drosophila melanogaster*, since the examples of homoeotics to be described shortly concern this species. However, many of the broad features of *Drosophila* development are shared by other insects, and indeed homoeotic mutations are known in a wide range of genera (see Table 10.1).

Table 10.1. Homoeotic mutations in insects other than *D. melanogaster*. From Ouweneel (1976)

Species	Mutation	Symbol	Locus	Alteration
D. ananassae	bithorax	bx	2	Haltere → wing
D. funebris	Aristapedia	?	autos.	Arista → tarsus
D. hydei	Antennapedia	se^{Anp}	2	Antenna → leg
D. pseudoobscura	aristapedia	ar	2–32 ±	Arista → tarsus
D. simulans	Ultrabithorax	Ubx	3–71 ±	Abdomen I + MT → MS
D. simulans	aristapedia	ap	3–75 ±	Arista → tarsus
D. subobscura	quasimodo	qm	E	Leg II,III → leg I
D. subobscura	macrohaltere	?	?	Haltere → wing (?)
Musca domestica	aristapedia	ar	5	Arista → tarsus
M. domestica	antennapedia	atp	?	Antenna → leg
Aedes albopictus	proboscipedia	prb	1–20 ±	Labella → tarsus; maxillary palpi → tarsus/antenna
Anopheles quadrimaculatus	supernumerary mouth part	?	?	Extra mouth part like that of 'proboscipedia'
Bombyx mori	supernumerary legs	E	6–0.0	Thoracic and extra abdominal legs on larval abdominal segments
Tribolium castaneum	antennapedia	ap	8	Antenna → leg
T. confusum	labiopedia	lp	1	Labial palpi → forelegs
Blatella germanica	Pro-wings	Pw	?	Pronotum → wings

A fertilized *Drosophila* egg develops into a syncytial blastoderm which, after the formation of cell membranes, undergoes a complex series of cell movements and divisions together with differentiation into a variety of cell types, the end-result of which is the production of a first-instar larva which breaks out of the thick chorionic membrane of the egg. The larvae go through two further instars before pupating and undergoing the metamorphosis which produces the adult insect or imago. The whole process from fertilization to eclosion takes about ten days when cultures are kept at 25 °C.

At a very early stage in this rapid developmental process, the cells from which adult tissues will be formed are segregated from the bulk of larval cells and are concentrated in a series of *imaginal discs*. Some of these can be recognized histologically even in newly hatched first-instar larvae (Poulson, 1950), though at this stage the discs consists of a relatively small number of cells. During larval development the discs remain distinct, and grow in size and cell number. For example, the wing disc has around 100 cells at the time of egg hatching, but has roughly 10 000 cells by the middle of the third larval instar (Garcia-Bellido *et al.*, 1973).

It is inappropriate here to go into a lot of detail about the imaginal discs themselves (see Poodry (1980) for a detailed description) but an important question that does need to be answered is: Which discs give rise to which parts of the adult? In answer to this question it must first be said that most investigations have concentrated on the origin of adult ectodermal structures; that is, the eyes, wings and cuticle. Although adult mesodermal musculature may also be derived from certain cells within the imaginal discs (Lawrence and Brower, 1982), less is known of the origin of such tissues and I will concentrate here on ectodermal structures.

There is a series of discs giving rise to various parts of the head and thorax, and each one is named according to the main adult structure(s) to which it gives rise; examples are the eye-antenna disc and the wing disc. Abdominal segments of the imago are formed from a series of bodies known as histoblasts which, along with the imaginal discs, are shown in Figure 10.6. Genetic analyses of development, including studies of the homoeotic mutations, have concentrated on head and thoracic structures. Accordingly, nothing further will be said here about the abdominal histoblasts.

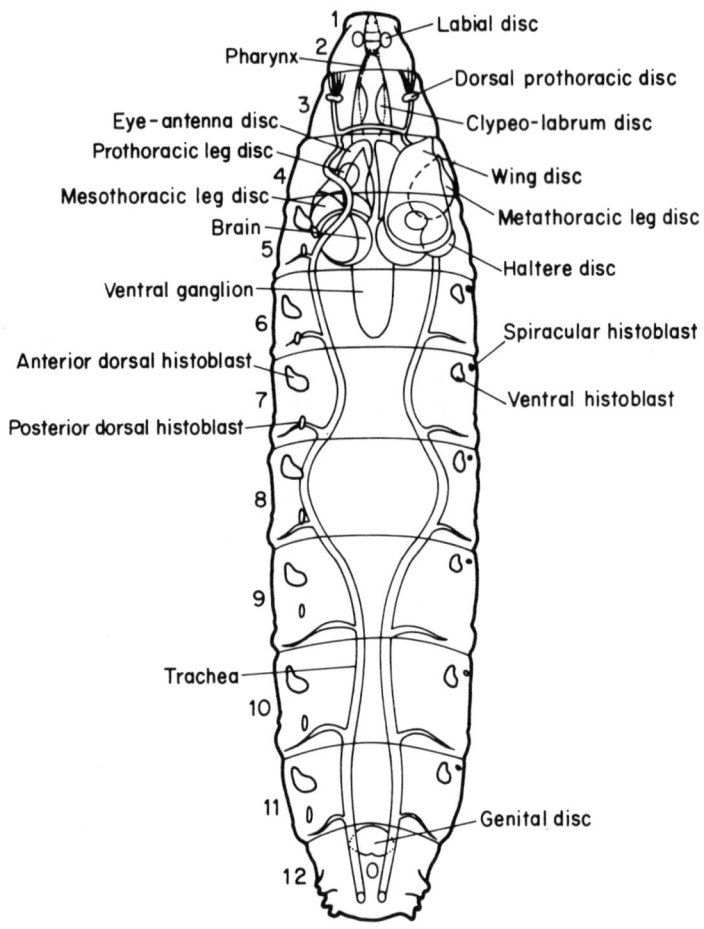

Figure 10.6. Approximate positions of imaginal discs and abdominal histoblasts in the larva of *D. melanogaster*. From Poodry (1980).

As imaginal discs grow in size during the larval phase of development, the cells within them come, as they multiply, to be restricted to a gradually narrower fate. That is, the extent of a particular adult segment to which any cell's descendants can contribute diminishes. One of the main methods used to establish this fact is the induction of somatic recombination by X-irradiation of embryos or larvae. If mitotic recombination is induced in a stock that is heterozygous for some recessive mutation such as multiple wing hairs (*mwh*) then the descendants of a cell in which the recombinant genotype *mwh/mwh* is produced are visibly distinguishable in the adult from the other cells (see Figure 10.7). The area covered by the mutant patch or *clone* of cells shows the extent of the adult epidermis that is formed by the descendants of any single disc cell present at the time the irradiation took place. Comparing different individuals subjected to the same treatment shows whether cell lineages are fixed in development. For example, if a particular section of wing is totally occupied by a mutant clone in one adult fly but only partially in another, cell lineages are clearly indeterminate. This is indeed the case in *Drosophila*, as shown in Figure 10.8a where partially overlapping clones can be seen.

By varying the phase of development at which irradiation takes place, clones of varying size are produced. Mutant clones can even be made to extend over structures normally produced by two or more imaginal discs, providing irradiation is carried out prior to the formation of the discs concerned. For example, irradiation at 3 h from fertilization produces mutant clones which extend over parts of wing and leg, whereas if the treatment is delayed until the 7–10 h stage, no such clones are found (Wieschaus and Gehring, 1976).

In order to determine the *potential* rather than the actual extent of coverage of a particular clone, a method known as the 'Minute technique' has been used (see Garcia-Bellido *et al.* 1973; Morata and Ripoll, 1975). The dominant Minute mutation (M) causes a lowered mitotic rate. Thus somatic recombination to wild-type M^+/M^+ cells in a heterozygous M/M^+ background produces clones of cells which grow much more rapidly than their neighbours. If the M mutation is linked to a visible marker such as *mwh*, then the faster-growing clone can be visually identified because the same recombinational event producing M^+/M^+ also produces *mwh/mwh* from *mwh/+*.

Use of the Minute technique in analysing wing development has shown that, although cell lineage is largely indeterminate, there are certain fixed boundaries which cells never cross. These boundaries often go unnoticed when the clones are small simply because a mutant clone may be formed too far from a boundary for its shape to be affected by it. The regions delimited by such boundaries are known as compartments (Garcia-Bellido *et al.*, 1973). The anterior and posterior compartments of the *Drosophila* wing, together with their separating boundary, are shown in Figure 10.8b.

The whole developmental process leading from imaginal discs to adult structures is in fact compartmentalized in the same way as the wing. Colour diagrams of the main compartments are given by Garcia-Bellido *et al.* (1979), and the central issues in the 'compartment hypothesis' are discussed by Lawrence and Morata (1976). The two important aspects of compartments for our present purposes are as follows. First, as development proceeds, a stepwise process of subcompartmentalization occurs. For example, the wing disc's

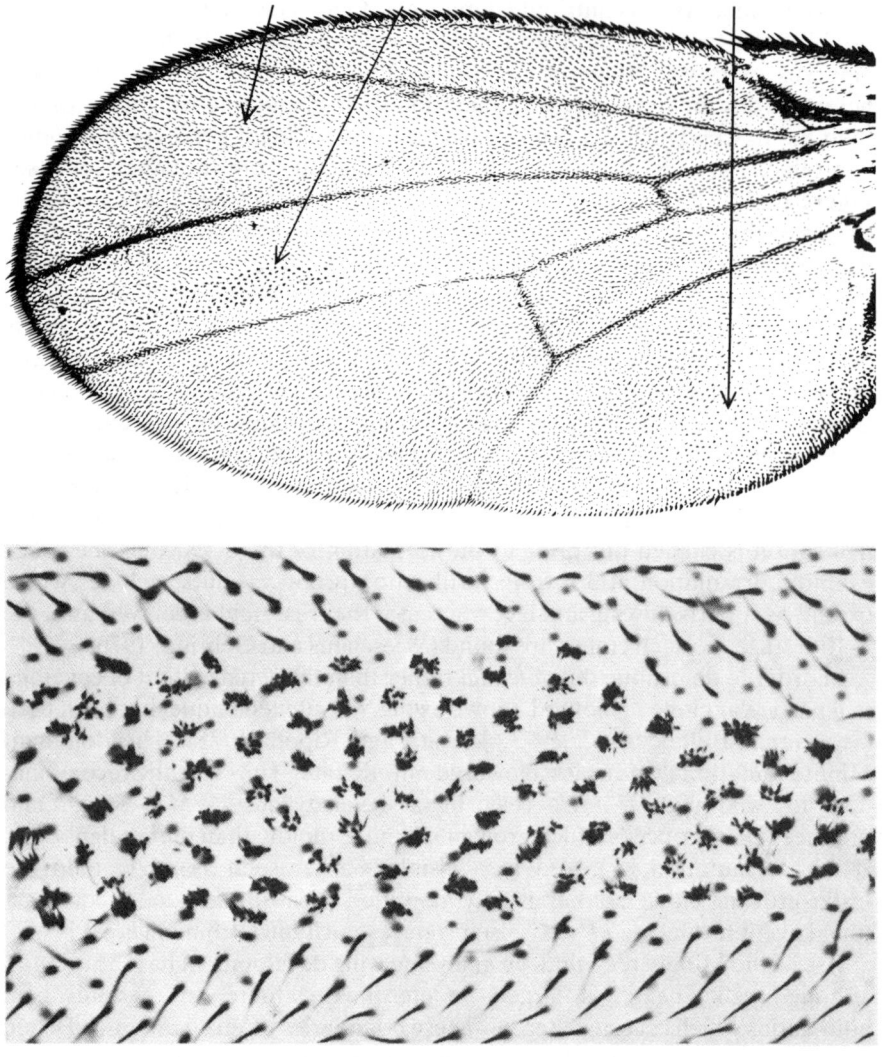

Figure 10.7. Clone of wing cells with multiple hairs (top, central arrow) in a wing whose other cells have only single hairs (flanking arrows). The lower section is a magnified view of the marked clone and the area immediately around it. From 'Compartments in Animal Development', by A. Garcia-Bellido et al. Copyright © 1979 by Scientific American, Inc. All rights reserved.

anterior and posterior compartments, themselves established by 10 h after fertilization, both become divided into four smaller compartments delimited by fixed boundary lines at about 40 h. Second, compartments are precisely the regions of the adult body affected by the homoeotic mutations. These mutations thus reveal the existence of a group of D-genes which have been called *selector genes* (Garcia-Bellido, 1975) which, when functioning normally,

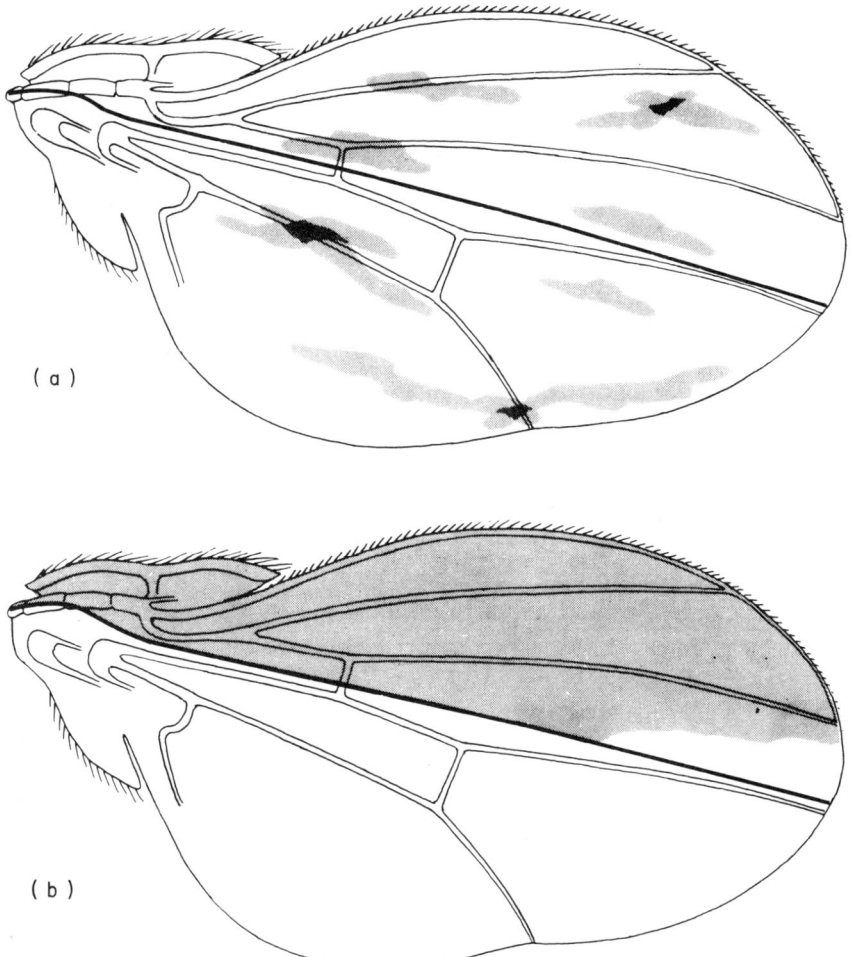

Figure 10.8. (a) Marked clones induced in different flies overlap partially when plotted on a single wing, showing small-scale indeterminacy of cell lineage.
(b) Large marked clone produced by the Minute technique (shaded) reveals a compartment boundary (heavy line) running longitudinally through the middle of the wing. From 'Compartments in Animal Development', by A. Garcia-Bellido et al. Copyright © 1979 by Scientific American, Inc. All rights reserved.

cause selection of the appropriate developmental pathway for a particular compartment.

An example of a series of homoeotic mutations which is particularly relevant here, since the possible evolutionary significance of the loci involved and others like them will be considered in the following chapter, is the bithorax series in *D. melanogaster* studied by Lewis (1951, 1963, 1964, 1978). This series, which includes the mutations bithorax (*bx*), Contrabithorax (*Cbx*), Ultrabithorax (*Ubx*), bithoraxoid (*bxd*) and postbithorax (*pbx*), is known as a pseudoallelic series because, although the mutations all map to a very small area (position 58.8 on chromosome 3), they are not coincident but fall into a very closely linked sequence in the order in which they are listed above. In addition to their chromosomal locations, the dominance or otherwise of the various bithorax mutations is known, this being indicated by the initial letter of the individual mutations. As can be seen, Contra- and Ultra-bithorax are dominant, while the rest are recessive. All the mutations show an ordinary Mendelian pattern of inheritance, and not the delayed-effect type illustrated by the shell-coiling *D*-genes of *Lymnaea* and *Partula*, though interestingly the switching on or off of the bithorax series appears to be controlled by an earlier-acting *D*-gene (extra sex combs or *esc*: see Struhl, 1981) which does exhibit the delayed-effect pattern indicative of maternal action.

The phenotypic effects of mutations in the bithorax complex are rather drastic, and are illustrated diagrammatically in Figure 10.9. It can be seen that each mutation alters the development of one or more segments or their major compartments so that their fate is switched to that which is appropriate for some other segment or compartment. The effects of two separate mutations in the complex are additive; thus a fly homozygous for *bx* and for *pbx* has the entire haltere transformed into a wing.

As the foregoing account shows, a considerable amount is known about the homoeotics. However, the question of precisely how the genes produce their morphogenetic effects is, as in the case of the shell-coiling *D*-gene of gastropods, unanswered. Again, it is difficult to see the connection between the genetic aspects of development and the morphogen/positional information system in which the genes must somehow be involved.

Despite this gap in our understanding of homoeotic gene action, one thing is certain: these genes do not produce an enzyme that sets up a gradient of a diffusible substance which in turn governs cell differentiation in the way that the morphogen in the French flag model does. The effect of the homoeotics is on pattern formation, not on cell differentiation. The arrays of cell types of wing and haltere, or of leg and antenna, are broadly similar; it is their spatial arrangement that differs. Also, the action of homoeotic genes is *cell autonomous*; that is, the gene's effects are restricted to those cells in which it is switched on. This has been shown by studies on various homoeotics (e.g. engrailed: Garcia-Bellido and Santamaria, 1972) in which marked clones of homoeotic tissue are produced in a wild-type background. In these mosaics the genetically marked patches of tissue and the homoeotically transformed patches correspond, indicating a lack of diffusible effects accompanying the homoeotic transformation.

Although the mode of developmental control effected by the selector genes of which the homoeotics are mutant forms is not yet clear except in the rather negative sense that we can state some ways in which these genes do not act, it is nevertheless instructive to consider how different selector genes interact to determine the developmental pathways adopted by compartments. The gradual process of subcompartmentalization that occurs as development

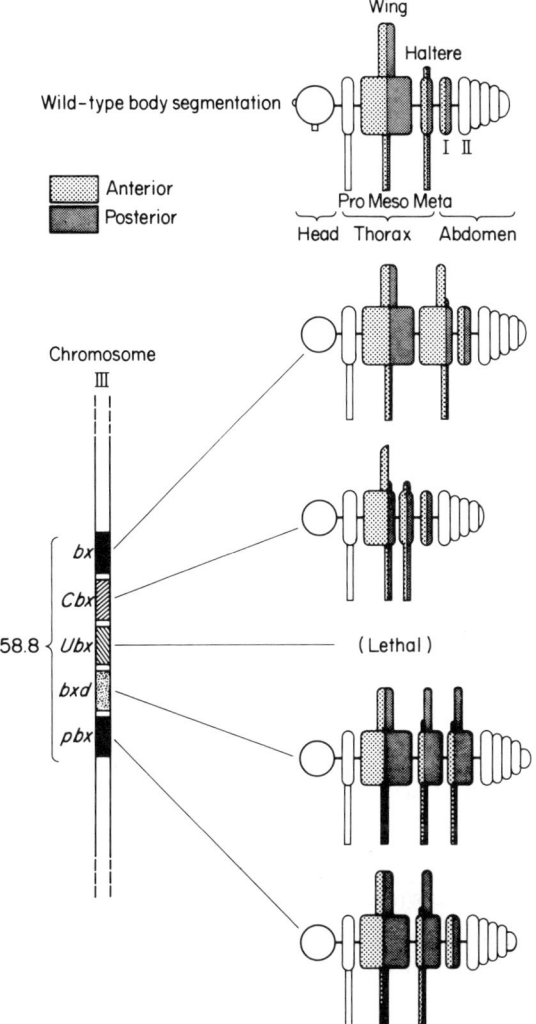

Figure 10.9. Diagrammatic respresentation of the segmental morphology of wild-type *D. melanogaster* (top) and of flies exhibiting four types of bithorax mutation. The chromosomal locations of the genes concerned are also shown in a simplified and diagrammatic manner. From Lewis (1963).

proceeds appears to be controlled by a series of these genes acting one after the other. The stages of compartmentalization of the mesothorax are shown in Figure 10.10a, while Figure 10.10b indicates the genes that are involved in some of the 'decisions' as to which compartment is specified. It has been suggested (Garcia-Bellido et al., 1979) that each compartment specified by a series of selector genes can be thought of as having a binary code, each bit of which represents a particular gene being switched on or off (Figure 10.11)

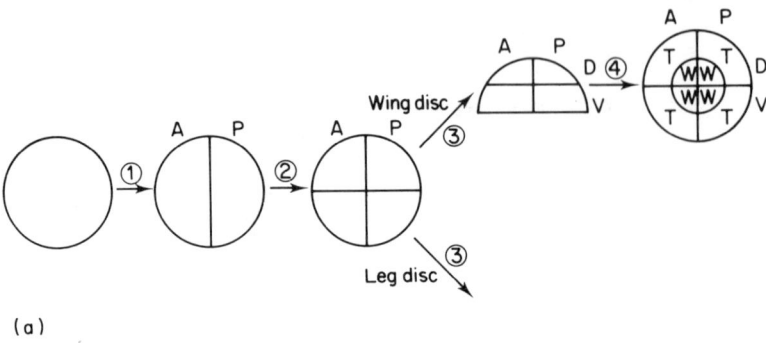

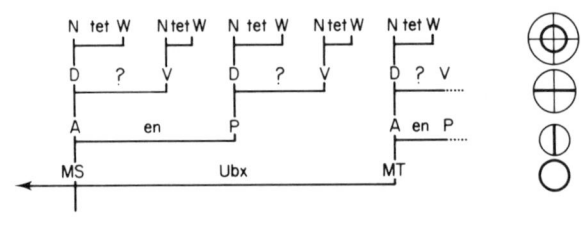

Figure 10.10. (a) Stages in the progressive subcompartmentalization of the wing disc. A = anterior; P = posterior; D = dorsal; V = ventral; T = thorax (notum); W = wing.
(b) Role of selector genes in the subcompartmentalization process. MS = mesothorax; MT = metathorax; Ubx = Ultrabithorax; en = engrailed; tet = tetraltera. From (a) Morata and Lawrence (1977), and (b) Garcia-Bellido (1975).

It is clear that, since different homoeotics affect different steps in a time series of developmental commitments, different selector genes become operative at different stages. The earliest-acting are presumably those with a maternal pattern of inheritance such as *esc*, whose normal functioning is necessary for the control of the bithorax complex (Struhl, 1981). The activation of bithorax thus follows that of *esc* and is itself followed by engrailed (the anterior/posterior controller) and so on. Once operational, many selector genes apparently need to function until the late third-instar larval period if development is to proceed correctly. The evidence for this is reviewed by Morata and Lawrence (1977); this article is also a good starting-point for anyone wishing to follow up the compartment/homoeotic story in more detail than has been allowed by the necessarily brief treatment given here. Morata and Lawrence (1977) also discuss the question of whether compartmentalization of development occurs in groups outside the arthropods, for example the vertebrates. This question has also been examined by Garcia-Bellido *et al.*

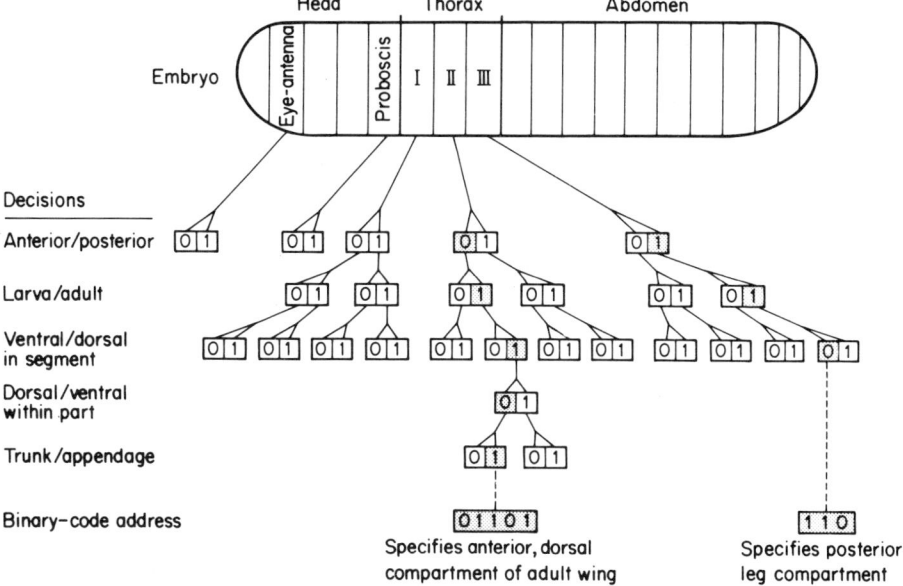

Figure 10.11. A view of the compartmentalization process as a series of binary decisions representing off (0)/on (1) states of selector genes. From 'Compartments in Animal Development', by A. Garcia-Bellido et al. Copyright © 1979 by Scientific American, Inc. All rights reserved.

(1979) and by Stewart and Hunt (1982), but as yet no clear answer to it is available.

10.5 A GENERAL FRAMEWORK FOR D-GENE ACTION

So far I have distinguished between the morphologically important category of D-genes and the alternative categories of R- and S-genes, have given a series of examples of D-genes, and have gone into some detail on two of these. However, an understanding of morphological evolution in general requires not just a recognition of the existence of D-genes and an intimate knowledge of isolated examples, but, in addition, it requires that we have a general scheme of D-gene action incorporating the interrelationships between the activities of different D-genes during the developmental process. Although the molecular details of such interrelationships are largely unknown, this may not present an insurmountable problem. It is worth recalling Waddington's (1975) comment that for evolutionary purposes we need to have a conceptual framework for development, not a molecular catalogue. In this section I attempt to provide such a framework; one which is basically a gene-orientated version of the morphogenetic tree concept put forward in chapter 9. The main proposals relating to this framework are as follows:

1. Genes with major effects on development such as those described in the two previous sections normally begin to act at an early developmental stage.
2. There is a relatively small number of these genes.
3. As development proceeds, the number of D-genes which affect the phenotype of each successive stage increases.
4. Paralleling this increase in the number of D-genes is a decrease in their average magnitude of phenotypic effect.
5. This process culminates with the adult phenotype being finalized by a large number of D-genes many of which have negligible effects and whose main functions are not necessarily morphogenetic; these are the polygenes of quantitative genetics.
6. The development of an organism with a complex life-cycle involving, for example, alternation of generations or metamorphosis, can be seen as two (or in some rare cases more than two) interconnected phases within each of which points 1–5 apply.

This suggested outline of the genetic control of development, which is illustrated diagrammatically in Figure 10.12, forms the basis for the evolutionary discussion that follows in chapters 11, 12, 13, and 14.3. Its validity must therefore be examined in some detail.

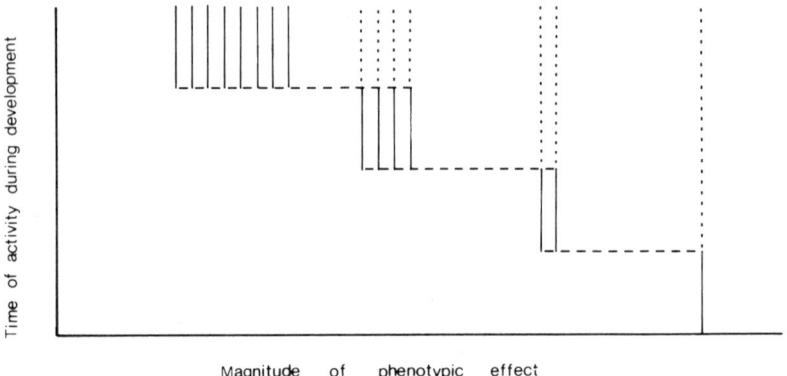

Figure 10.12. A gene-orientated version of the morphogenetic tree, showing a decreasing average magnitude of effect of active D-genes, and an increase in their number, as development proceeds. Solid vertical lines show periods of activity. Dashed horizontal lines show causal interconnections, i.e. the direct or indirect switching on of later-acting D-genes by earlier-acting ones. In reality the situation must be much less regular than shown here. Among other complexities, some early genes would be expected to persist in activity after they have switched on other genes. This is indicated by the dotted vertical lines.

The evidence for early activity of D-genes with major phenotypic effects has already been considered, in the case of selected examples, in the previous two sections. All the genes described therein were seen to act at a very early stage of development, with some of them, namely the shell-coiling genes of *Lymnaea* and *Partula* and the *esc* gene in *Drosophila*, indicating through their patterns of delayed inheritance that their products are required during oogenesis and take effect shortly after fertilization. In fact, we would expect most *D*-genes with large phenotypic effects to begin activity early in development from *a priori* reasoning. The development of an adult organism from an egg (or from a series of imaginal discs) is largely achieved by cell proliferation and differentiation. The 'opposite' processes of cell destruction and de-differentiation play a relatively minor role and indeed there are many cases where differentiation is irreversible. Thus at a late stage in development a mutant gene can achieve a lesser amount of phenotypic modification than at an earlier stage simply because more of the developing organism is already constructed. It is important to note here that the 'magnitude of effect of a gene' refers to the average magnitude of phenotypic effect of mutations of that gene. This point will be discussed further in section 12.2.

The number of early-acting *D*-genes with large effects on development is clearly unknown even in any one organism, and definitive comments in this area must await the results of further studies in developmental genetics. Workers on the genetic control of development in *Drosophila*, which is undoubtedly the best known system, have stressed the potential ability of a few key selector genes (Garcia-Bellido, 1975) to organize complex morphogenetic processes. Thus at least the suggestion made here that there are few such *D*-genes is in accord with current thinking on the development of *Drosophila*. Also, the small number of genes involved in making key morphogenetic decisions is a central theme in the recent review of developmental-genetic aspects of evolution by Raff and Kaufman (1983).

That later phenotypic stages are affected in some way or other by a larger number of *D*-genes than are earlier stages is inevitable. It may be true that the number of *D*-genes actually *active* increases through development. This would certainly occur in a morphogenetic tree system (see Figure 9.3) if there is a constant ratio of *D*-genes to morphogens; and I have discussed elsewhere (Arthur, 1982d) some lines of evidence which support this assertion of a gradually increasing number of active *D*-genes. However, even if such an increase does not occur, the number of *D*-genes that affect, directly or indirectly, the phenotype at time t_j in development must exceed the number that affect the phenotype at some earlier time t_i, simply because of the carry-over of effects from earlier stages. Such indirect effects of *D*-genes on phenotypic stages occurring after the switching off of the gene concerned can occur either by genetic or phenotypic means. An example of the former is the affect of the allelic state of the *esc* locus on larval stages of *Drosophila* occurring after the period of activity of the locus itself, because of its switching effect on the bithorax complex (see Struhl (1981) and previous section). Turning to phenotypic carry-over effects, a *D*-gene mutation halving the number of cells in a rudimentary organ will have an effect late in development after it is switched off if, for example, a hormone or other morphogen causes the cells of the initial rudiment to divide at a fixed rate for a fixed period. Because of these two kinds of carry-over effect of early-acting genes on late development, it is almost certainly true that later phenotypic stages are affected by more D-genes than earlier ones, regardless of how the number of active D-genes alters through development — though this number may well increase itself. However, assuming that the number of active D-genes does indeed increase during development, there must also be a 'terminal decrease' in this number because, towards the end of development, all D-genes then active will not be switched off simultaneously, as suggested in the rather simplified picture given in Figure 10.12.

Points 4 and 5 above imply that the number of D-genes with individually small phenotypic effects increases during development. This is a fairly contentious point for which, as far as I am aware, there is little evidence. Clearly, the number of polygenes affecting the overall adult form of an organism such as *Drosophila* or the mouse is very large. Estimates of the number of loci contributing to quantitative variation in any one character (wing length, number of bristles, etc.) are usually in the 10–100 range (see Falconer, 1960, 1981) and there are many such quantitative variables in any adult organism, albeit their genetic bases must overlap to some extent. What is not clear is whether most of these D-genes are late-acting ones, as proposed here. They are certainly not all involved in the very final stages of the determination of the adult form since polygenic control of components of size and shape of immature stages is well known (see chapters 3, 4). However, the immature mice and other organisms used in such studies are at a very late stage of development compared with the stages at which major D-genes like the homoeotics begin to take effect. It would be of interest to know whether organisms at the blastula/blastoderm or gastrula stages exhibit polygenic variation; on the model of development put forward here they should do so to a much lower degree than the corresponding adults.

The validity of the final point (number 6 above) depends on the extent to which the different stages of a complex life-cycle are separate developmental systems. There can be little doubt that they are indeed independent developmentally if the different stages are physically separate, as in the case of the sporophyte and gametophyte stages of ferns or the different stages within a complex parasitic life-cycle, e.g. that of the cestodes, which inhabit different species of hosts. In *Drosophila*, despite the close physical association of the two developing systems — those of larva and adult — which is apparent from the location of the imaginal discs within the developing larva, it is clear that the developmental activities occurring in the two systems are well segregated. Thus, even when spatial separation of the different stages does not occur, there still may be a good case for considering the different stages as only loosely interconnected.

A point that has not been made explicit so far is that the generally greater effects of early-acting D-genes include two components: a greater effect on an individual phenotypic character, and a greater number of characters affected.

This point may be exemplified by considering shell shape in *Lymnaea*. If we take the variable mentioned in chapter 2 of 'number of degrees turned through', the gene for sinistrality alters this character of the shell from $-360W°$ to $+360W°$, W being the number of whorls which, though highly variable in *L. peregra*, is typically around five (Ellis, 1969). In contrast, a polygene causing shell growth to slow slightly during a late developmental phase might affect this character by a few degrees — perhaps in single figures — as opposed to $7200°$. Also, while the gene for sinistrality clearly re-orientates all somatic structures, a very slight change in the rate of accretion of shell during a late developmental stage can presumably occur without alterations to most of the soft tissues. Early-acting *D*-genes may be severely constrained in their evolution either because of their greater effects on individual characters or because of the fact that they alter many characters and must do so in a co-ordinated way if the organism is to survive at all. These and related evolutionary issues will be discussed in the following two chapters.

10.6 THE MOLECULAR BASIS OF *D*-GENES

It was noted when *D*-genes were introduced in section 10.2 that this category is heterogeneous in the sense that some genes within it probably produce enzymes, with others in some way regulating the rate of production of enzymes at neighbouring or distant loci, and yet others perhaps performing some other functions. Those *D*-genes that are essentially structural genes producing enzymes which are involved in the setting up of a morphogenetic gradient are likely to be composed of single-copy DNA, that is a sequence of the order of 10^3–10^4 bases, including coding sections and sections whose corresponding pieces of heterogeneous nuclear RNA are removed before translation, which occurs once, or at most a few times, per haploid genome.

There may well be *D*-genes which control the activities of batches of other genes in some way that does not involve the production of a protein. One possibility here is that moderately repetitive DNA, that is sequences present a few hundred times per haploid genome, is involved in morphogenetic activity. Transcripts from this kind of DNA appear not to be translated, but they exhibit tissue-specific patterns in abundance in sea-urchins (Scheller *et al.*, 1978), which suggests a possible role in cell differentiation and morphogenesis (Davidson and Britten, 1979). However, much remains to be done before the precise role of this class of DNA is established. Highly repetitive DNA, which consists of sequences of a handful of bases repeated thousands of times per genome, is unlikely to be involved in morphogenesis. At present it appears from its concentration at centromeres that its role is probably a structural rather than a genetic one.

Two other recently discovered genomic entities deserve brief mention. Pseudogenes, best known in the mammalian globin gamily, appear to be functionally redundant and thus seem unlikely to be involved in gene control or morphogenesis. Transposable elements, on the other hand, most certainly are relevant from a developmental viewpoint. These elements were originally discovered in maize because of their effect on pigmentation of the endosperm (see McClintock, 1951). More recently, several types of transposable elements have been discovered in the *Drosophila* genome. Of particular importance here is the fact that all bithorax alleles examined in a recent study by Bender *et al* (1983) were due to the insertion of a mobile element (either the 'gypsy' or the '412' element) into a certain region of the bithorax complex. However, while this observation confirms that insertion of a transposable element can cause a homoeotic mutation, there is no necessary link between the two. Bender *et al.* also showed that all postbithorax alleles studied were due to deletions, while *Ubx* mutations could occur as a result of a variety of types of molecular change. These observations lend weight to the idea that interference with a morphogenetic gene's normal activities is the cause of mutant phenotypes, and that exactly how this interference is achieved at the molecular level is quite variable. In the case of a gene making an enzyme that has a morphogenetic role, it seems reasonable to suppose that insertion of a transposable element, deletion of a large section of the gene, and a point mutation causing an amino-acid substitution in the active site, are all capable, together with several other types of molecular event, of causing a radically altered pattern of development.

The uncertainties surrounding the molecular nature of *D*-genes and the probable heterogeneity of this genic category at the molecular level are unlikely to prevent useful discussion of morphological evolution. The evolutionary schemes put forward in the following chapters are heavily dependent on the proposed framework for genetic involvement in development given in the previous section (Figure 10.12), but they are relatively immune to unexpected discoveries — such as a morphogenetic role for pseudogenes — that might occur at the molecular level.

10.7 SUMMARY

Although the way in which morphogenetic genes produce their effects is largely unknown, many such genes have been identified, their patterns of inheritance established, and their phenotypic characteristics thoroughly described. These genes, which affect the spatial arrangement of developing systems, are referred to as *D*-genes, and they are contrasted with conventional regulatory or *R*-genes which control the rate of production of proteins which lack morphogenetic effects. Two examples of *D*-genes, those governing the direction of shell coiling in gastropods and the pattern of compartmentalization in insects, are described in some detail. A scheme is proposed wherein all *D*-genes, including those of major effect and those like the polygenes whose effects are relatively small, interact in a tree-like pattern. This parallels the scheme of morphogen interactions put forward in chapter 9. Various lines of evidence supporting the suggested framework are discussed, and the main gaps in this evidence noted.

Finally, the likelihood that *D*-genes constitute a heterogeneous group at the molecular level is stressed; but it is also stressed that this probable heterogeneity need not prevent useful discussion of *D*-gene evolution.

Chapter 11

The Evolution of Major Genes and the Origin of Higher Taxa

"The paleontologist has more reason to believe in a qualitative distinction between macro-evolution and mega-evolution than in one between micro-evolution and macro-evolution."

G. G. Simpson, 1944.

11.1 INTRODUCTION: TWO TYPES OF ORIGIN

One of the most fundamental questions in evolutionary biology is how the various major body plans which we recognize, for example those of the insects, vertebrates, gastropods or gymnosperms, became established. Neo-Darwinism provides a reasonably satisfactory explanation of the adaptive radiations that have occurred within such groups, producing what might be described as variations on a theme. However, it is by no means clear that the same process of natural selection on polygenic variation that is responsible at least for intraspecific and intrageneric evolutionary divergence is the sole or even the predominant process giving rise to higher taxa such as orders, classes or phyla. The alternative, of course, is some form of saltational evolution where a radically new phenotype is produced by a single, or at least relatively few, fixations of mutations with major effects. The proposal that new *higher groups* sometimes originate in this way is, as noted earlier, quite a separate issue from that of whether new *species* usually arise abruptly, which has been dealt with in chapter 8. These two possible kinds of saltational evolution certainly do not stand or fall together. Indeed the view taken here is that punctuated equilibria are largely illusory whereas the possibility that new body plans originate, at least in some cases, by a more drastic form of saltational change than envisaged by the punctuationalists, needs to be taken seriously.

The evolutionary literature of the century and a quarter since the publication of *On the Origin of Species* is strewn with suggestions that saltational evolution, including the saltational origin of higher taxa, has indeed occurred. Less than two months after the publication of the first edition of 'The Origin', T. H. Huxley wrote to Darwin saying that "you have loaded yourself with an unnecessary difficulty in adopting *Natura non facit saltum* so unreservedly" (see F. Darwin, 1887). Exactly what sort of saltations Huxley had in mind is not clear, but the same cannot be said of D'Arcy Thompson (1942) whose words on the subject (see quotation in section 9.1) clearly indicate that he had a saltational origin in mind for higher taxa such as classes and phyla. More recently, Lawrence and Morata (1976), in a discussion of the developmental compartments of *Drosophila* and the selector genes which control them, comment of such genes that "small changes in the activity of their products could have a large effect on the morphology of the fly, suggesting an opportunity for rapid evolutionary changes." Finally, A. Huxley (1982) stresses the distinction between punctuations and larger saltational changes, and says of the latter: "It may yet turn out that these are important."

Despite the numerous suggestions of saltational evolution, this proposed process has largely failed to gain acceptance in evolutionary theory. The reason for this is probably the converse of the reason for the acceptance, after much heated debate, of Darwinian natural selection; that is, there has so far been no generally applicable *mechanism* outlined for saltational evolution. The main problem here is that although mutations are known which lead to radically altered phenotypes (see chapter 10) some of which are quite viable in their own right, the fitness of such variants relative to other members of the population is usually considerably reduced so that, in a sense, mutation and selection oppose each other. Thus at the population level no evolutionary change occurs except for the elimination of the new variant. This contrasts with the slight, successive, quantitative variations proposed by Darwin, where mutation and selection are 'compatible' in the sense that many of the slight phenotypic changes caused ultimately by mutations of the polygenes are selectively advantageous.

Although the lack of a satisfactory mechanism for saltational evolution has led to its dismissal by many evolutionists, it is questionable whether in fact there is a satisfactory non-saltational or neo-Darwinian mechanism for the origin of new body plans. Many texts on evolution by neo-Darwinian authors do not even consider in any detail how new forms arise. Huxley's (1942) *Evolution, The Modern Synthesis*, for example, lacks a chapter on this fundamental question. This can hardly be because the question is regarded as unimportant, and must presumably be because the authors concerned take the view that no special explanation is required; that is, that given some appropriate ancestral form, two or more classes or phyla can be established from this ancestor by a long series of minor quantitative changes. There are, however, some cases in which evolutionary divergence cannot occur in this way. Also, quite apart from individual instances of large evolutionary changes which defy an explanation in terms of accumulation of many smaller changes, the extrapolation of selection on intraspecific polygenic variation to the origin of major taxonomic groups is quite a conceptual leap, and one which we should not be too ready to make. In the following sections, therefore, I will not only examine a possible mode of origin of higher taxa through saltational evolution (11.3), but will also consider in some detail the neo-Darwinian alternative and its associated problems (11.2)

11.2 NEO-DARWINIAN ORIGINS

General outline

It is difficult to give a clear exposition of the neo-Darwinists' view of the origin of higher taxa because of their tendency, already noted, not to deal with this major evolutionary problem as a distinct issue. However, Mayr (1963, chapter 19) devotes sections to 'The origin of a new type' and 'The origin of higher categories' which effectively summarize the neo-Darwinian position. What follows is based largely on Mayr's account, and I have taken several quotations from this account to illustrate just how extreme the neo-Darwinian view can be. It should be noted that the view presented here differs from the more moderate version of neo-Darwinism given in section 1.5. Mayr's 'extreme' neo-Darwinism is less acceptable than its more moderate counterpart, but it serves here in the development of the argument. The first of these quotes, with my italics, is as follows: "*all* evolution is due to the accumulation of small genetic changes, guided by natural selection". It is in fact clear, because he goes on to contrast this view with that of evolutionists believing in phenotypic saltations, that Mayr is actually referring to small *phenotypic* changes rather than small genetic ones. This is worth noting because, in the heated controversies that surround all proposals of saltational evolution, there has been a tendency to confuse the magnitude of effect of a mutation at genic, chromosomal, phenotypic and population levels. I have already emphasized in chapter 1 the non-correspondence between the degrees of effect of a mutation at any two of these levels.

The belief that all evolutionary changes are of the quantitative, polygenic kind known to occur within present-day species (see chapter 5) leads to the view, again expressed by Mayr (1963), that "Every species is an incipient new genus, every genus an incipient new family, and so forth." In this view, the speciation events leading to species from which new classes or phyla arise would not be recognizably different from any others at the time they occur, and are only distinct in retrospect. Hence the tendency, in books on evolution written by neo-Darwinists, to deal only with the cycle of quantitative intraspecific variation, selection and speciation described in section 5.1. A proliferating series of such cycles, none of which is in any way unique, is thought to be responsible for the entire spectrum of living and extinct species.

Along with this denial of any role for genes of large effect goes an over-emphasis on the channelling of evolution by ecological, as opposed to developmental, forces. Mayr attributes considerable importance to major habitat shifts, such as water to land, and states: "Each of the successful branches of the animal kingdom, for instance the insects, the tetrapods, or the birds is the product of such a shift." Although all evolution takes place in, and is to some extent directed by, environments, as reflected by Hutchinson's (1965) clever title *The Ecological Theater and the Evolutionary Play*, the correspondence between habitats and taxa is not as clear-cut as Mayr implies. His choice of examples certainly backs up his point, but the choice of different examples can seriously alter the conclusion. Going 'upwards', for example, arthropods and vertebrates are notable for their diversity of major habitats, as are molluscs and angiosperms. Going downwards beneath the level of classes we find, for example, the pulmonate molluscs of whose two major suborders one is largely terrestrial, the other largely freshwater. A further important point here is that within any of the groups cited by Mayr, the aerial/land/water habitat types are about all that the constituent species have in common ecologically. In fact, a group such as the insects is enormously ecologically diverse. It is only from the rather blunt view of ecology often adopted by evolutionists that a major taxon can be seen as even reasonably homogeneous. An alternative view of a higher taxon is that it is a series of ecologically *diversified* species produced by polygenic modification of a basic theme which itself, though ecologically constrained in that it may be of survival value in only one major class of environments, is specified by a relatively small number of genes with large phenotypic effects which to some extent restricts developmentally the range of quantitative variants that may be produced. These opposing views will be discussed further, below, in the context of the question of whether the lack of a major taxonomic group 'in between' the vertebrates and echinoderms is due to developmental-genetic or ecological constraints.

Problems

As stressed in the previous section, there has been a tendency in evolutionary debate for the problems of saltational origins of major taxa to be highlighted and those of neo-Darwinian origins to be neglected. I intend, here, to redress this balance and to pinpoint what seem to be the three major weaknesses in the extreme neo-Darwinian view.

1. What happens to major D-genes?

We have already looked at a number of examples of D-genes with major phenotypic effects. Let us now consider the insect homoeotics, whose evolutionary role is both more controversial and potentially more important than that of the gastropod shell-coiling gene. The question that arises is: What has happened to the loci at which homoeotic mutations occur, during arthropodan evolution? The extreme neo-Darwinian view is that these genes have not evolved, because the aberrant phenotypes resulting from mutations in them are accompanied by severe fitness-depression. While it is indeed difficult to see how evolutionary saltations can take hold at the population level (but see next section on saltational origins), we need to ask what the alternatives are.

That is, if these genes do not evolve, what happens to them? There are only three possibilities: the loci concerned may be lost from the genome, retained but permanently inactivated, for example, as pseudogenes, or retained in a functional state but with their activities over-ridden in some way by batteries of genes with relatively minor effects. The first two of these possibilities seem very remote for the following reason. Although the developmental genetics of orders of insects outside the Diptera are relatively poorly known, it seems reasonable to suppose that the basic process of genetically controlled compartmentalization described in section 10.4 for *Drosophila* is prevalent in most if not all insect groups. Certainly, the imaginal discs in which the compartmentalization occurs are found in a wide range of insects (Poulson, 1950). If loci with major effects on key steps in this process are inactivated or lost during evolution, then new loci with the same general function of each lost locus must somehow be acquired. That evolution proceeds in this way cannot be ruled out but it does seem exceedingly unlikely. We are left, then, with the possibility of functional but unevolved homoeotics and other major D-genes whose effects have been altered by batteries of minor D-genes, the evolution of which does not present a serious problem. To some extent, this sort of thing — modification of the effects of small numbers of major D-genes by the activities of many minor ones — must occur, and indeed all quantitative changes can be thought of as examples of just such a process. However, if the scheme of genetic involvement in development proposed in section 10.5 is correct, the D-genes whose evolution is allowable within a strict neo-Darwinian framework are predominantly late-acting genes whose ability to modify the effect of an earlier-acting homoeotic will be limited and whose ability to switch completely the developmental pathway specified by an earlier-acting gene can be doubted altogether. Thus we are left in the uncomfortable position that, while evolution of D-genes with major phenotypic effects is difficult to envisage, so is a lack of such evolution.

2. *What about fitness-depression in phenocopies?*

Clearly, some saltational evolution at the phenotypic level must be admitted even by extreme neo-Darwinists if only because certain phenotypic structures vary in a binary or integer manner. If the direction of shell coiling in gastropods or the number of segments in an annelid or athropod alters in evolution, as they obviously do, phenotypic saltations are not only possible but inevitable. However, some neo-Darwinists argue that such changes are produced by polygenic modification according to the threshold model (section 3.5) and are not 'genuine' saltations produced by mutations of large phenotypic effect. The problem here is that the fitness depression widely cited as a major difficulty in accepting the evolution of genes with large effect is in fact a property of aberrant *phenotypes*, and not just of those produced in a single-gene manner. No one would claim that Waddington's (1956) ether-induced phenocopies of bithorax mutants were as fit as, or fitter than, wild-type *D. melanogaster*.

Equally, a sinistral snail produced as an ontogenetic accident in a population of dextrals will have just as much difficulty in mating with its neighbours as will one produced by the shell-coiling D-gene described in section 10.3. (This mating effect of the direction of shell coiling will be discussed more fully in section 11.4.) Thus we end up being unable to see how qualitatively variant phenotypes have evolved, despite knowing that they must have done so on many occasions, and we have no additional difficulty in understanding the evolution of such phenotypes in the specific cases in which they are produced by single mutations.

3. *Why are there major gaps in the evolutionary tree?*

One consequence of the 'universal quantitative variation' scheme of extreme neo-Darwinism is that any species can be progressively modified in any direction and that whether one species is modified in a particular way depends on the availability of empty niche space in the sense of Crozier (1974) rather than on any kind of developmental constraints on the direction that evolutionary changes can take. However, we need to ask whether this system of exclusively ecological channelling of evolution is compatible with the shape of overall evolutionary tree in so far as we are able to discern it.

It is my view that these two things are not as compatible as they first seem. The problem is that many ecological processes operate very much at a species-specific level. The extinction of a species due to competitive exclusion or its non-production in the first place due to a lack of unutilized niche space will usually affect one species, or perhaps a small group of species (see chapter 14), at a time. The lack of any major taxonomic group morphologically intermediate between vertebrates and echinoderms is difficult to explain in this way. Admittedly, the destruction or alteration of a whole class of environments might cause widespread extinction, and such events may conceivably be the cause of known mass extinctions such as that of the dinosaurs, though this particular example is still poorly understood. At any rate, the largest gaps between major branches of the evolutionary tree appear to be due to non-production, rather than extinction, of intermediate taxonomic groups. This is much more difficult to explain in ecological terms.

The key question in this area of the compatibility of evolutionary theories and evolutionary trees is this: What shape of tree would be *predicted* from each theory if no palaeontological or comparative neontological data were available? Now while the known shape of the overall evolutionary tree is not incompatible with an evolutionary process driven entirely by selection on quantitative variation coupled with speciations and extinctions — if they were incompatible, the more extreme versions of neo-Darwinism would have disappeared long ago — it is not the most likely, and hence not the predicted kind of tree, given this sort of evolutionary mechanism. On the other hand, it is readily predicted on the basis of a more complex form of evolution in which polygenic modification and saltational change both occur. I will return to discuss this point in more detail after considering how new groups could originate through saltational changes.

11.3 SALTATIONAL ORIGINS, AND THE ROLES OF w- AND n-SELECTION

An important distinction

As Berry (1982) has pointed out, attacks on neo-Darwinism have a tendency to re-surface after being refuted, and it is clearly undesirable, in any branch of science, to have recurrences of a particular controversy if, at any stage, it has been satisfactorily resolved. In this context, it is necessary to stress that the partially saltational theory of evolution developed herein is quite distinct from the theories of de Vries (1905), Goldschmidt (1940) and Eldredge and Gould (1972). The difference is an important one and needs to be made abundantly clear. Basically, all of the above-mentioned authors formulated schemes of evolution in which the main discontinuity of mechanism was between microevolutionary (i.e. intraspecific) change and the production of new species and genera through speciation events, the latter process being referred to as macroevolution. As noted by Simpson (1944), the occurrence of a mechanisitc discontinuity at this level, and the predominance of some non-polygenic mechanism for the production of new species, would mean that "the innumerable studies of micro-evolution would become relatively unimportant and would have minor value in the study of evolution as a whole".

In contrast to the previously formulated saltational theories of evolution, the view taken here is as follows:

1. Most speciation events, and indeed the evolutionary production of most new genera and families, can be seen as a straightforward consequence of the action of natural selection on quantitative, polygenic variation within a single species, coupled with the action of one or more isolating mechanisms.
2. Some, and perhaps most, classes and phyla originate through one or more speciation events involving mutation to, and fixation of, new alleles of major D-genes with large phenotypic effects.

If this view is correct, then not only are microevolutionary studies important, but indeed they relate to the predominant mode of evolutionary change as measured by the number of species produced. Saltational changes are regarded as important not because they are common — which seems most unlikely — but because when they do occur they open up a whole new body plan for modification through quantitative variation. Also, the mechanisms by which these occasional saltational changes may occur (see below) are entirely selective. For these reasons I see the theory outlined in this book as being an expansion, rather than a replacement, of neo-Darwinism, though how many neo-Darwinists will accept it remains to be seen.

Along with D'Arcy Thompson (1942), Simpson (1944) was one of the first people specifically to propose that a discontinuity of mechanism might exist not between microevolution and macroevolution but between macroevolution and the origin of higher categories such as orders, classes and phyla, which he calls mega-evolution (see quotation at the start of this chapter). I wish, here, to complement Simpson's point that palaeontologists have some reason to expect a macro-/mega-evolutionary discontinuity by suggesting that neontologists too have more reason to suspect the existence of such a discontinuity than they have to accept the lower-level one proposed in the theory of punctuated equilibrium.

Having clarified the extent to which saltational change is considered here to contribute to evolution, and distinguished in this context between the ideas discussed in this book and those of earlier proponents of saltations, I will say little more about the work of these earlier authors. De Vries's 'macromutations' turned out anyhow to be chromosomal variants which, it now seems, have no general relationship with morphology (see chapter 6). The work of Goldschmidt, however, will receive some further attention because of his suggestions relating to the evolution of homoeotic genes; but this will be deferred until section 11.4. The most pressing need at this stage is to consider precisely how saltational evolution could occur.

Possible mechanisms

Minor phenomena such as seasonally oscillating gene frequencies aside, any evolutionary change requires the introduction of a new variant and its fixation in a population. In the case of quantitative microevolutionary changes in morphology, these two stages of the overall process are normally accomplished by mutation and selection, respectively. We now need to enquire whether mutation and selection will also suffice to explain saltational evolution.

The initial appearance of a new variant is no more of, and possible less of, a problem in saltational evolution than in microevolution. As seen in chapter 10, many cases are known where a mutation of a particular D-gene produces a new morphological variant, and indeed the developmental sequences leading to the adult phenotype have often been well described. This is a considerable advance on our understanding of mutations of polygenes, which is restricted to a knowledge that such mutations occur. Moreover, the occurrence of mutations of D-genes with major phenotypic effects is not restricted to laboratory systems. Some species are naturally polymorphic for such mutations. The cases of *Lymnaea* and *Partula* have already been discussed; another example is the lateral plate polymorphism of the stickleback, *Gasterosteus aculeatus*, described by Bell (1981). Also, many of the major D-gene mutations of *D. melanogaster* outlined by Lindsley and Grell (1968) are spontaneous rather than mutagen-induced and so will occur with low but non-zero frequency in natural populations.

It is clear, then, that no novel mechanism need be invoked to explain the introduction of a 'saltational variant'. The problem occurs at the level of the population rather than at that of the individual, and arises when we enquire whether selection can be so easily extended as mutation from microevolution to saltational changes. It is of interest to note that the proponents of saltational evolutionary theories have mostly been developmentalists, whose prime concern is of course with individuals, and palaeontologists, who often work with detailed morphological descriptions of fragmentary samples sometimes containing only one or a few intact fossils, where the individual by default rather than intent is again the focus of attention. Neo-Darwinists, on the other hand, are mostly to be found among those groups of biologists for whom the population is of paramount importance.

Because of the relative ease of understanding the introduction of qualitatively distinct variants, the remaining discussion of possible mechanisms of saltational evolution will be centred at the population level. The problem here is what happens to a saltational variant after it first appears in a population through mutation. There are two ways in which such a variant might be increased in frequency, though neither of them seems likely to be very widespread. Their rarity is not a problem in that it is not supposed that saltational changes are common anyway, but it does raise problems in relation to testability. The two possible mechanisms are as follows.

1. w-Selection

It is abundantly clear from a variety of passages scattered through *On the Origin of Species* that Darwin thought of natural selection as acting through competition; that is, the fitness of any individual was measured not in any absolute sense but relative to the other members of the population in which it found itself. It is equally clear, from several statements in A. R. Wallace's paper to the Linnean Society in 1858 (see Berry, 1982, for quotations), that he too thought of selection as acting in a competitive manner. Because our usual measure of relative fitness is the cross-product ratio w (see chapter 1), I will refer to this kind of selection as w-selection. (The contrasting process of n-selection will be discussed shortly.)

We now need to ask: Could the frequency of a new, qualitatively distinct variant be increased through w-selection? Some of the most apparently persuasive evidence against the common occurrence of such a process comes from *Drosophila* laboratories. Reasonably large populations of a variety of species, particularly *D. melanogaster* and *D. pseudoobscura*, have been kept for various lengths of time, in a variety of laboratories, starting around 1920. Yet despite this, none of the many morphological mutants that have been competed with wild-type stocks have 'won' as measured by an increase in frequency. In addition, no new distinct morphological mutants have occurred and increased to fixation in any wild-type population cage, despite the fact that billions of individuals must have been reared over the last 60 years or so.

There are three reasons why we should be cautious in drawing a general conclusion from these observations. First, although considerable numbers of individuals have indeed been reared in *Drosophila* laboratories, their total number is minute when compared to the number of living organisms that have existed in the biosphere since its origin. If we are dealing with a very rare process, the *Drosophila* work may not be a sufficient sample within which to detect it. Second, the probability of a randomly chosen morphological mutant being selectively advantageous may depend on the starting point; that is, the particular body plan of the species being investigated. Thus not only may a particular genus such as *Drosophila* be unrepresentative simply because it only is a single genus, but there may be an underlying trend in evolution towards the production of less grossly modifiable body plans. In this context it should be noted that bithorax mutations of *Drosophila* are in a sense retrogressive mutations of a higher dipteran. In order to investigate the possible involvement of the homoeotic genes in insect evolution it would be better to have information on the fitness of 'monothorax' mutations in primitive four-winged flies.

The third reason why the *Drosophila* work may be misleading concerns the nature of the 'habitat'. The environmental conditions in which *Drosophila* populations are kept are rather standardized among different laboratories, and are kept very constant over time at any one laboratory. (It is perhaps also worth noting that these standard conditions were arrived at largely because of their suitability to wild-type flies.) The likelihood of saltational mutants being selectively advantageous (i.e. favoured by conventional w-selection) is considerably increased by two sorts of imbalance. One is of a coadaptational kind where the imbalance is between different components of the same organism; this can arise through n-selection as will be discussed shortly. The other, which is relevant here, is an imbalance that develops between the current morphology and the environment. This second kind of imbalance can occur either through invasion of a new habitat type by migration or through drastic change (for example, climatic) in the environment already inhabited. In either case, the further the current morphotype is from the optimal morphotype for the environment concerned, the greater the magnitude of effect a mutation may have and still be selectively favoured. This argument was formulated by Fisher (1930) and will be discussed in more detail in chapter 12. The important point to note here is that the standard laboratory environment for *Drosophila* precludes habitat changes of either of the two types discussed, and consequently prohibits the occurrence of imbalances between environment and morphology which can give rise to selectively driven phenotypic saltations.

In conclusion, it can only be said at present that the increase in frequency of highly mutant morphotypes under w-selection is neither a common, nor even a moderately rare, event. Whether it has occurred on a very rare basis or not at all is uncertain, though the former alternative seems more likely, providing the mutations concerned are not too drastic in terms of the type of phenotypic revision they produce. It is worth noting, in this context, that a conventional (w-selective) mechanism has been proposed for the evolution of the direction of shell coiling in gastropods (see section 11.4). Thus w-selection may be responsible for occasional, 'minor' saltations. If this process does indeed occur, it provides a conventional mechanism for saltational evolution at the population level. It seems likely that, in these occasional instances of w-selection for a new and distinct variant, the selective coefficients in favour of the new morphotypes are almost as large as those that normally favour the wild-type against such mutants. The mutant phenotypes involved are most unlikely ever to be selectively netural, or even nearly so.

2. *n-Selection*

(a) Introduction

Current population genetics theory (with the exception of neutralist models) is basically a formalization and articulation of Darwin's paradigm of the 'survival of the fittest'; and in this formalization, the central variable, fitness, is, as noted earlier, often measured by the cross-product ratio, w. Using this measure of fitness, and assuming the absence of balancing mechanisms, such as frequency-dependent selection, a new variant (B) will spread through a population of A if, and only if, $w_B > 1$. Now this *relative* fitness, w, is both comparative and competitive. It is comparative in the sense that the way the measurement is made ensures that we are always comparing a variant with some other one. It is competitive in the sense that a less fit variant may not be excluded from a population unless that population is subject to intraspecific competition; that is, it begins to encounter the problem of a limiting resource, whether food or space. This point is related to the one made by Ford (1971), who noted that mutant forms are favoured during times of population growth. The reason for this is, of course, that even if a mutant has $w_B < 1$, its absolute *numbers* can increase concomitantly with its decrease in frequency in a growing population.

As mentioned in Chapter 5, the only three attributes that a group of entities must possess in order that they are subjected to selection are reproduction, variation and inheritance. We can now add to this and say that for fixation of a new type through selection we need, in addition, not only mortality of individuals but also limitation of populations. The latter will not always occur through competition in the strict ecological sense of the word (see Bakker, 1961), since in a few cases population limitation may occur through other mechanisms such as predation; however, self-limitation through competition is probably predominant.

We now ask the question: Can a non-comparative and non-competitive form of selection be envisaged, where the criterion for establishment of a new variant is something other than $w > 1$? The answer to this question is an emphatic 'yes', and the form of selection of this kind that I now wish to outline can be called *n*-selection. The *n* comes from *n*et reproductive rate (R_0) which, as explained below, is the key variable in the assessment of fitness in this 'new' form of selection. (This choice of name seems better than R-selection, since the latter phrase might be confused, at least in spoken communication, with MacArthur and Wilson's (1967) related but not identical concept of r-selection.) The questions of whether n-selection will actually occur in nature, and if so, how often, and under what circumstances, will be discussed shortly. Before such issues are addressed, however, it is necessary to make clear exactly what is meant by n-selection, and in what way the net reproductive rate is involved in the criterion for success.

To clarify this matter, it helps us to use a 'gedanken experiment' (or thought experiment). Let us suppose that a series of population cages of *Drosophila melanogaster* is set up, and into each is placed a propagule of, say, 100 fertilized eggs of a particular genotype. Each cage is inoculated with a different genotype, but each genotype is a homozygote for a mutant allele that affects morphogenesis in some fairly substantial way. (The problem of heterozygotes and the dominance or otherwise of the mutations will be considered later.) We then ask the question: Which of the genotypes will, in these pure cultures, establish persistent and presumably stable populations? It is easy to find examples of mutations for which the outcome of such a culture is known. The vestigial-winged homozygote *vg/vg* would establish a stable population; the larval lethals studied by Nüsslein-Volhard and Wieschaus (1980) clearly would not. Although a few genotypes would no doubt give indeterminate results in that some replicate propagules would establish populations and others would not, due to uncontrolled environmental variables or variation in the genetic background, the vast majority of morphogenetic mutants could be classified as definite successes or failures. The criterion of success in such an experiment is that, *at low density*, $R_0 > 1$. (It is assumed that the cages have sufficient resources so that $N = 100$ of any life-stage is substantially below carrying capacity.) In the case of a population of asexual organisms with no age structure, this simply means that when resources are plentiful the average number of offspring per individual exceeds 1.0. The meaning of R_0 in more complex populations, and its relation to life-table theory, was briefly outlined in chapter 1, and is dealt with in detail by a number of authors, including Krebs (1972).

It is clear that in our overall gedanken experiment a form of selection has occurred: some variants are fitter than others (as measured by R_0), and this variation in fitness leads to some variants being, to use Darwin's words, naturally selected. The process going on within any one cage of our experiment is less easily describable as selection — rather, it is a form of 'screening' of one particular variant on the basis of its R_0 value. But this is no way detracts from the selective nature of the overall process which, it should be noted, would still prevail if the experiment was staggered; i.e. one mutant being tested for R_0 at a time. This form of selection, then, in which R_0 rather than w is the criterion of success or failure, may be described as n-selection. It should be noted that in n-selection the term fixation has no meaning; we talk instead of the *establishment* or otherwise of a mutant population.

It must be stressed at this stage that all the previously formulated subdivisions of 'selection', with the possible exception of Wallace's (1968, 1981) hard selection, are subdivisions within my category of w-selection. n-Selection as proposed here is a new concept or, more accurately, a process which has not previously been considered as selection at all. The relationship between MacArthur and Wilson's (1967) r-selection (see also Anderson and King, 1970; King and Anderson, 1971) and my n-selection is a subtle one: r-selection is w-selection for variants with high R_0 values in a situation where all of the variants are likely to have $R_0 > 1$ at low density; n-selection is a selective process actually based on R_0 and involving the separation of variants with (at low density) $R_0 > 1$ from those with $R_0 < 1$.

It should be readily apparent that n-selection is a much less rigorous form of screening than w-selection. If our gedanken experiment had involved mutant propagules sympatric with each other and with the wild-type, rather than allopatric propagules, the outcome would, under normal laboratory conditions, have been fixation of the wild-type and extinction of all the other variants. Thus the different outcomes of w-selection and n-selection, and particularly *the power of n-selection to permit the establishment of a drastically altered morphotype with a lowered degree of coadaptation*, cannot be doubted. What most certainly can be doubted is whether conditions ever exist in nature that allow the fitness of a new variant to be assessed only in terms of R_0; that is, to be subjected only to n-selection. This problem will be examined in (b) and (c) below.

(b) Conditions under which R_0 rather than w is the appropriate measure of fitness

In the discussion above, the distinction between the ecological situations in which n-selection and w-selection operated was simply that between allopatry and sympatry. That is, if two variants were together in the same population then w was the appropriate measure of fitness, whereas in pure culture R_0 was appropriate. However, the question of whether, *in general*, the fitness of one type of organism should be measured comparatively to that of some other organism does not simply revolve around whether or not they are sympatric. We do not, for example, measure the fitness of a *Drosophila* variant in a natural population relative to that of a sympatric spider or oak tree. While this fact is so obvious as to make the very possibility seem ridiculous, the question of the necessary and sufficient conditions for w rather than R_0 to be the appropriate measure of fitness is not a simple one, and it has not received the attention it deserves. We will now take steps to remedy this.

Initially, four phenomena seem relevant to this question: interbreeding, competition, sympatry and descent/degree of relationship. We can, however, dispose of the last two of these fairly easily. Identity of descent ultimately applies to any two living organisms. The degree of genetic identity between two individuals is a continuous variable, and there is no clear point anywhere on this scale at which to switch from one form of fitness measurement to the other.

There are therefore two possibilities: either we accept some arbitrary degree of identity by descent as a necessary condition for the comparative measurement of fitness or we reject considerations about descent altogether. From both a theoretical viewpoint and from a practical one (in which the measurement of degree of identity can be problematical) the latter alternative — rejection of considerations of descent — is clearly preferable.

Turning to sympatry, the case for rejecting this as a sufficient condition for measuring the fitness of two organisms comparatively is equally strong. Sympatry was only important in our gedanken experiment in a second-hand way, through its effect on competition: allopatry means no competition, sympatry means *possible* competition. The reason that the fitness of a naturally occurring vestigial-winged *D. melanogaster* (for example) has to be measured relative to that of coexisting wild-type *D. melanogaster* but not relative to that of coexisting spiders or trees inhabiting the same environmental patch is simply that the resources used by *D. melanogaster*, trees and spiders do not overlap, and they thus do not compete. It should be noted here that lack of interbreeding does not remove the need to measure fitness by w: highly mutant *D. melanogaster* that would pass the test of n-selection (i.e. $R_0 > 1$ at low density) are nevertheless driven to extinction in mixed culture with wild-type *D. simulans*, despite a complete lack of interbreeding in the experiments concerned (Arthur, 1980a,b). In fact, w is a measure of interspecific competitive ability as well as of intraspecific fitness (Arthur, 1982a) and indeed when experiments involving mixed cultures of haploid variants are performed, it is equally possible to consider the system as one of polymorphism of interspecific competition (Levin, 1972).

In the process of rejecting sympatry as a sufficient condition for the relative measurement of fitness, we have, simultaneously, established the importance of competition. What remains to be discussed is whether the ability of two variants to interbreed is relevant to the way in which we should measure their fitness. This is actually a more difficult question than implied by the apparent dismissal — in the last paragraph — of interbreeding as being important, and I will return to it shortly. However, we can at least dispense with the problem of interbreeding in asexually reproducing organisms. In their case we may now say that the fitness of one variant should be measured relative to that of another *if and only if the two variants compete for limited resources.*

Returning to sexually reproducing organisms and the question of whether or not the ability of two variants to interbreed is important, one thing at least has already been established. If two variants compete but do not interbreed, then w rather than R_0 is the appropriate fitness measure. Thus we can say that interbreeding is not a *necessary* condition for the comparative measurement of fitness. When two variants compete *and* interbreed, then we necessarily measure their fitness by w because of their competition, and the fact that they interbreed as well is irrelevant. The remaining question that needs to be addressed is: when two variants interbreed but do not compete, is w or R_0 the more appropriate measure of their fitness?

Two rather different ecological situations need to be considered here: populations whose constituent variants are temporarily not competing because of population growth/unutilized resources, in which situation the lack of competition applies to any inter-organism comparison, regardless of their genetics; and the situation in which two genetic variants do not compete because they utilize different limiting resources (see Levene, 1953). In both situations it seems preferable to use w rather than R_0 if only because the meaning of R_0 becomes obscure if it is applied to a subgroup within an overall, interbreeding, population. Also, in cases of population growth, resources must ultimately become limiting, which strengthens the case for w. However, while w seems a preferable measure of fitness for the *evolutionist* to use in this situation, it is important to note that *nature* is, essentially, using both measures of fitness simultaneously. This is reflected, in a growing population, by the increase in density and decline in frequency of a mutant with $w < 1$, $R_0 > 1$. This last point serves to emphasize the fact that, although most of the discussion here has been phrased in terms of what measure the evolutionist should use, it is the natural 'measurement' of fitness that really matters.

Having arrived at a fairly clear decision as to the conditions under which fitness should be measured comparatively by w, and the alternative conditions under which R_0 is the appropriate measure (see Table 11.1 for a summary), it is now necessary to enquire about the likelihood of the latter conditions (cases 4 and 5 in the table) being realized in nature.

Table 11.1. Summary of conditions under which it is appropriate to measure fitness by w or by R_0

Situation	Allopatric (A) or sympatric (S)	Competing (C) or not competing (N)	Interbreeding (B) or not (N)	Appropriate measure
1	S	C	B	w
2	S	C	N	w
3*	S	N	B	w
4	S	N	N	R_0
5**	A	N	N	R_0

*This case is rather complex: see text.
**The situation in our gedanken experiment. *Note* that in this case the variants are not actually competing or interbreeding because they are allopatric; but they are potentially so in that they would interbreed and compete if made sympatric.

(c)   On the likelihood of n-selection in nature

The best candidate for a situation in which n-selection would operate (i.e. the variant concerned would be screened only in terms of its R_0) might at first seem to be small, isolated, newly founded mutant populations (see Table 11.1, row 5). These populations would be originated in those few rare cases where mutational and migrational events coincide, and might consist of a single group of siblings or even, in the case of asexuals or self-fertilizing hermaphrodites, a single individual. However, such a scheme for the long-term establishment of a new morphotype suffers a serious problem: the newly founded mutant population (itself a very unlikely event) will be susceptible to invasion from without, by migrant wild-type individuals, and from within, by back-mutations. Only if the new population remains isolated until a new coadapted gene complex is built up through 'internal w-selection' on modifiers of the new major D-gene will it stand any chance of remaining in existence. Since this build-up of a new coadapted gene complex is a slow process operating in evolutionary time, whereas the migration of wild-type individuals is an orders-of-magnitude more rapid one operating in ecological time, the chances of long-term survival of the mutant morphotype are very close to zero. The discrepancy between the ecological and evolutionary time-scales is very marked in this particular type of situation because, if the mutant population is established from a small propagule of mutant individuals, it will have gone through a severe bottleneck and will be dependent on mutations of modifiers occurring before its degree of coadaptation can be improved by selection. Thus even the occasional establishment of a new form through n-selection operating in temporary ecological isolates composed exclusively of that form can be doubted. This problem of reinvasion outpacing 're-coadaptation' has not been sufficiently stressed by other authors who have suggested processes of, essentially, n-selection in spatial isolates. For example, Stanley (1979) is, I think, too optimistic when he states that "it is within a weakly populated ecologic island, whether large or small, that the chances of immediate expansion and *long-term survival* of aberrant populations are greatly increased" (my italics).

The problem of eventual re-invasion, and hence re-introduction of w-selection involving the new mutant and the original form, applies only in this particular regime — ecological isolates — with which previous authors have been preoccupied. There is, however, a second situation in which n-selection will operate (Table 11.1, row 4) which is not subject to this problem and in which the newly established variant is protected for a much longer period from eradication through w-selection.

If the process discussed above is described as n-selection by isolation in space, the alternative form of n-selection, upon which I now wish to focus, may be described as 'isolation in hyperspace' (a reference to Hutchinson's (1965) multidimensional niche). In other words, competition between mutant and 'wild-type' is avoided due to the mutant's pattern of resource utilization being shifted far enough from that of the wild-type that the two forms do not compete when sympatric. Clearly, while the probability of success of a mutant under w-selection decreases with increasing magnitude of phenotypic effect (Fisher, 1930; see also chapter 12), the probability of a mutant being subjected to 'sympatric n-selection' because of non-competition with the wild-type form varies in the opposite direction. That is, only mutations of very large effect indeed will produce this release from w-selection. It is therefore likely that these mutations involve early-acting D-genes with major effects on morphogenesis, such as the insect homoeotics (see section 10.4). This contrasts with the allopatric n-selection situation discussed earlier in that we were in that case implicitly assuming that the mutant form — though much more distinct than a polygenic variant — was nevertheless sufficiently similar to its wild-type progenitor that the two forms would compete if they became sympatric.

Although, in a sexual species, a group of mutant individuals ecologically disjunct from the wild-type but not initially reproductively isolated might rapidly evolve such isolation through a rather drastic version of the Wallace effect, such situations are difficult to deal with in detail and I will restrict the discussion here to cases where reproductive isolation is produced simultaneously with ecological (but not spatial) isolation, and is caused by the same mutation. Such a situation is not unreasonable in principle — though it will of course be extremely infrequent — because, not only are most major-effect mutations highly pleiotropic, altering a number of characters, but also any one character may be involved in the determination of a resource-utilization pattern as well as in determining the reproductive compatibility of the affected individual. Clearly, the morphology of the feeding apparatus will usually have little direct reproductive effect. However, appendages such as legs and wings may well be important in mating as well as in locomotion to particular resource patches. A drastic effect on such structures could thus conceivably cause both ecological and reproductive isolation, even if no other morphological characters were altered by the mutation concerned, though, as already noted, most mutations of early-acting D-genes will in fact alter many characters together.

Given that mutations as described above will occasionally occur, and recalling that in asexually reproducing organisms the mutation need only produce ecological isolation, under what conditions will a mutant population be successfully established? There are two aspects of this problem: first, under what conditions can a group of mutant individuals (i.e. a mutant propagule) be derived from a single mutant allele?; and second, under what conditions will such a propagule give rise to a larger population and to an effectively permanent evolutionary line? I will deal with these two aspects of the problem in turn.

The problem of propagule formation was essentially by-passed in our gedanken experiment because this stage of the process was actually carried out by the experimenter. This problem, however, which certainly cannot be by-passed in natural populations, is not as severe as it might at first seem. The crucial points here are (a) whether the mutation concerned is dominant or recessive, (b) in which type of cell the mutation occurs, and (c) whether or not the locus affected has a maternal type of action, as outlined in section 10.3. We will now consider the possible fates of mutations which vary in these characteristics. In all cases it will be assumed that we are dealing with a diploid sexual organism. (Note that in an asexual haploid a mutant allele *is* a mutant propagule.) It will also be assumed that the mutation concerned involves an early-acting D-gene with major morphological effects, sufficient to ensure ecological and reproductive isolation from the wild-type progenitor.

Dominant germ-cell mutations This is the least likely kind of mutation to become established as a propagule. The result of a fertilization involving a gamete carrying such a mutation is the production of a heterozygous and phenotypically mutant individual which, regardless of the likelihood of its own survival, is an evolutionary dead-end because of its inability to interbreed. The only way in which a propagule could be established is through the occurrence of two similar mutations in the same generation — a very unlikely (though not impossible) event.

Recessive germ-cell mutations If a recessive mutation occurs in an egg cell or a spermatozoan, a fertilization involving the mutant gamete and a normal one will produce a heterozygous, non-mutant individual. Until the first homozygotes are produced, the frequency of such a gene will be subject only to drift. While in the majority of cases drift will lead to extinction of the mutant allele (see Fisher, 1930), given enough repeats of this situation the mutant allele will increase in frequency in a few of them. In such a population, once the frequency of the new allele has risen far enough, homozygotes will begin to appear. If two or more appear in the same generation then we have a propagule from which a mutant population may or may not be established, depending on whether or not $R_0 > 1$; that is, on whether or not it passes the test of n-selection.

Dominant mutations in germ-cell precursors While it is an elementary and often-noted fact (see chapter 1) that mutation in somatic cells are evolutionarily irrelevant, and are thus fundamentally distinct from germ-line mutations, it has been less often stressed that precisely where in the germ-line a mutation occurs is also of considerable evolutionary significance. In most organisms, the cells of the germ-line are separated from somatic cells at a very early stage in development (for example, the 'pole cells' of *Drosophila* are distinct from the soma prior to the blastoderm stage: Sonnenblick, 1950). The number of cells originating the germ line is always small whereas the number of gametes ultimately produced in the adult is usually very large. A mutation occurring in an early germ-line cell will thus be represented in a large fraction of the gametes: the earlier the mutation, the higher the fraction. If the mutation concerned is dominant, then, assuming the individual in which it occurs leaves a number of offspring, a mutant propagule is immediately established. Among the relatively few authors who have clearly stated this possibility are Sinnott *et al.* (1958) who put it as follows: "Mutations in cells ancestral to the gametes, such as spermatogonia, may affect several or many gametes, resulting in the appearance of clusters of mutant individuals in the progeny." Thus we have here a method of producing mutant propagules which is not dependent on the vagaries of genetic drift.

Recessive mutations in germ-cell precursors If a mutation occurring in an early germ-line cell is recessive rather than dominant, then the main difference, from a population viewpoint, is that the mutant propagule appears one generation later. In the first progeny generation we will have a cluster of heterozygotes;

when these mate a smaller group of homozygous, mutant individuals will be produced. The only exception to this delaying effect of recessivity is systems of self-fertilization in which, as in the case of a dominant mutation, it is possible for a group of phenotypically mutant individuals to appear in the first progeny generation.

Mutations in loci with maternal action All the foregoing comments relate to the 'usual' sort of locus where an individual's phenotype is determined by its own genotype. However, a number of loci are known, in a variety of organisms, where the individual's phenotype is determined instead by its mother's genotype (see sections 10.3, 10.4). In such cases, a single dominant mutation (or the existence of a single individual with a homozygous recessive mutant genotype) can give rise to a propagule of phenotypically mutant individuals. Although the fate of such a propagule is a complex matter and one which has not yet received much attention from evolutionary theorists, it may be worth further study in the future since loci exhibiting maternal action often control key early morphogenetic processes and are thus subject to mutations with very large phenotypic effects; i.e. precisely the kind necessary to give rise to an organism that will be subjected to n-selection rather than w-selection.

In conclusion, there are several ways in which a propagule of individuals phenotypically mutant for the effects of a major D-gene mutation may be established. None of them is likely to be a common event, but then it is not proposed that such propagules are *often* involved in evolution — rather it is the *importance* of such rare evolutionary events that is postulated; so the rarity of mutant propagules is not a problem except, as noted earlier, in relation to testability.

We now turn to the fate of a mutant propagule, once it has been established: will it subsequently become extinct, or will it survive, be progressively improved by w-selection on genes which modify the new mutant phenotype, and establish a more-or-less permanent evolutionary line and, ultimately, evolutionary 'branch'? At one level, the answer to this question is remarkably simple. If the propagule passes the test of n-selection (i.e. if it has $R_0 > 1$ at low density) then it will survive; otherwise it will not. Since R_0 depends on both l_x and m_x of the life table (see section 1.4), and hence on both survival and reproduction, these are the two phenomena that are 'assessed' in n-selection, just as they are in w-selection. Thus we are simply stating that if the individuals comprising the mutant propagule can survive long enough to reproduce, can actually reproduce, and can produce offspring at a rate sufficient that, initially, $R_0 > 1$, then the beginnings of a new evolutionary line have been accomplished.

This issue becomes rather more complex if we enquire into what factors (other than the magnitude of a mutation's effect, which has already been noted) affect the likelihood of a mutant propagule having its fitness assessed in terms of R_0 rather than w, and the likelihood of R_0 being greater than 1 when this is indeed the variable being assessed. Consideration of these factors is, as will become apparent, informative, and strengthens the case for n-selection as a means of establishment of major new body plans. The two factors that seem most important in this respect are sex and the presence of species additional to the one whose evolution is under consideration. I will deal with these separately, taking 'additional species' first.

As already noted, the fitnesses of competing organisms are measured relative to each other whether or not they interbreed; that is, whether or not they are in the same species. Thus we need to know not only whether a mutation takes the organism in which it occurs out of ecological range of its parental wild-type variant (i.e. whether their multidmensional limiting resource-utilization curves become disjunct) but also whether the new variant overlaps in ecological requirements with some other already existing species. If it does overlap, then w-selection will occur, in the form of interspecific competition, but if not then the new variant will be subject only to the test of n-selection.

The main factor influencing the likelihood of a given mutant 'jumping' into an already occupied area of ecospace is the diversity of taxa existing prior to the occurrence of the mutation. The higher this diversity, the lower the probability that the new mutant will be screened only by n-selection and not by w-selection. Because of this effect, n-selection would be expected to decrease in commonness as the evolution of multicellular forms proceeds, and it is conceivable that a point will be reached where diversity is such that the probability of n-selection occurring is negligible. In this context, it is worth noting that all animal phyla, with the possible exception of the vertebrates, arose in early metazoan evolution, before 500 MYBP (see, for example, Frazzetta, 1975).

It should be noted that whether or not *any* two forms overlap in actual resource utilization is dependent on the extent to which the environment in which they co-occur allows potentially differing patterns of resource utilization to manifest themselves. Two variants that would, in a heterogeneous environment, utilize completely different resources, can be forced to compete if the environment is relatively homogeneous. Indeed, in work on interspecific competition, the contrast between frequent competitive exclusion in laboratory environments and frequent coexistence in natural ones has been attributed to the induction or intensification of competition in simple laboratory systems (MacArthur, 1972). This distorting effect of 'homogeneous' environments applies equally to the outcome of mixed cultures of single-species variants, and it would be interesting, in this context, to know whether major morphological mutants would persist in mixed culture with wild-type organisms in artificial environments allowing multidimensional ecological separation. Such a persistence is at least more likely than the spread of such mutants through w-selection in competitive conditions, and an experimental approach to this problem is easily within reach. A recent experiment by Jones and Probert (1980) has shown that an allele (affecting pigmentation rather than structure) that is deleterious and rapidly removed by selection in relatively homogeneous environments can persist in a more heterogeneous environment. This different outcome of selection in homogeneous and heterogeneous environments is a result of reduction of intergenotypic competition by differential habitat choice in the latter type of environment. In fact, w-selection and n-selection as I have described them can be thought of as occurring at the two extremes of 'degree of habitat choice', with a variety of intermediate situations being possible.

I will come back to the issue of competition between a saltational variant and other species in chapter 12. It is now necessary to turn our attention to that other factor which influences the probability of establishment of a new form through n-selection, namely sex. Two points can be made here. First, since the conditions necessary for a large-effect mutation to produce a propagule with $R_0 > 1$ are much more restrictive in sexual than asexual 'species' due to the necessity of reproductive compatibility among different members of the mutant propagule in the former case, n-selection should be a more common process in asexually reproducing organisms. It should be recalled at this point that we are dealing, throughout, with mutations which affect the between-cell component of development (i.e. pattern formation/morphogenesis), so all organisms under discussion, whether hypothetical or actual, are necessarily multicellular ones. Thus, if there was a phase between the origin of the eukaryotes at around 1.4 BYBP (Schopf and Oehler, 1976) and the origin of eukaryotic sex at around 1 BYBP (Maynard Smith, 1978) (both of these estimates are very rough ones) during which asexual but multicellular forms were common, then n-selection of highly mutant morphotypes would have been a more frequent event than at any later stage of evolution. Second, within sexual species, complex patterns of courtship behaviour, especially those where 'receptivity' based on visual information is important, must make interbreeding more difficult between the organisms of a mutant propagule. This fact also suggests that successful establishment of a mutant propagule should be more likely during earlier evolution; and also makes it particularly unlikely that successful evolution by 'sympatric n-selection' will occur in the vertebrates.

In conclusion, then, an asexual (or hermaphrodite) multicellular organism subject to major morphogenetic mutations and inhabiting a taxon-sparse early environment might be expected to give rise, *relatively* frequently, to mutant forms which establish themselves as populations because they can survive well enough and reproduce fast enough that $R_0 > 1$. At the other extreme, a sexually reproducing population with a complex courtship pattern and inhabiting a species-diverse environment would give rise very rarely if at all to mutant populations established through 'sympatric n-selection'. Thus this proposed evolutionary mechanism has characteristics such that it should decline in frequency of occurrence as evolution proceeds. This is a requirement of any mechanism involved specifically in the establishment of radically new multicellular body plans, since these nearly all originated in the Pre-Cambrian, during early multicellular evolution.

New mutant forms established through sympatric n-selection (which, it should be noted, immediately constitute new populations, new species and new body plans) will be protected from invasion from without, unless a similar type has already been produced. They are also protected from invasion from within; a back-mutant will simply revert to the alternative non-interbreeding and non-competing population. Later back-mutants will be very unfit after a revised co-adapted gene complex has been built up around the new major mutation. It may be, as noted earlier, that a mutant of the bithorax series, or the leg-for-antenna mutant antennipedia, or indeed any of the homoeotics, are the reappearance by back-mutation of old major D-gene alleles which are now less fit than their later replacements, but which in an earlier genetic and phenotypic background were the fittest alleles for the D-loci concerned.

Finally, it must be stressed that the process of n-selection would not involve the production of a perfectly coadapted form or one which was perfectly 'pre-adapted' to some particular ecological niche, as Goldschmidt (1940, 1955) considered his 'hopeful monsters' to be. The new morphotype may in fact be very inefficient in many respects; the initial requirement of n-selection is only that it survives and reproduces in the absence of competition.

(d) The interaction between n-selection and w-selection

Two important distinctions need to be made between the w-selective origins of occasional, 'conventionally fit' saltational variants such as sinistral gastropods (see section 11.4) and the proposed origin, by n-selection, of most other very drastically altered variants. The first distinction is that, because of the way in which n-selection operates, it will usually produce a variant with a *reduced* level of coadaptation. The second is that, while w-selection can provide a self-contained explanation of the origin of a new major variant in those few cases where such variants are indeed fitter (in the w sense) than the original morphotype, n-selection can not. These two points are related: n-selection may establish a body plan capable of surviving and reproducing, but it is not likely to be a very efficient one. However, like any other organism, one produced by an n selective saltation will be continually subject to w-selective modification. Thus the evolution of a new body plan may be initiated by n-selection but it will be completed by w-selection. Moreover, such events may be characterized by a particular pattern of w-selection, with the magnitude of effect of mutations that are fixed in the population, and that modify the basic mutant form, gradually declining as the time since the initial saltation elapses and the actual body plan converges with the 'best' version of that plan. This can be pictured as coadaptation taking place by mutations sweeping upwards through morphogenetic trees of the sort illustrated in Figure 10.12. This rather speculative picture will be returned to in chapters 12 and 14.

The shape of the mega-evolutionary tree

In the artificial selection experiments of quantitative genetics, evolutionary change is often represented (see, for example, Figure 4.6) in graphical form with a simple, unidimensional measure of morphology as the y-axis and time as the x-axis. When we 'ascend' to macroevolution and the theory of punctuated equilibrium, the axes have switched and morphology is now measured by some complex and unspecified variable on the x-axis (see Figure 8.1). In mega-evolution the situation is even more complex if only because the morphological changes involved are, whether they occur abruptly or gradually, enormous, and so the interpretation of the x-axis, if such can be said to exist, becomes even more problematic. If we take a proposed tree of metazoan mega-evolution (see, for example, Barnes, 1980), then it is possible to take a point on the x-axis which, if projected upwards, is crossed by several lineages, indicating that an ancestral chordate, for example, repeats, millions of years later, the body plan of an early sponge. Such an interpretation is clearly ludicrous and no doubt none of the authors presenting these evolutionary trees intended them to be interpreted in this manner. Nevertheless, divergence of two lines in the x-dimension is taken to indicate increasing morphological difference between the groups involved, so this axis clearly does represent morphology in some way.

This problem of the interpretation of mega-evolutionary trees would presumably be removed if we could draw, or even envisage, their multidimensional equivalents. However, we cannot easily do so and the problem thus remains. I have pointed it out because I now wish to contrast three types of trees and it is necessary to ensure that criticism of the tree approach in general is not confused with the central issue, which is the question of which type of tree would be predicted from an assumption of a particular kind of evolutionary mechanism. The three alternative trees discussed below all suffer from the problem of an implicitly multidimensional x-axis, so none is preferable to any other on this basis.

There are at least three distinct ways in which an organism might be constructed genetically: by a small set of D-genes with major morphogenetic effects; by a large number of polygenic D-genes each with very minor morphogenetic influences; or by a mixture of these two cateogires with, in addition, a range of D-genes with intermediate effects. It is now clear beyond all reasonable doubt, from the investigations of developmental and quantitative geneticists, that most multicellular organisms, and probably all of them, are constructed in this final manner (Figure 10.12 shows one proposed morphogenetic system of this third 'mixed' type). However, it is instructive to consider the most likely shape of mega-evolutionary tree under the assumption of each of these three kinds of morphogenetic process, and also assuming, following Fisher (1930), that the rate of evolution of a gene declines as its magnitude of effect increases.

We would expect a wholly polygenic morphogenetic process to lead to the easy and frequent modification of any body plan in all ecologically favoured directions. The alternative extreme kind of morphogenetic process would produce a body plan which, if it evolved at all, would do so in occasional large steps. A combined form of morphogenesis would lead to frequent multiple polygenic modifications of basic themes which themselves make only very occasional transitions. These three types of evolutionary trees are illustrated in Figure 11.1. It is clear that the type III tree represents the closest approximation to the actual pattern of morphological mega-evolution.

The point of this exercise is that, although we know that organisms are constructed by D-genes of very varied magnitudes of effect, the extreme neo-Darwinian approach (section 11.2; cf. more moderate version of neo-Darwinism given in section 1.5) is to regard only the polygenes as being capable of evolution in natural populations. Thus this approach effectively 'writes off' the major D-genes from an evolutionary viewpoint, which means that the most readily predictable kind of evolutionary tree is the type I. Clearly, this type of tree is not a good approximation to the facts. Nor, for that matter, is the type II tree that would result from an extreme saltational view of evolution which denies any importance, outside of intraspecific variation, to the polygenes.

One reason why the more extreme shades of neo-Darwinism survive despite this apparent incompatibility is that the gaps between major branches of the mega-evolutionary tree are explained in terms of ecological rather than developmental constraints. As I have already stated, this is difficult to accept since an ecological reason for the non-production of a large taxon intermediate between the embryologically related vertebrates and echinoderms is hard to imagine. This evolutionary gap is much more easily explained in terms of the limited range of quantitative modifications of the basic body plan produced by a set of early-acting D-genes with major effects, coupled with the rarity of evolutionary events in that key set of D-genes itself. The very fact that phyla are recognized on morphological and especially embryological grounds, and are heterogeneous ecologically, argues for the latter explanation.

To summarize, then:

1. The shape of a mega-evolutionary tree is dependent on the kind of morphogenetic system of D-genes underlying the construction of the organisms whose evolution is being considered, and on the extent to which each of these D-genes is capable of variation in natural populations.

2. The shape of the actual mega-evolutionary tree of the metazoans and metaphytes is incompatible with a purely saltational theory of evolution, explicable but with some difficulty on an extreme neo-Darwinian theory, and readily explicable on the basis of a mixed saltational/neo-Darwinist scheme in which quantitative change predominates but saltational changes occur occasionally and are responsible for the production of new body plans.

11.4 EXAMPLES OF SALTATIONAL EVOLUTION

Shell coiling in gastropods

One reason why I used the shell-coiling D-gene of *Lymnaea* and *Partula* as an example of an early-acting gene with a major morphogenetic effect (section

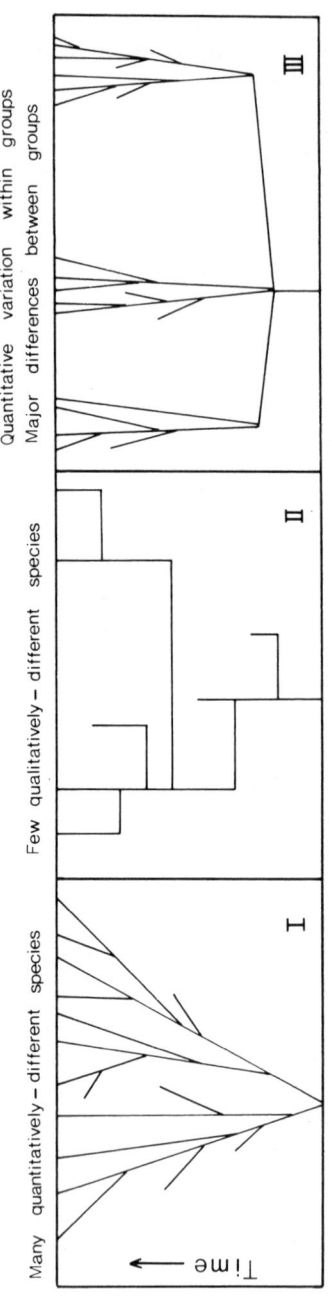

Figure 11.1. Three types of mega-evolutionary trees illustrating the predicted pattern of evolution of organisms whose morphogenesis is exclusively under polygenic control (I), exclusively under the control of a few key genes with major effects (II), or determined by a range of different kinds of *D*-genes with varied magnitudes of phenotypic effect as illustrated in Figure 10.12 (III).

10.3) is that this gene, or equivalents of it in other genera, has almost certainly been involved in a recurring kind of saltational evolution in the gastropods. It is impossible, here, to discuss or even to mention all of the cases in gastropod evolution where a sinistral/dextral saltation has occurred since there are more than 75 000 living species in this group and around 15 000 fossil forms (Barnes, 1980). I will restrict myself to a very brief consideration of these phenotypic saltations in *Partula*, where the selective mechanism has been investigated, and in the terrestrial stylommatophoran pulmonates of the British Isles and North-western Europe. This latter group (which will be discussed first) is easily dealt with since a recent text (Kerney and Cameron, 1979) provides excellent descriptions, both verbal and pictorial, of the species concerned.

Excluding families whose only members are 'greenhouse aliens', there are 22 families of this group of molluscs in the area of Europe which Kerney and Cameron cover. The number of species per family ranges from one (Pyramidulidae, Subulinidae, Bradybaenidae) to 69 (Helicidae). Four of the 22 families (Arionidae, Milacidae, Limacidae, Testacellidae) are slugs and so do not really concern us here. The distribution of the sinistral body plan in the remaining 18 families is shown in Table 11.2. It is immediately apparent that the 33 sinistral species are distributed over these 18 families in a highly non-random manner.

Table 11.2. Distribution of sinistral species in 18 families of North-western European terrestrial pulmonates. Data from Kerney and Cameron (1979)

Family	Number of genera	Number of species	Number of sinistral species
Succineidae	3	5	–
Cochlicopidae	2	4	–
Pyramidulidae	1	1	–
Vertiginidae	3	22	2
Orculidae	2	5	–
Chondrinidae	4	19	–
Pupillidae	4	8	–
Vallonidae	5	10	–
Enidae	4	5	1
Enodontidae	3	5	–
Vitrinidae	5	11	–
Zonitidae	8	28	–
Euconulidae	1	2	–
Ferussaciidae	2	3	–
Subulinidae	1	1	–
Clausiliidae	14	30	30
Bradybaenidae	1	1	–
Helicidae	28	69	–

This sort of distribution would be expected of a gene involved in the origin of a higher taxon. However, it must be said that, while sinistrality appears to be a basic feature of the family Clausiliidae, this characteristic has not generated an entire order, or ever suborder, of gastropods. As regards the number of reversals of symmetry in the evolution of the North-western European stylommatophorans, the simplest hypothesis is that there have been three such reversals; though, clearly, considering the evolution of a geographically restricted group whose families have member species outside the area concerned is unsatisfactory. In general terms, though, the choice of the particular group of gastropods examined in Table 11.2 is not too misleading. Sinistral snails are not restricted to North-western Europe, as indicated by the sinistral form of *Partula suturalis*; nor are they restricted to terrestrial species, as the rare sinistral variant of *L. peregra* and completely sinistral freshwater species such as *Physa fontinalis* show.

A search for the selective mechanism(s) causing evolutionary shifts from dextrality to sinistrality or vice versa is most easily made in species that are genuinely polymorphic for this characteristic. It is not surprising, therefore, to find that a thorough study of the selective pressures acting on the direction of coiling has been conducted in the polymorphic species *Partula suturalis* (Johnson, 1982). As will be seen, a switch in the direction of coiling is explicable in terms of conventional w-selection, and indeed a specific selective mechanism — reproductive character displacement (see section 6.5) — appears to be involved.

Although the initial direction of coiling in *P. suturalis* is not known, Clarke and Murray (1969) have suggested that this species was originally sinistral, with dextrality evolving subsequently in certain populations. Most populations of *P. suturalis* are in fact monomorphic for either sinistral or dextral shells, with these two types of population being separated by relatively narrow, polymorphic, 'hybrid zones'. A clue to the evolution of dextral populations of *P. suturalis* is provided by the fact that these coincide geographically with populations of the uniformly sinistral species *P. mooreana, P. tohiveana* and *P. olympia*. If snails with opposite directions of coiling have difficulty in mating, as indeed they do (see below), and if interspecific hybrids in the *P. suturalis* complex are relatively unfit, then selection might favour dextral *P. suturalis* in areas where the sinistral species are common, because of the reduced frequency of interspecific mating characterizing the dextral variant.

A recent study by Johnson (1982) has confirmed the relative reproductive incompatibility of snails with opposite directions of coiling. In a no-choice mating experiment involving 'same-coil' and 'mixed-coil' pairs, the latter showed an equal amount of courtship behaviour, but a significantly reduced frequency of successful copulation as a result of anatomical 'mis-match'. Johnson (1982) followed up this observation by collecting offspring from the snails involved in the same-coil and mixed-coil pairs. Over a period of between one and two years, the number of offspring produced from the mixed-coil pairs was considerably (about 28%) lower than the equivalent number for the same-coil pairs.

One potential problem for an explanation of the evolution of dextrality in *P. suturalis* in terms of reproductive character displacement is that, within any sinistral population of this species, the original dextral mutants will be at a disadvantage due to a degree of reproductive incompatibility with the rest of their own population. (*P. suturalis* can self-fertilize but does not usually do so;

Murray and Clarke, 1976.) However, the relative magnitudes of this selective pressure (which is an inverse frequency-dependent one) and the counteracting pressure stemming from the disadvantage of interspecific mating, depends on the relative abundances of *P. suturalis* and the coexisting sinistral species. As Johnson (1982) has pointed out, the ecological situation most likely to lead to evolution of dextrality in *P. suturalis* is where this species is rare relative to a sympatric sinistral species; in this situation, selection against sinistral *P. suturalis*, because of their involvement in interspecific mating, will be strongest.

Although we have, then, a feasible mechanism for the evolutionary shift from sinistrality to dextrality in *P. suturalis*, it is by no means certain that a similar mechanism (i.e. reproductive character displacement) was involved in the origin of groups, such as the Clausiliidae, that are now uniform for a coil opposite in direction to that of their nearest relatives. It is certainly *possible* that the same mechanism was involved, but it is unlikely that we will ever be able to draw a firm conclusion on this matter, because the groups concerned originated so long ago that an accurate reconstruction of the ecological situations in which these sinistral-dextral shifts took place is no longer possible.

Although the *D*-gene controlling the direction of shell coiling has a taxonomic distribution that indicates that it is sometimes involved in the origin of new families, although it is rarely polymorphic within species which suggests the evolutionary constraint to be expected of a gene that creates a new body plan, and although a selective mechanism underlying its spread can be envisaged, the evolutionary significance of this gene is in fact rather limited, for two distinct reasons. First, because sinistral and dextral snails are simple mirror images of each other, a mutation leading from one to the other does not produce a severe imbalance between the early-acting *D*-gene and its modifiers which, as we will see later, may be an important step in the formation of *radically* new body plans. Second, the gastropoda is the only group in which the sort of reversal of symmetry this gene causes is possible. Although some other early-acting molluscan *D*-genes with major morphogenetic effects may have contributed to the divergence of taxa higher than families, for example the divergence of gastropods and bivalves, such suggestions are highly speculative. Even an explanation of the origin of the Clausiliidae in terms of a mutation in a single gene controlling the direction of shell coiling is speculative since it is conceivable, though perhaps unlikely, that shell coiling in the original members of this group had a polygenic/threshold basis. To investigate a case of the evolution of major *D*-genes with wider significance we turn to the insect compartments and the genes that control their formation.

Insect compartments

The insects are far more numerous in species than the rest of the animal kingdom as a whole. Thus the evolutionary diversification of insect form represents a substantial part of morphological evolution, and any processes involved in this diversification are of considerable importance.

We have already discussed compartmentalization of development in *Drosophila* and the selector genes which control it. *D*-Genes of this kind are known in a good many insects other than *D. melanogaster*, and some examples of the homoeotic mutations that reveal their existence are given in Table 10.1. The question that needs to be answered is whether, and to what extent, homoeotic mutations have been involved in insect evolution in general and in the origin of orders such as the Diptera in particular. The possibility of such involvement has been discussed by Goldschmidt (1940, 1952, 1953, 1955) and, more recently but much less extensively, by Garcia-Bellido (1975), Lawrence and Morata (1976) and Lewis (1978). The question is not, of course, whether phenotypic saltations have occurred in insect evolution; this can be taken for granted. As noted earlier, characters which vary in an integer manner, such as numbers of legs or segments, must necessarily evolve in a saltational manner at the phenotypic level. The main questions are whether these phenotypic saltations are produced by a small number of major mutations rather than by polygenic variation acting in conjunction with thresholds, and what kind of selective processes are involved.

As regards extrapolation from the intraspecific and generally non-naturally occurring variation in homoeotic genes that is well known, to the evolutionary production of the insects and their constituent orders which is at issue, Lewis (1978) has this to say: "Flies almost certainly evolved from insects with four wings instead of two and insects are believed to have come from arthropod forms with many legs instead of six. During the evolution of the fly, two major groups of genes must have evolved: 'leg-suppressing' genes which removed legs from abdominal segments of millipede-like ancestors followed by 'haltere-promoting' genes which suppressed the second pair of wings of four-winged ancestors. If evolution indeed proceeded in this way, then mutations in the latter group of genes should produce four-winged flies and mutations in the former group, flies with extra legs. In *Drosophila*, not only have both types of mutations been observed, they have been shown to involve a single cluster of pseudoallelic genes known as the bithorax complex".

In postulating that the origin of the insects as a whole, and the origins of particular orders of insects, have involved alterations in the selector genes, Lewis is merely repeating the suggestions of Goldschmidt (1940, 1952, 1953, 1955). Although, interestingly, Lewis has not been taken to task for his suggestions, Goldschmidt certainly was and it is pertinent to enquire why his work was largely rejected. The main reasons seem to have been these:

1. Like Eldredge and Gould (1972), Goldschmidt drew his line of mechanistic discontinuity between micro- and macroevolution rather than between the latter and mega-evolution. For example, he states (1940): "Microevolution does not lead beyond the confines of the species, and the typical products of microevolution, the geographical races, are not

incipient species. There is no such thing as incipient species. Species and higher categories originate in single macroevolutionary steps as completely new genetic systems." This rather extreme view led to an equally extreme reaction and counter-view as exemplified by Mayr's (1963) comments given earlier (section 11.2).

2. As regards the genetic basis of mutant phenotypes (both homoeotics and others), Goldschmidt totally rejected any explanation in terms of genic mutation, and talked instead of 'chromosomal repatterning' and 'systemic mutation'. He goes into considerable detail on these proposed processes in his books of 1940 and 1955, but the key point is that he equates mutations that are large in the genetic sense (i.e. chromosomal) with those that have large phenotypic effects, As we have seen, there is no such *general* link; whether a particular mutation has a large phenotypic effect depends on the role of the 'normal' allele's product in morphogenesis.

3. The proposed process of chromosomal repatterning, which involves changes in chromosome structure such as inversions, coupled with position effects on the inverted gene sequences, was thought to evolve, in a population, in a series of genomic stages. None of these except the last, which produced a new, fully integrated 'systemic mutant' was manifested phenotypically. The intermediate chromosomal stages, which would have resulted in the production of uncoadapted phenotypes *if expressed*, were mysteriously 'silent' at the phenotypic level. This suggests a link between phenotypic expression of a gene (or chromosomal sequence) and the fitness of the expressed phenotype, such that the genes or chromosomes 'know' when the phenotype they encode will be ecologically fit. While such a process would be of enormous evolutionary importance if it existed, and indeed would transform our whole view of evolution, there is no known mechanism through which this process could occur.

4. Goldschmidt regarded the known homoeotic phenotypes as reasonably well coadapted because of, among other things, their bilateral symmetry. Thus instead of considering them as distinctly 'un-coadapted', compared with the wild-type (the view taken here), he thought of them as more or less equivalent to the wild-type in this respect, and as much more co-adapted than the hypothetical intermediate types which, as noted above, he imagined to exist only in encoded form in the genome. Because of their supposedly high degree of coadaptation, and also the 'pre-adaptation' which he imagined them to have for a particular ecological niche, Goldschmidt saw the homoeotic mutants as roughly equivalent in fitness to the wild-type in that, while they were selected against in standard laboratory conditions, they would be favoured over the wild-type in certain (largely unspecified) environments. That is, they could increase in frequency through conventional w-selection, given the right circumstances; a point also briefly made by Wright (1950).

5. When Goldschmidt (1940) put together his main treatise on long-term evolution, no-one had sufficiently emphasized the difficulties of the alternative view to evolution of the selector genes, namely the view that they have *not* evolved. This was, I think, a consequence of the homoeotics being seen as 'mutations', in some general sense, rather than as *mutant alleles* of particular loci. At that stage the process of compartmentalization was not understood, the existence of selector genes controlling this process (of which the homoeotics represent mutant alleles) had not been proposed, and these genes were not seen as comprising some of the most fundamental supporting links in a morphogenetic tree system (see chapter 10 for details of all these phenomena and hypotheses). It is easier to imagine a phenotype of unknown genetic origin as being 'non-contributory' to evolution than it is to envisage a particular *locus* as never evolving, when the process it controls clearly *is* evolving. It was thus easier, in the 1940s, to reject Goldschmidt's proposal that homoeotics were involved in evolution (regardless of precise mechanisms), without proposing a constructive alternative, than it is today.

These five reasons for the rejection of Goldschmidt's work are of three fundamentally different kinds. The first of the five involves criticism of his whole view of evolution; the second, third and fourth involve criticism of his proposed *mechanisms*, both at the individual and population levels; and the final one involves criticism of his idea that the homoeotics have contributed to evolution, regardless of mechanism. Also, some of the five reasons for rejection still stand (1–3), while others are not relevant if n-selection is invoked (4) or are no longer tenable (5). Does this mean, perhaps, that we should accept some of Goldschmidt's proposals and reject others? It is my view that this is the only reasonable course of action. It now seems difficult to reject Goldschmidt's proposals totally, and equally difficult to accept them completely.

Although no definitive statement can yet be made about the evolutionary role of the homoeotics, the following represents a 'compound hypothesis', parts of which (see chapter 15) are testable by experiment. First, homoeotic phenotypes are usually (but not always) the result of possession of mutant alleles at D-loci (the selector genes) which have a key role in morphogenesis. Second, those loci are subject to very occasional evolutionary change. Third, when such changes do occur (at least those involving functionally distinct alleles; see next chapter), phenotypic saltations take place. Fourth, at the population level, the mutant morphotypes may in exceptional circumstances spread by w-selection, but will more often establish themselves through n-selection. Finally, such a process is orders-of-magnitude less common than conventional speciation (see chapter 6), but when it does occur, will lead, in conjunction with gradual w-selective modification of the mutant morphotype, to a new body plan. This, then, entails acceptance of Goldschmidt's proposal that some homoeotic mutations are involved in evolution, rejection of his view that most speciations are saltational, and rejection of his views on mechanisms, with the proviso that the occasional spread of a homoeotic mutant under w-selection cannot be ruled out. This proviso is necessary because environments can be envisaged where normal w-type fitness relationships are reversed; an example would be the fixation of a saltationally transformed wingless form of an insect in a population inhabiting a windswept plain. This overall hypothesis is more conventional than Goldschmidt's in the sense that it is not incompatible with moderate versions of neo-Darwinism (see section 1.5) though it certainly is incompatible with the 'extreme' version (section 11.2).

A final cautionary comment should perhaps be added in order to prevent misinterpretation of the above hypothesis. The known homoeotic mutations of *D. melanogaster* such as bithorax and postbithorax have *not* contributed to evolution, since four-winged taxa have not, apparently, evolved from a dipteran body plan. The proposal being made here is rather that the underlying selector genes *have* evolved, through unknown homoeotic mutations including some of large effect, and that these mutations need not initially have been conventionally fit since they may have become established through n-selection.

11.5 SUMMARY

There are two alternative ways in which higher taxa such as orders, classes and phyla may arise in evolution: through prolonged, slow, quantitative divergence coupled with extinction of intermediate forms; or through some kind of saltational change. Both of these types of origin are discussed and the problems of the quantitative type are noted. Possible mechanisms of saltational evolution are considered, with emphasis being given to the means by which mutant morphotypes may spread at the population level. It is proposed that very long-term evolution is a mixture of saltational and gradual change, with the latter predominating but with the former being responsible for the origin of many new body plans. This mixed kind of evolution appears to be more compatible with the known pattern of mega-evolution than either an extreme neo-Darwinian, or a purely saltational, evolutionary process. Finally, examples are given of some particular situations where phenotypic saltations necessarily occur and are almost certainly underlain by mutations of early-acting D-genes with large morphogenetic effects.

Chapter 12

Variation in Evolutionary Rates

"Monstrosities cannot be separated by any clear line of distinction from mere variations."

Charles Darwin, 1859.

12.1 INTRODUCTION

Although the origin of higher taxa is a topic of obvious interest whereas the phrase 'variation in evolutionary rates' carries much less impact, saltational origins of higher taxa as proposed in the previous chapter can be seen as merely one component of a general pattern of evolutionary change which can be outlined in terms of differences in the evolutionary rates of different groups of D-genes and a consequently variable rate of evolution of morphological characters. The purpose of the present chapter, which is essentially a revision and extension of the ideas I put forward in a recent article (Arthur, 1982d), is to describe this overall pattern, which subsumes not only the saltational changes considered in chapter 11, but also the microevolutionary changes discussed in chapter 5.

It is necessary at the outset to distinguish both between different kinds of evolutionary rates and between different sources of rate variation. Taking kinds of rate first, many quite distinct variables, ranging from Δq to taxon-turnover rates, have been used to measure how fast evolution occurs in particular situations. The choice of variable depends on: (a) whether we wish to concentrate on the genotype or the phenotype; (b) whether, in the case of phenotypic evolution, the variants involved are discrete or continuous; (c) over how long a time period a rate is to be calculated; and (d) whether the focus of attention is morphological or taxonomic.

For the most part I intend to describe here only those kinds of measurement of evolutionary rate that will be made use of in sections 12.2 and 12.3. Discussions of other types of evolutionary rate measurement can be found in Haldane (1949), Simpson (1953) and Arthur (1982d) among many others. However, the recent theory of species selection (Stanley, 1975, 1979) deserves brief mention here along with the kinds of rates — namely rates of speciation and extinction — to which it relates. Stanley points out that speciation and extinction are to the species what birth and death are to the individual. In the same way that individuals leaving more progeny increase in frequency in a population, so do taxa with high net speciation rates increase in their number of species relative to others. While it is undoubtedly true that this is the case, we need to ask whether such a process is important in morphological evolution. In fact, species selection cannot produce new morphological characters, as already pointed out by Maynard Smith (see Gould, 1982); all it can do is to alter the relative frequency of different kinds of species already in existence whose morphology differs because of mutation and individual selection occurring *within particular lineages*. Thus species selection is a secondary process important in determining the relative species diversity of different taxa, but not itself responsible for the origin of new body plans or adaptation to particular ecological niches.

Turning to the variables that are of interest in the context of morphological, rather than 'taxonomic', evolution, we have the following:

1. The rate of occurrence of mutations

Taking any particular *D*-gene, we may distinguish between four kinds of mutation rates. First, the rate, θ', of production of altered DNA base sequences in the whole population of germ cells of a particular generation. Second, the rate, θ, of production of altered base sequences in functional gametes; that is, those contributing to the next generation. In a stable population of diploids in which the average member produces m gametes, the relationship between these two rates, when each is expressed as number of mutations per unit time, is

$$\theta = 2\theta'/m \tag{24}$$

Not all changes in base sequence necessarily alter the activity of the gene product. Indeed, in the case of a *D*-gene that makes a protein, not even all amino-acid substitutions necessarily alter the morphogenetic process controlled by that protein. Thus it is useful to define μ' and μ, the rates of production of *functionally distinct alleles*, the former relating to the whole population of germ cells, the latter to those that will contribute to the subsequent generation. Clearly $\mu' < \theta'$ and $\mu < \theta$. Also, the relationship between μ' and μ is the same as that between θ' and θ, so we have

$$\mu = 2\mu'/m \tag{25}$$

Of the four measures of the mutation rate, μ is the most relevant to discussions of morphological evolution and will be employed in the following section.

2. The rate of fixation

Corresponding to θ and μ there are two distinct fixation rates. We will consider here only the rate of fixation of functionally distinct alleles, since only this rate of fixation is important for morphological evolution. The overall rate of such fixations over a long period of evolutionary time is dependent on μ, assuming that a fixed (and very small) proportion of mutations appearing in the population are selectively advantageous in the case of any particular locus.

3. The rate of change of a metric character

Turning from the D-genes themselves to the morphological characters they produce, the simplest form of measurement is a unidimensional one such as body length, wing length, and so on. Rates of evolution of a simple metric character of this kind are normally expressed relative to the initial value of that character. Haldane (1949) has suggested that such evolutionary rates should be measured in the form

$$(\ln \bar{x}_1 - \ln \bar{x}_0)/t \tag{26}$$

An alternative form, again giving the rate of change relative to the initial value, is

$$[(\bar{x}_1 - \bar{x}_0)/\bar{x}_0]/t \tag{27}$$

These measures of the rate of evolution of a metric character can be used regardless of whether the character concerned is altered in a quantitative, polygenic way or through fixation of a large-effect mutation. However, in the latter case, our simple linear measurement excludes much information on what is in fact a multidimensional morphological change. Even if a metric character changes only through quantitative modification, simple linear measurements omit any information on changes in correlated characters. The situation can be improved slightly by applying expressions (26) and (27) to some ratio x/y rather than to a single measurement, x, but there are still strict limits to the amount of phenotypic information that can be included. This has led some authors to devise more complex morphological measurements.

4. Rates of change of character complexes

One of the most severe problems encountered in attempts to discuss mega-evolution in a quantitative manner is that there is no satisfactory way in which the large-scale phenotypic changes that occur (whether gradually or saltationally) can be measured. One possibility is to have parallel scales of some sort for each of a series of characters and to sum the scores for each character to obtain a compound index for each individual. This sort of approach has been used by Simpson (1953) to investigate the rate of evolution of the lungfish studied by Westoll (1949). Attempts such as this, however, all involve subjective weighting of one character against another. Even giving equal weight to each of a series of characters is to make assumptions about their relative importance. Also, this approach is not applicable to very long-term evolution in which new characters appear and old ones disappear. Because of these problems, I will not attempt to devise any quantitative measure of the rate of evolution of overall morphology. Rather, this kind of evolution, together with associated questions of coadaptation, will be discussed in a qualitative manner after the simpler kind of morphological evolutionary rates represented by expressions (26) and (27) have been considered quantitatively.

Turning from measurement of rates to sources of rate variation, the major distinction that needs to be made is between ecologically induced and developmentally induced changes in evolutionary rate. Clearly, in a lineage undergoing quantitative modification, the rate of change as measured by expression (27) will depend to a large extent on the rate of ecological change or, in the case of a population migrating into a new environment, on the degree to which its morphological characteristics differ from the optima for survival in that environment. It seems likely that in microevolution, and even in macroevolution, this ecological influence on the rate of evolution is predominant. However, if we consider evolutionary changes occurring over a long enough period that rare saltational changes as well as the commoner quantitative ones occur, then some of the rate variation will be due to the differing magnitudes of phenotypic effect of different D-gene mutations. Thus in mega-evolution there are two sources of rate variation: developmental-genetic and ecological. The discussion in the following two sections concentrates on the first of these. This is because it is difficult to theorize in general in a non-trivial way about ecologically induced variation in evolutionary rates; it is better to investigate particular instances of this through case studies of natural populations. In contrast, it is easier — though there are still considerable problems — to construct a general theory of developmentally induced variation in evolutionary rates, and one which is less trivial than its ecological counterpart that rapid evolution is associated with rapid ecological change.

12.2 DEVELOPMENTALLY INDUCED RATE VARIATION: GENETIC ASPECTS

Some introductory remarks

For any locus, the probability of a fixation occurring in a given population over a particular period of time is the product of the simple probability of a mutation occurring and the conditional probability (see Arthurs, 1965) of mutations that occur eventually becoming fixed. Thus we can write

$$p(\text{Fixation}) = \mu \cdot p(\text{Fixation} | \text{Occurrence}) \tag{28}$$

This probability that a mutation will be fixed, given that it has occurred in the population, is a positive function of its probability of being selectively advantageous, $p(A)$, that is,

$$p(\text{Fixation} | \text{Occurrence}) = +f[p(A)] \tag{29}$$

There is not strict equality between these two probabilities for a number of reasons, the most important of which is that many selectively advantageous mutations are lost, while at low frequency, through genetic drift (see Fisher, 1930, chapter 4). Thus the probability of fixation is likely to be much lower than the probability that a mutation is selectively advantageous. Nevertheless, the positive correlation between these two variables indicated in expression (29) holds. As should be apparent from the use of 'fixation' above. I will be considering only w-selection in this section. In the next (12.3) I will examine complications in the argument resulting from occasional cases of establishment of a major mutant by n-selection. Also, where 'selection' and 'fitness' are used without qualification, the w sense should be inferred.

Relative evolutionary rates of different groups of D-genes

In general, mutations in early-acting D-genes will both alter a larger number of characters and, in the case of any one of these, alter it by a larger amount, than mutations of later-acting D-genes. This correspondence between, on the one hand, magnitude and number of effects and, on the other, the time at which a particular D-gene first affects morphogenesis, is not a perfect one, and exceptions to it may easily be envisaged. For example, a gene altering in some way the activity of a hormone involved in termination of growth at reproductive maturity will simultaneously affect the size of many morphological characters. Also, there must presumably be at least *some* early-acting D-genes with relatively minor effects (contrary to the rather simplified picture given in Figure 10.12). However, the relationship proposed here is not intended to be an absolute one. What is suggested is that the *average* number of characters affected and the *average* magnitude of effect declines as we consider mutations of D-genes coming into effect at progressively later stages of development.

As well as the relationship between magnitude of effect and time of earliest activity, a further relationship is important to a consideration of the relative rates of evolution of different groups of D-gene. This is the relationship between magnitude of effect and numbers of loci. It was asserted in chapter 10 (Figure 10.12) that there were fewer early-acting D-genes of large effect than later-acting polygenes with individually negligible effect. This appears likely from a comparison of the results of studies by quantitative and developmental geneticists. However, this relationship should be regarded as a hypothesis, not a proven fact. If it is correct, then we have a three-way relationship between magnitude of effect, time of earliest morphogenetic activity, and number of loci.

It should be noted that I am assuming, here, that particular *D-loci* are typified by *mutations* which are themselves not too varied in magnitude of morphogenetic effect, at least compared to variation among mutations occurring at different loci. This might at first seem odd since there are certain 'non-morphological' loci, such as those producing the α and β chains of haemoglobin, where some mutations produce drastic effects (e.g. sickle cell anaemia) and others negligible ones (Harris, 1975). However, it should be recalled that we are dealing here only with *functionally distinct alleles* that do affect morphogenesis and that the sort of morphogenetic process controlled by major D-genes such as the bithorax complex often have only a limited range of outcomes all of which are very different from each other, whereas mutations in polygenes produce only a range of small, difficult-to-distinguish, variations.

We now consider the consequences of the proposed three-way relationship between magnitude of effect, number of loci, and time of earliest morphogenetic activity, for the evolutionary rates of different groups of D-genes. We will first examine the two components of the fixation rate, as given in expression (28), separately, and then put these two components together at a later stage. Throughout the following discussion two groups of D-genes will be considered: those beginning to take effect at time i in development, and those commencing activity at some later stage (time j). This makes a quantitative treatment simpler, but it should not be allowed to obscure the fact that in reality development is a continuous process, not one divisible into a number of discrete stages.

1. The rate of occurrence of mutations

Providing that the mutability of individual D-genes does not vary in a systematic way according to their time of initial morphogenetic activity, the rate of production of mutations in the *groups* of D-genes acting at different times in development will depend largely on the number of genes in each. Thus considering the per-group mutation rates M_i and M_j, the inequality

$$M_j > M_i \qquad (30)$$

will hold as long as the average per-locus mutation rates do not vary in such a manner as to outweigh the difference in the number of loci. That is, expression (30) holds as long as

$$\frac{n_j}{n_i} > \frac{\bar{\mu}_i}{\bar{\mu}_j}$$

where n is the number of D-genes coming into effect at the appropriate time (i or j).

2. The rate of fixation of mutations already in the population

(a) The influence of 'number of characters affected'

Here I intend to show why, at least under certain assumptions, the probability of a mutation being selectively advantageous (which, it should be recalled, is positively correlated with its probability of fixation) is reduced as the number of separate characters it affects increases The main assumption that will be made is a certain kind of independence of phenotypic effects. I will make the simplest possible comparison, namely a mutation affecting one phenotypic character *versus* a mutation affecting two characters, and will assume that in the latter the probability that one character is improved by the mutation is independent of whether the other is improved or worsened by the same mutation. Later, in section 12.3, I will examine the consequences of violating this assumption.

Let the probability of a mutation affecting a particular character producing an improved version of that character be $p(a)$, and the equivalent probability for a second (also non-pleiotropic) mutation affecting a different character be $p(b)$. Considering now a third mutation which affects both of the two characters, under the assumption of independent effects the three possible outcomes and their associated probabilities are as follows:

1. Both characters improved: $p(a)p(b)$
2. One character improved, the other worsened: $\{p(a)[1 - p(b)]\} + \{p(b)[1 - p(a)]\}$
3. Both characters worsened: $[1 - p(a)][1 - p(b)]$

The first of these phenotypes will be selected for, and the third will be selected against. Because the distribution of selective coefficients is recognized to be asymmetrical in the sense that the average coefficient acting against mutant forms is greater in absolute value than that acting in their favour, less than half (and probably much less) of type 2 mutant phenotypes will be selectively advantageous. This means that the number of pleiotropic mutations that will be selectively favoured is less than the number of single-effect mutations, because

$$p(a)p(b) + c\{p(a)[1 - p(b)] + p(b)[1 - p(a)]\} < \frac{p(a) + p(b)}{2} \tag{32}$$

as long as the fraction c is less than 0.5, as we have already seen that it will be. This argument can be extended to more complicated comparisons to show that, in general, the more phenotypic characters that are affected, the lower will be the probability of a mutation being selectively advantageous and hence of it being fixed in the population.

(b) The influence of 'magnitude of effect'

An alternative to considering how a varying number of effects alters the chances of success of a new mutation is to consider how, with a fixed number of effects, the average *magnitude* of effect influences the probability of a mutation being selectively advantageous. This question has been discussed, though not in a developmental context, by Fisher (1930) and I will deal with

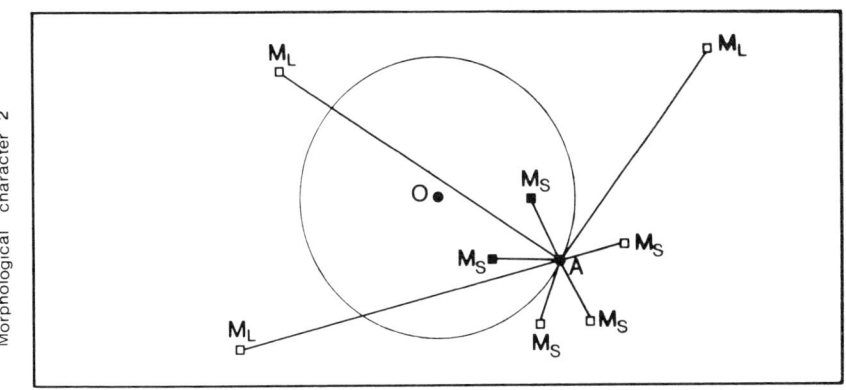

Figure 12.1. The influence of a mutation's magnitude of phenotypic effect on its probability of fixation. M_S, small-effect mutation; M_L, large-effect mutation; O = optimum phenotype. (■), selectively advantageous mutations; (□), deleterious mutations.

the special case of mutations affecting two phenotypic characters so that Fisher's main argument can be represented diagrammatically (see Figure 12.1).

The essence of the argument is as follows. If the type of variant predominating in a population at a particular time is represented by a point A, then unless our population is perfectly adapted to its environment, which is unlikely, the point A will be separated by some finite distance from a second point O which represents the optimal values of the two phenotypic characters that form the axes of the diagram. A mutation in the A-phenotype will produce a new variant represented by a point (M) somewhere else in our two-dimensional space. Such mutants will be selectively favoured over A-variants if the distance of the new point M from the optimum, OM, is less than the distance OA. It is clear that nearly half of a series of very small-effect mutations result in OM < OA. In contrast, for a series of very large mutations where the distance AM (which Fisher calls r) exceeds the diameter of the circle around O on which A lies, the fraction of all mutations that is favoured over A-variants is zero. In general, the larger the magnitude of effect, the lower the probability of fixation.

Although Fisher (1930) did not consider the influence, on the probability of fixation, of the number of characters affected in isolation from the magnitude of effect, he did in fact combine these two aspects of the problem by noting that the decrease in the probability of a mutation being selectively favoured as AM increases becomes more rapid as the number of phenotypic dimensions rises.

3. Synthesis

It follows from the argument advanced so far in this section that the number of mutations occurring in later-acting groups of *D*-genes will exceed the number occurring in earlier-acting groups; and also that the probability of mutations in later-acting genes being selectively favoured is higher than that for genes which take effect at an earlier stage in development. Thus, from expressions (28) and (29) it is necessarily true that the later the stage at which a group of *D*-genes begins to affect morphogenesis, the higher the frequency of fixations of functionally distinct alleles of genes within it.

12.3 DEVELOPMENTALLY INDUCED RATE VARIATION: MORPHOLOGICAL ASPECTS

Simple metric characters

Even without the additional complexity of varying rates of ecological change, a morphological character whose value is the result of the activities of several *D*-genes with differing magnitudes of effect will show a definite pattern of rate variation in long-term evolution due to the infrequent occurrence and fixation of mutations with large effect. This is illustrated in Figure 12.2, which shows the results of a computer simulation in which a morphological character, *y*, is allowed to evolve over 250 arbitrary time periods, each of which can be thought of as consisting of many generations. The program upon which this simulation was based incorporates a battery of 11 *D*-genes, one of which was designated 'early-acting', the rest as 'late-acting'. The per-locus mutation rates were equal over all 11 loci; the probability of fixation of a mutation at a late locus was twice as high as that for an early one, but the phenotypic effect was only 1/10th as great. Both lineages — which differ only in the times at which mutations occur via a random number generator — show a variable rate of morphological evolution resulting from variation in the rate of fixation and magnitudes of effect of mutations in the underlying *D*-genes.

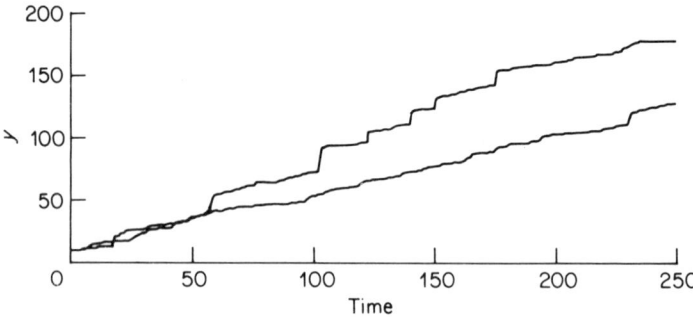

Figure 12.2. Simulation of the evolution of a metric character, *y*, in two 'replicate' lineages. The starting value of *y* was 10 units; and only mutations increasing the value were permitted. Time is in arbitrary units.

Whole organisms

Contrary to what might be supposed by a newcomer to biology initially exposed to an evolutionary literature consisting largely of studies of single loci or individual metric characters, an organism is neither a random collection of genes nor a series of linear measurements whose optima bear no relation to each other. Instead, as we all know but sometimes temporarily forget, a multicellular organism is a highly integrated entity whose survival depends on the compatibility of its various morphological, physiological and behavioural characteristics. Four main things are demanded by natural selection of these integrated entities: development, maintenance of the adult form for a limited period, the ability to interbreed during this period with at least some members of the same species, and sufficient effectiveness in a particular ecological role to ensure an adequate supply of energy to power the other three requirements. All of these necessitate morphological integration of the whole organism as well as certain values of particular morphological characters. Yet thus far in our discussion from chapter 1 onwards this integration has received less attention than it deserves.

The main reason for the relative lack of attention to the integrative aspects of organisms both here and throughout the evolutionary literature is that although this topic is of recognized importance it is extremely difficult to measure the 'degree of integration' of an organism. In this area we are still in what D'Arcy Thompson (1942) calls the "phase of mere description". The following considerations are thus, regrettably, rather descriptive, but they may be useful nevertheless in curbing the excesses that result from the explicit and implicit assumptions of the abstract arguments put forward in the foregoing discussion.

In fact, both the argument based on the number of characters affected by a mutation and that based on the magnitude of effect contain invalid assumptions which cause them to underestimate the frequency of evolutionary events involving early-acting D-genes with major phenotypic effects. The pattern of evolution suggested by the arguments (see right-hand curve of Figure 10.2) consequently needs some modification. I will now deal with these hidden assumptions, treating the two lines of argument separately.

The main problem with the argument based on number of phenotypic effects is the assumption of independence of two or more effects of a mutation in the sense that whether one is an improvement over the original type is taken not to affect whether the other(s) also represent improvements. We need to ask to what extent the prediction of slower rates of evolution of early-acting D-genes is affected by the violations of this assumption that certainly occur.

One possibility is that the prediction is unaffected because the argument can be rephrased in terms of character clusters. An example will help to illustrate what is meant by this. A vestigial-winged mutant of *D. melanogaster* appearing in a natural population possesses several morphological characters whose values are different from those of the co-occurring wild-type flies. The length of the right-hand wing, for example, is reduced, and this is a decided worsening in relation to the effect of this character on fitness. The direction of effect of *vg* on the length of the left-hand wing and on fitness as affected by this character is entirely dependent upon, and corresponds to, the direction of effect on the right-hand wing. On the other hand, the effect of *vg* on fitness in relation to optimum weight is unpredictable. It is likely that, on average, a vestigial-winged fly is very slightly lighter than one with normal wings. In a natural population, the average body weight at any moment in time is likely to be slightly displaced away from the optimum value of this character. Assuming that the difference in weight between normal and vestigial-winged flies is very small, then there is a 50% chance of improvement of this character as regards its effect on overall fitness. Of course, such an improvement will be irrelevant compared to the much larger detrimental effect on the wings and so the mutant will be rapidly removed from the population by natural selection. Nevertheless, this example serves to illustrate that while *some* of the phenotypic effects of a *D*-gene are co-determined as regards their direction of effect on fitness, others are essentially independent of each other. Thus the quantitative argument outlined earlier in terms of number of characters could be re-phrased in terms of character clusters, the clusters being defined so that they are more or less independent of each other even though the characters within each cluster are highly interdependent.

There are two complications that this revised argument still does not overcome, and these are probably of considerable importance in affecting the rates of evolution of some particular groups of *D*-genes. First, *D*-genes acting at the very beginning of development affect *all* characters together and, as a result of this, produce a much more integrated phenotype and have a much higher probability of being selectively advantageous than would be predicted by the earlier argument. For example, the shell-coiling gene in *Lymnaea* rotates all morphological characters rather than just some of them. Most mutations 'attempting' to do the latter would almost certainly be early-stage lethals. Thus we would expect that, although the earliest-acting *D*-genes probably have a much lower chance of evolving than very late-acting polygenes, they will evolve faster than moderately early-acting *D*-genes which will usually cause some degree of incompatibility between affected and unaffected characters.

Second, within that group of *D*-genes whose mutations usually produce some degree of mutual incompatibility of characters, there may be some genes that are involved in a relatively independent branch of the morphogenetic tree, the alteration of which, without change in the rest of development, is not too problematic. Such D-genes would be expected to evolve more rapidly than others even if the effects of mutations in them are saltational at the phenotypic level. The obvious example of such a branch is the 'development' of external pigmentation. Darwin (1859) noted the rapidity of evolution of colour patterns and referred to colour as "that most fleeting of characters". Moreover, most of the studies of ecological genetics which involve genes of large phenotypic effect are based on patterns of pigmentation. Even Mayr (1963) who puts forward the extreme neo-Darwinian view that *all* evolution is through genes of small effect, puts in a footnote to contradict himself and admit that the evolution of industrial melanism in moths takes place through a saltational mutation. We now know of numerous cases of transient and stable polymorphism of genes affecting pigmentation in butterflies, moths, snails, ladybirds and other organisms, many of which are discussed by Clarke *et al.* (1978; snails), Bishop and Cook (1980; moths and ladybirds), and Turner (1977; butterflies). The high frequency of occurrence of such polymorphisms indicates that the genes controlling external pigmentation are much more free to evolve than those controlling structure. This is at least partially due to their relatively weak developmental constraints. Mutations in genes controlling pigmentation interfere little if at all with developing *structures* since they involve no protrusions into, or constrictions of, these structures; nor do they necessitate re-routing of blood or nerve supplies or other major overhauls that would be required by mutations altering structural components. Although colour pattern is not part of morphology in the strict sense of the word (see section 1.1), nevertheless it is a pattern in essentially two-dimensional space and some of the genes controlling it are *D*-genes. (This is, incidentally, one of the few cases in which pattern formation and morphogenesis can be clearly distinguished.)

Turning now to Fisher's (1930) argument based on magnitude of effect, from which he concludes that the chance of a mutation with such a large effect that OM > OA being selectively advantageous is zero (see Figure 12.1), we again find a hidden assumption that causes the argument to be misleading. This is the assumption that selection always operates in a competitive manner, i.e. that it takes the form of w-selection. However, this is only likely to be true within a certain distance of the point A. If a mutation takes a new variant phenotype a great distance from A, its fitness (w) relative to A becomes irrelevant, and its absolute fitness (R_0) becomes all-important. Thus, while the probability of a mutation becoming established decreases initially as its magnitude of effect rises from being negligible, it will eventually increase again once the mutant form is outside of the sphere of influence of the A-variant and the mutant's success or otherwise is determined by n-selection. (Note that this point applies also to the argument based on the number of characters affected.)

There is a further complication here relating to the presence, in the environment concerned, of species additional to the one whose evolution is under consideration. Interspecific competition is, as noted earlier, a form of w-selection. Thus, while a saltational mutant has some chance of success in a taxon-sparse community, it has almost no chance of becoming established in a highly diversified one. This is illustrated in Figure 12.3, which shows two extremes of already existing diversity and their effect upon the fraction of saltational mutants of the A species that will be subjected to w-selection. Because of this effect from competing species, it is to be expected that

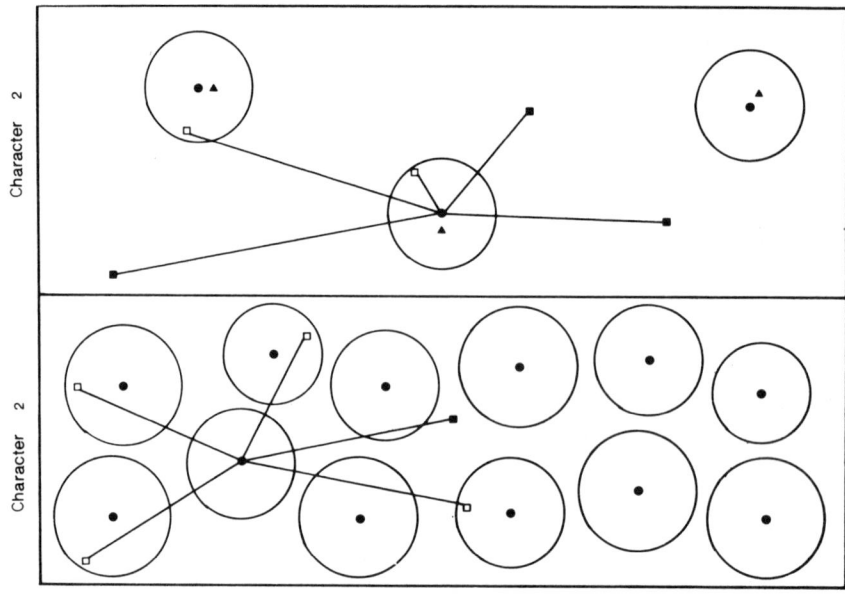

Figure 12.3. The effect of prior evolutionary production of taxa on the chances of success of a saltational mutant. *Top*: community of three species each representing a higher taxon. *Bottom*: community of many species, each again representing a higher taxon. Circles are centred on the points representing the original form of each species (A in Figure 12.1); these circles represented the areas of 'morphospace' within which a new variant undergoes w-selection with the form in the centre of the circle. (□), unsuccessful saltational variant removed by w-selection. (■), saltational variant subject only to n-selection, of which a proportion will survive. (Triangles (▲) represent the O-points of Figure 12.1; these are not shown in the lower half of the figure.)

saltational origins of major taxa through n-selection will be restricted to the early stages of the evolution of multicellular organisms.

The qualifications made above to the previous theoretical arguments about the evolutionary rates of D-genes suggest an alteration to the right-hand curve in Figure 10.2. This alteration is incorporated into Figure 12.3, which is thus a revised proposal of how the probability of a mutant becoming established depends on its magnitude of effect. Two final points should be noted: first, that 'establishment' is now used instead of 'fixation' since the latter term has no meaning in contexts of n-selection; and second, that some D-genes controlling relatively independent branches of the morphogenetic tree will depart from the pattern represented in Figure 12.4.

12.4 THE MORPHOGENETIC TREE AND THE MEGA-EVOLUTIONARY TREE

It will be clear by now that one of the main proposals being made in this book is

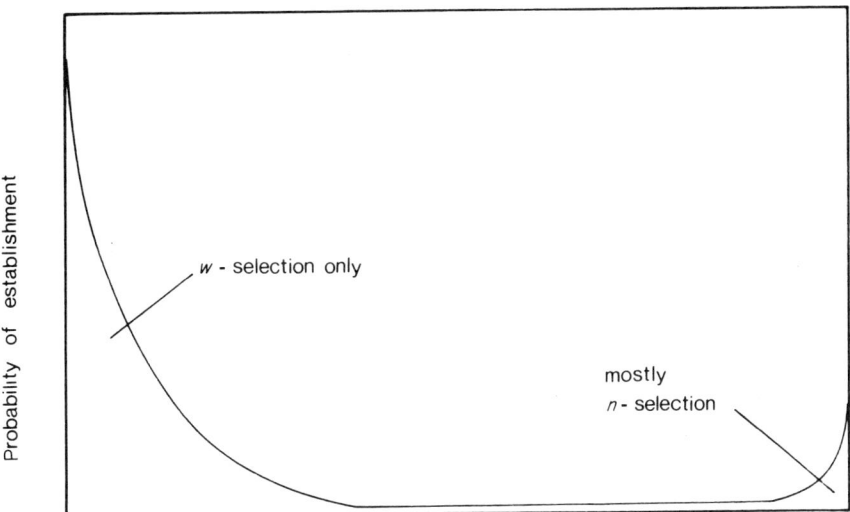

Figure 12.4. Proposed relationship between the probability of establishment of a mutation and its magnitude of effect, taking into account arguments (see text) not considered in Figure 10.2. Note that the interpretation of the rising probability of establishment at the right-hand end of the x-axis depends on how this axis is measured. If it takes the form of 'number of characters affected', then this rise is due to the increased w-type fitness of mutant phenotypes where all characters are affected and there is little or no coadaptational imbalance (e.g. a sinistral snail). If, however, the x-axis takes the form of 'overall magnitude of effect on a particular character complex', then the increasing probability of establishment at the right-hand side reflects the increasing likelihood of the phenotype concerned being subject only to n-selection.

that the shape of the mega-evolutionary tree (see, for example, Figure 12.5) is to some extent determined by the structure of the morphogenetic tree, and in particular by the number of D-genes, their magnitude of phenotypic effect, their temporal distribution through development, and the relationship between these three phenomena. The main point of correspondence between the two trees lies in the small number of early saltational evolutionary changes corresponding to the small number of, and severe selective constraints on, early-acting D-genes of major effect; and the large number of quantitative evolutionary changes corresponding to the large number of polygenes and their higher probability of undergoing mutation to a selectively favoured allele. This overall view of morphological evolution is compatible with both forms of neo-Darwinism outlined in section 1.5, and can account for the following phenomena:

1. The relatively close resemblance of the early developmental stages of different types of organism.

2. The lack of recent origins of phyla and classes.

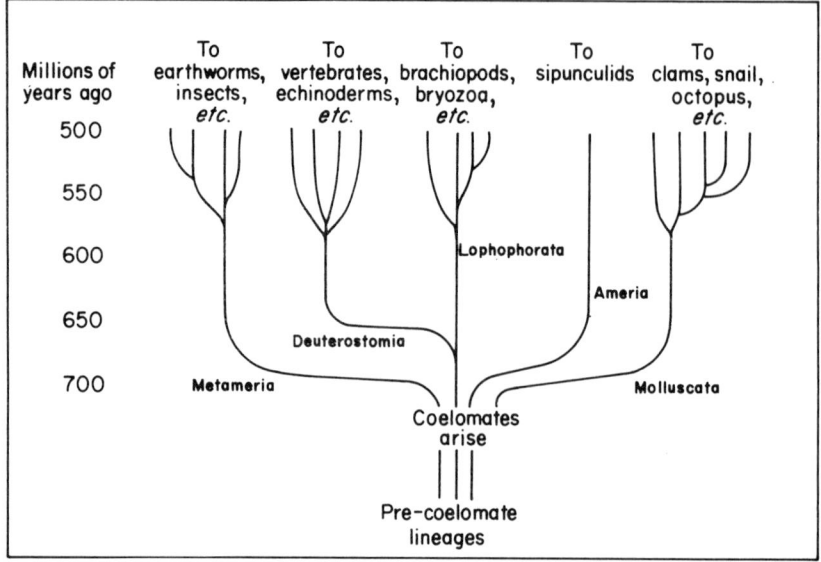

Figure 12.5. A proposed mega-evolutionary tree for coelomate animals, illustrating the occurrence of a few rapid divergences in early evolution and the subsequent occurrence of more numerous but less significant divergences. From Ayala and Valentine (1979).

3. Major 'morphological gaps' between major taxa.

4. The apparent possession, by organisms from different orders or classes, of different alleles at loci with a major effect on morphogenesis.

Of course, the fact that a theory can account for a series of observations does not necessarily mean that it is the correct explanation for these observations, but as yet no alternative evolutionary theory appears to be able to explain all of the above. Included in this 'alternative theory' category are the extreme version of neo-Darwinism (section 11.2), and the theories of de Vries (1905), Goldschmidt (1940) and Eldredge and Gould (1972) which claim that the *majority of species* are produced in a saltational manner.

As regards the form of neo-Darwinism outlined in section 1.5 whose proponents believe that morphological evolution *usually* occurs through selection acting on minor quantitative phenotypic variations, my view is that this is correct but that the concentration on what usually happens has prevented a search for important but infrequent evolutionary events and their underlying mechanisms. In doing so it has also prevented a search for some sort of overall pattern of evolution that includes both common and rare types of change. As MacArthur (1972) has said, "to do science is to search for repeated patterns", and clearly the phenomena explained by any branch of science are restricted to those for which an overall pattern can be discerned. Thus the neo-Darwinism of section 1.5 is not incorrect; it is simply restricted because of its concentration on the commonest form of evolution. I view the ideas put forward herein as an attempt to expand this form of neo-Darwinism, not to replace it, and it is worth noting in this context that all variants are considered here to arise by mutation and to become established either by w-selection or by n-selection. No Lamarckian, neutralist, or other non-Darwinian form of evolutionary change is involved, with the minor exception of the initial spread of some recessive morphogenetic mutations through drift while in the heterozygous condition.

12.5 SUMMARY

Evolutionary rates can be measured by any of a large number of variables, the choice of variables depending on the length of evolutionary time under consideration and on whether the focus of attention is genes, morphology or taxa. Several variables used to measure rates are described, and are then employed in a theoretical discussion of the evolutionary rates of D-genes and the morphological characters they control. It is concluded that, in general, there is a tendency for early-acting D-genes to have larger and more numerous effects than later-acting ones, and consequently to be more constrained in their evolution. This results in an erratic pattern of morphological evolution, with periods of slow, quantitative change being occasionally interrupted by the establishment of a new mutation of an early-acting gene of major morphogenetic effect. Complications to this general pattern are noted; these result from a higher-than-predicted rate of evolution of genes acting at the very beginning of development or controlling a relatively independent section of the developmental process, such as pigmentation. Finally, it is emphasized that the view of evolution put forward in this and the previous chapter can be seen as a proposal that the shape of the mega-evolutionary tree is to some extent determined by the structure of the morphogenetic tree; and that such a proposal is not incompatible with neo-Darwinism except in its most extreme form.

Chapter 13

The Evolution of Morphological Complexity

"The main task of any theory of evolution is to explain adaptive complexity."

J. Maynard Smith, 1972.

13.1 INTRODUCTION

We now return to discuss a major phenomenon of morphological evolution mentioned briefly in chapter 1, namely its tendency to proceed in the direction of increasingly complex forms. This is not, of course, an exclusive tendency and examples can be found where particular taxa appear to go against the general trend. However, the predominance of evolutionary change in the direction of increased complexity is obvious and has been noted by various authors who have highlighted either complexity itself or associated properties such as 'organization'. For example, Fisher (1930) states that "evolutionary changes are generally recognized as producing progressively higher organization in the organic world". Huxley (1942) notes the trend towards increasing complexity and attempts to use it in defining evolutionary progress. Maynard Smith (1972; see above quotation) is clearly taking the increase in complexity for granted and emphasizing the need for an explanation of why it has taken place.

In this chapter I will attempt to clarify what the major question is within the area of the evolution of morphological complexity, and will put forward the proposal that complexity increases not because it is intrinsically advantageous but rather because it is often impossible for natural selection to produce a simpler organism that retains its original functions. Before going into the evolution of complexity, however, it is necessary to consider what is meant by the term 'complexity' itself.

13.2 WHAT IS COMPLEXITY?

Overall phenotypic complexity includes complexity of structure, function and behaviour; and structural complexity is a concept which can be applied to any of several levels of structure from molecular to organismic. We will be concerned here only with morphological complexity; that is, the degree of large-scale structural complexity of an organism. This aspect of complexity relates, then, to numbers and kinds of cells, tissues, organ systems and body segments, but not to any subcellular levels of organization. We now need to ask whether a precise definition of morphological complexity can be formulated. This is a difficult problem and it seems unlikely that any one definition will be universally applicable. Saunders and Ho (1976) have not been quite cautious enough when they assert that one advantage of considering complexity rather than some other evolutionary variable is that "there is no difficulty in defining it". These authors note von Neumann's (1966) definition of the complexity of an automaton as the number of components it contains, and go on to suggest that for biological systems it is preferable to define complexity as the number of different types of component.

At the level of 'pure definition', this seems a reasonable suggestion. However, as noted in section 1.3 in the context of defining morphological evolution itself, a definition must be workable in that it must readily connect with some scale of measurement. Saunders and Ho's (1976) definition does not do this. If we wanted to compare, for example, the relative morphological complexity of a trilobite and a fruitfly, the definition 'number of different types of components' is insufficient to enable us to choose which measurements to make.

The main problem here is that metazoans and metaphytes are constructed not of one series of components but of several such series which are related but which do not necessarily behave in the same way during evolution. In particular, it is necessary to separate the four levels of components mentioned earlier, namely cells, tissues, organs and body segments. As the number of cell types increases, so does the number of tissue types, since each tissue is essentially defined by the cell types it contains. However, the number of body segments or of types of body segment in an arthropod, for example, has no simple relationship with the number of cell types. Thus we are faced with the problem that, in a comparison between two forms, measurements based on one level of components might suggest that form A is more complex than form B while measurements based on another level might suggest the reverse. Since there is no satisfactory way of weighting the different aspects of complexity, it is only meaningful to speak of one form as being more morphologically complex than another if all components of complexity vary in the same direction. Thus we can say that both a man and a millipede are more complex than a sponge, but which of men or millipedes is more morphologically complex is a more difficult question.

It may make the study of morphological complexity more manageable to concentrate on one particular series of components at a time; and it seems likely that which evolutionary time-scale is the focus of attention determines which series of components should be concentrated upon. For example, the number of cell types clearly rises in mega-evolution but, as has already been noted, men and chimpanzees differ hardly at all in the array of cell types they possess; indeed, within the mammals as a whole there is little variation in this array. Within a class or phylum, therefore, it is more useful to consider numbers of types of body segments or some other 'higher' series of components than cells. This also has the advantage of enabling fossils to be included in any survey of morphological complexity, since body segments are often preserved whereas individual cells are usually not.

13.3 ALTERNATIVE THEORIES OF THE EVOLUTIONARY INCREASE IN MORPHOLOGICAL COMPLEXITY

Three explanations have been given as to why morphological complexity increases in evolution. Theses are not mutually exclusive, but it simplifies matters to introduce them as if they were. The reason why the theories are not distinct alternatives to each other is discussed later, and this helps to clarify the main question that needs to be answered.

1. The 'nowhere to go but up' theory

It is clear that protozoans are less morphologically complex than metazoans and also that the latter arose from the former during early evolution. Assuming that after the origin of cellular organisms subcellular ones such as coacervates (see Rose (1966) for a brief description) were no longer viable except as parasites — which are necessarily limited in number relative to non-parasitic forms — then the increase in complexity that occurred with the origins of the metazoa and metaphyta can be thought of as occurring simply because no counter-balancing decrease was possible. This 'nowhere to go but up' theory has been briefly outlined by Maynard Smith (1972) who describes it as "obvious and uninteresting". I would like to give one reason why it is uninteresting and also to point out a limitation to its applicability.

The theory is uninteresting because it provides no mechanism for the increases in complexity that occur and, associated with this, it does not explain why morphological complexity does not remain the same. After all, although non-parasitic subcellular forms may now be prohibited, unicellular organisms, both prokaryotic and eukaryotic, are abundant. Thus, although a decrease in average morphological complexity can be ruled out, a lack of change cannot, and we need to know why average complexity has risen rather than remaining constant.

The main limitation to this theory is that its value diminishes as complexity increases. Once multicellular forms have arisen, increases and decreases in complexity are both possible. Thus we have moved from the situation where there really is nowhere to go but up, to the less extreme situation where the probability distribution may still be skewed in favour of more complex forms, but the probability of simpler ones evolving is not zero. In this context it is interesting to note that no-one has, as far as I am aware, proposed a multicellular origin of any major unicellular taxon. If a lack of such origins is indeed the case, then we need to know what prohibits them, and the 'nowhere to go but up' theory cannot provide an answer to this question.

2. The 'advantages of complexity' theory

One possible explanation for rises in morphological complexity in evolution is that mutations increasing complexity bestow new functions which, at least in some cases, substantially increase the fitness of the organisms concerned and are consequently fixed in the population. Gould (1980) seems to be implying this when he states that, with regard to complexity, "the evolutionary transition from any level to the next occurs more than once; the advantages of increased complexity are such that many independent lines converge upon the few possible solutions". The general impression that such statements give is that certain advantages emanate specifically from complex phenotypes and that these advantages could not be obtained if a revised, but simpler, phenotype were to be produced. The problem with this 'advantages of complexity' theory is that the explanation it gives is an *a posteriori* one. Certainly, many complex phenotypes have homeostatic capabilities and other functions which are less well developed in relatively simple organisms. However, while this fact may help to explain the maintenance of complex forms once they have arisen, a problem occurs if the increase in complexity does not take place in a single step. If it occurs in a long series a small steps, and major new functions do not become available until a late stage in this series, how do the initial steps occur? No answer to this question is provided by the 'advantages of complexity' theory. Thus to assess this theory further we need to ask whether complexity tends to increase in a saltational or gradual manner; this issue will be considered in section 13.4.

3. The 'maintenance of functions' theory

At this point it is worth pausing to consider what we are really trying to explain. As already noted, some evolutionary changes lead to decreases in morphological complexity, others to increases, and others to a level of complexity that is unchanged. The trend that is observed in evolution can be summarized in the following rather abstract way. If some compound scale of complexity could be devised upon which any species could have its value specified, then we would expect that, considering a well-spaced series of points throughout evolution (say 0.25 BY apart), we would see the species of median complexity rise (by variable amounts) when each point is compared with the previous one. This must be due to the ratio of changes that increase complexity to those that decrease it being in excess of 1, unless, of course, changes increasing complexity are individually large and those decreasing it individually small, which seems unlikely. The relative frequency of evolutionary changes which leave the level of complexity unaltered is irrelevant.

Given that most species are well adapted to their environment in that all their components help to ensure survival in it, or that they are genetically linked to ones that do, mutations causing a decrease in complexity (i.e. removal of a component) have an almost zero chance of being selectively advantageous. Those that increase complexity, on the other hand, *retain the original components and their functions*. What they add may be negligible and its initial selective advantage slight but, since it is not opposed by selection against the mutation concerned because of an adverse effect on the original functions, even a very slight advantage may fix this mutation in the population. Because the additional component must have some slight selective advantage to ensure its fixation other than by drift, this theory is difficult to distinguish from the previous one. They are perhaps best separated on the basis of whether a revised and simpler phenotype *could* perform as well as a complexified one. The previous theory asserts that it could not, whereas the present theory argues that the difficulty lies in producing a simpler, revised phenotype by a series of steps some of which involve the removal of functioning components, and makes no comment on whether such a simplified phenotype would be successful if it could somehow be artificially produced. Also, as I have stated it, the 'advantages of complexity' theory is not an explanation of why there should be a *net* change in morphological complexity, but rather why in some cases complexity will increase, which is not really the main issue.

13.4 *D*-GENES, COMPLEXITY AND SALTATIONS

The whole area of the evolution of morphological complexity is a confusing one as the difficulty of coming up with clear alternative theories shows. In the present section I will attempt to reduce this confusion by re-stating the problem in terms of the relative frequency of fixation of the following three categories of *D*-gene mutations:

1. *Reductive mutations*. These remove all or part of some component(s) of an organism and do not add anything in its place.
2. *Supplementary mutations*. These add one or more components (of whatever sort or size) while removing nothing, and so are the opposite of reductive mutations.
3. *Status-quo mutations*. These alter morphogenesis in some way without having any net effect on morphological complexity.

Examples of these three categories, respectively, are: eyeless in *D. melanogaster* in which the eyes are reduced or absent (Lindsley and Grell, 1968); multiple wing hairs in *D. melanogaster* where, as the name suggests, additional hairs appear on the wings (see chapter 10); and symmetry-altering mutations such as that causing sinistrality in *Lymnaea peregra*. In general, all mutations of D-genes that are not of the status-quo type must be reductive or supplementary, albeit some will have a *net* effect in one of these two directions rather than an exclusive one.

Having distinguished between these three kinds of mutation, it is now clear that the statement that average morphological complexity increases during evolution is equivalent to the statement that there is an excess of fixations of supplementary mutations over fixations of reductive ones, provided that the average degree of change in complexity caused by supplementary mutations does not exceed that caused by reductive ones. The rate of fixations of status-quo mutations does not affect the argument.

An excess of fixations of supplementary mutations is by no means inevitable. Although it is reasonable to expect the probability of a supplementary mutation being selectively advantageous to be higher than the equivalent probability for a reductive one, we need to take into account the relative mutation rates also; see expressions (28) and (29). The direction of change in morphological complexity is unpredictable because it depends on the relative magnitudes of

$$\mu_r p(A)_r \quad \text{and} \quad \mu_s p(A)_s$$

and, if $\mu_r > \mu_s$ as may well be the case, then we need to assess the relative magnitudes of the counteracting inequalities in μ and $p(A)$ which is clearly an impossible task. It could be argued that for a well-adapted organism $p(A)_r = 0$, and if this is so then evolution will proceed in the direction of increased morphological complexity, forced that way by a sort of ratchet mechanism analogous to Muller's ratchet acting on the genetic load of asexual organisms (see Maynard Smith, 1978). However, there must be some situations, such as species making major environmental transitions like from water to land, where the energetic and parasitic costs of a structure, coupled with its redundancy in the new environment, make some reductive mutations favourable; that is, $p(A)_r \neq 0$. Also, the removal of whole stages of the life-cycle is likely to be easier than the removal of one or more major components from a particular life-stage. Processes of this kind, such as neoteny, do seem to have had an important evolutionary role, and are discussed in detail by Gould (1977). In these and perhaps a few other situations, an excess of fixations of reductive mutations is quite feasible; in general, though, an excess of fixations of supplementary mutations is more likely, but it is not inevitable unless, most of the time, $p(A)_r = 0$.

Turning to the question of whether complexity increases gradually or saltationally, it is likely that many increases in complexity occur through a process of duplication followed by divergence of the duplicated components. This process has been postulated as occurring both at the morphological level (see, for example, Rensch, 1959) and at the genic level (Ohno, 1970). In this process the step generating the duplication is saltational while the divergence of replicated parts is gradual. If this form of evolutionary increase in morphological complexity is the usual one, then an explanation in terms of large selective advantages at every stage of the process is unlikely. Even if there are cases where morphological complexity increases in a totally saltational manner, the initially produced more complex form would, under the scheme put forward in chapter 11, be un-coadapted and only allowed to establish itself through n-selection. The idea of a saltationally complexified and perfectly pre-adapted form spreading through a population under strong w-selection in its favour is difficult to entertain.

13.5 FOUR MISCONCEPTIONS TO BE AVOIDED

In this section I will briefly discuss four misconceptions relating to the evolutionary increase in morphological complexity which either exist already or which might well be generated by the foregoing discussion if some cautionary comments were not appended.

First, it must be emphasized that neither of the first two theories described in section 13.3 provides a complete explanation for the *net* trends towards increased complexity. The 'nowhere to go but up' theory has nothing to say about why increases in complexity occur at all or about why the average level of complexity continues to rise once this level is such that the chances for decreases in complexity equal those for further increases. The 'advantages of complexity' theory, on the other hand, does propose an explanation of why increases in complexity sometimes occur but does not deal with the main issue of why complexity is more often advantageous than simplicity. Only the third theory of 'maintenance of functions' addresses this main issue and is applicable to all stages in the evolutionary increase in complexity rather than just to the initial ones. This third theory, in contrast to the second one, proposes that the difference in the average value of selective coefficients acting on supplementary and reductive mutations is largely due to their effect (or lack of it) on existing structures rather than to their effect (or lack of it) in adding new ones.

Second, Saunders and Ho (1976) appear to be putting forward a theory similar to the 'maintenance of functions' theory proposed here, as evidence by such comments as, "States of higher fitness and lesser complexity may well exist, though they are unlikely to be arrived at by piecemeal modification". However, these authors combine this view with another: "in our view there are two separate laws of evolution, survival of the fittest and increase in complexity". It is important to stress that the proposals embodied in these two quotations are independent. The suggestion that there are two separate laws of evolution seems misguided because one 'law' is a mechanism, the other a phenomenon produced as a spin-off from the action of that mechanism. Like many other authors adopting a developmental approach to evolution, Saunders and Ho (1976; see also Ho and Saunders, 1979) are more anti-Darwinian than is necessary.

Third, it appears that the general trend in the size and complexity of the genome is an upward one. However, a simple causative relationship between genomic size and morphological complexity should not be inferred. Alterations in morphological complexity can be caused by allelic changes at a single locus, while large alterations in the amount of DNA per haploid genome can occur without marked alteration in morphological complexity. For example, the haploid amount of DNA varies by up to 100-fold among amphibian species, while the average coelenterate differs from the average mammal by a factor of about ten (see Britten and Davidson, 1971). Thus, while some relationship may exist between the size or complexity of the genome and morphological complexity, it is certainly not a very straightforward one. (The only relationship that is discernible is one between the *minimum* haploid genome size of a taxon and its morphological complexity. See Lewin (1983, chapter 17) for a recent summary of the data.)

Finally, although several other authors, notably Medawar (1967), have already squashed this fourth potential misconception, it is worth noting that the overall trend towards increased morphological complexity in the biosphere does not constitute a violation of the second law of thermodynamics. Even if complexity can be equated with 'negative entropy' so that evolution is seen as producing a reduction in entropy in contrast to the universe as a whole where entropy increases, the biosphere belongs to a class of systems to which the second law of thermodynamics does not apply. Complete energetic systems consist of an energy source, an energy sink and some intermediate steady-state system in which, over a long enough period of time, the amounts of energy entering and leaving are equal. The second law of thermodynamics dictates that entropy must rise in complete systems, but does not dictate what happens in their steady-state components into which category the biosphere and its constituent ecosystems and organisms fit. As long as the rest of the universe increases in entropy sufficiently to outweigh the local decreases occurring as a result of evolution in this, and possibly other, biospheres, then there is no violation of the second law.

13.6 SUMMARY

Although there are difficulties involved in defining and measuring morphological complexity, these do not obscure the fact that the overall trend in long-term evolution is an upward one. Possible explanations of this trend are discussed and it is noted that the increase in average complexity may result not from the advantages of complexity *per se* but rather from the difficulty of producing, through natural selection, a simpler, revised phenotype which retains its original functions. Some potential misconceptions that may arise in the area of morphological complexity are discussed, including the incorrect suppositions that there is a straightforward relationship between the size of the genome and morphological complexity, and that the evolutionary increase in complexity contravenes the second law of thermodynamics.

Chapter 14

Limits to the Evolutionary Proliferation of Species and Body Plans

"Once you have got a theory which makes continued evolution possible the major problem is to provide a general theory of phenotypes."

C. H. Waddington, 1975.

14.1 INTRODUCTION: TWO TYPES OF LIMIT

It seems appropriate to end a book on evolution with a discussion of evolutionary limits. There are, however, many different kinds of limits and it would be pointless to try to make a few brief comments about all of these. Rather, I will restrict the discussion here to two types of limit which seem particularly relevant to those aspects of morphological evolution considered in previous chapters, especially chapters 5–7 and chapter 11. As it happens, one of these types of limit (the ecological one: see below) can be treated in some detail because of the existence of a body of ecological theory which is highly relevant from an evolutionary viewpoint. Unfortunately, we can as yet make very little headway in the second area of developmental limits, so the discussion of that topic (section 14.3) will be kept relatively brief. In the remainder of this introductory section I wish to make clear what is meant here by ecological and developmental limits.

The ecological limit that will be considered relates to the proliferation of species which are morphologically and ecologically very similar to each other. This corresponds roughly, but by no means exactly, to the species within a genus. What will be discussed (in section 14.2) is not limitations to the rate of proliferation, but rather the limit to the number of such closely related species that can coexist. The importance of this limit is that it determines when evolution within a particular group switches from an *expansive* phase, where the number of species is increasing, to a *steady-state* phase, where no further species can be produced unless speciation is accompanied by the extinction of an already existing species. This sort of ecological limit clearly relates to rather narrow taxonomic groups, and to micro- and macroevolution as opposed to mega-evolution. The mechanism by which the limit is brought about is interspecific competition. In both of these areas — scale and mechanism — this ecological limit differs from the developmental one discussed below.

The species diversity of a higher taxonomic category such as a class or phylum can be divided into two components: the average number of species per genus or per family which is subject to the ecological limit discussed above; and the number of different morphological 'branches' within the taxon concerned, which corresponds roughly to the number of orders or classes. The enormous number of species within the Insecta, for example, is contributed to both by some very diverse genera and families (*Drosophila* has more than 2000 species) and by a large number of orders (Diptera, Lepidoptera, Coleoptera, etc.) among which adult morphology and the pattern of development exhibit fairly major variations.

Although distantly related groups may compete with each other, as suggested by Darwin (1859) and recently confirmed by experimental evidence of competition between ants and rodents (Brown and Davidson, 1977), it is generally accepted that such competition is less widespread, and on average much less severe, than intrageneric competition. Thus it seems unlikely that the number of major morphological branches of a higher taxon is limited in a competitive manner, though those originating saltationally may find it difficult to get established in the face of even diffuse competition which, as noted in chapters 11 and 12, constitutes w-selection. Other sorts of ecological limit certainly do exist at this level. For example, insectivorous groups (of whatever taxon) could obviously not have evolved before the insects did and, were insects to become extinct, so would insectivores — either through genuine extinction or through Stanley's (1979) 'pseudoextinction', i.e. evolutionary modification into something else. The ecological constraints in long-term evolution are probably determined more by trophic relationships than by competition. These constraints are no doubt important but they are difficult to generalize about, and having pointed out their existence I will say no more about them. I will concentrate instead on the other sort of limit which is important in mega-evolution, namely the developmental one.

The question of how (and whether) developmental constraints have in some way channelled mega-evolution is a difficult one and, like many interesting questions about long-term evolution, one which cannot easily, if at all, be approached experimentally. The question within this rather broad area that will be considered in section 14.3 is whether some form of developmental constraint could explain the gaps in the mega-evolutionary tree mentioned in chapter 11, and, associated with this, whether the number of major groups within a class or phylum is somehow limited or constrained developmentally. As noted earlier, the discussion of this important problem will be kept rather brief since it is difficult at present to draw any firm conclusions in this area.

14.2 SPECIES PACKING AND ECOLOGICAL LIMITS

The body of ecological theory often referred to as species packing theory, and due largely to Robert MacArthur and his colleagues (see, for example, May and MacArthur, 1972), deals with the question of what conditions must be satisfied if two or more competing species are to be able to coexist. This is not in itself an evolutionary question — though it is closely associated with the evolutionary concept of character displacement discussed in chapter 7 — but is, rather, an ecological one and relates to the competitive processes occurring *in a particular community*. To relate species-packing theory to the problem of ecological limits to the evolutionary proliferation of species, we need to consider the extent of migration between communities. This is necessary since most species have populations distributed through many communities, and evolutionary processes involving them depend on more than just the ecological events occurring in any one of these. Surprisingly, although species packing theory itself is relatively well developed (but not so well tested: see below), its evolutionary implications have not received the attention they deserve. In the present section I will separate these two issues, first discussing species-packing theory itself and the problem of coexistence within a single community to which it relates, and then turning to the evolutionary implications of the theory and in particular its relevance in helping us to understand the ecological limits to micro- and macro-evolution.

Species-packing theory

Although Darwin's (1859) purely ecological contributions were largely eclipsed by his evolutionary concept of natural selection, *On the Origin of Species* contains many important ecological comments, particularly on competition, this process being in fact central to Darwin's view of evolution. Especially relevant in the present context is the following passage: "That the number of specific forms has not indefinitely increased, geology shows us plainly; and indeed we can see reason why they should not have thus increased, for the number of places in the polity of nature is not indefinitely great, — not that we have any means of knowing that any one region has as yet got its maximum of species." If we read 'niche' for 'place' and divide 'polity of nature' into within- and between-community components, then we arrive at a modern 'translation' of Darwin's comment. This might be put as follows: The number of niches in any one heterogeneous environment and the number of communities in the biosphere as a whole are both limited, and this will limit the number of species that can be produced, though it is not usually clear if the number of species in any particular community has reached its competitive limit. Darwin's own version is much more elegant, of course, but the 'translation' serves to emphasize that Darwin was a founder of species-packing theory in that he identified the main questions which it seeks to answer.

The niche concept has been rather troublesome in ecology, and one of MacArthur's main contributions was to dispense with the Eltonian biotic niche

(Elton, 1927) as well as with Hutchinson's (1965) elaborate hypervolume niche, and to concentrate on one quantitative variable in relation to which closely related species appear to differ in their utilization of common limiting resources. Although there are many problems associated with these unidimensional niches, or resource-utilization functions, they have allowed a graphical and mathematical approach to the problem of limits to niche overlap which has produced more quantitative and testable predictions than were previously possible.

The small groups of species upon which species-packing theory concentrates are *guilds*; that is, groups of species using a particular resource in a similar, but not quite identical, way (see Root, 1967; Jaksic, 1981). This will often, but not always, mean a congeneric or confamilial group, though a guild may also consist of more than one such group. I will for simplicity assume here that we are dealing with a guild which consists of a single group of congeners.

Taking one such group, the degree to which the various species concerned overlap in their limiting resource (which can be food or space but is usually assumed to be the former) can be represented in the form of a resource spectrum with a series of resource-utilization curves superimposed upon it. This representation (Figure 14.1) is a multispecies version of the situation depicted in Figure 7.1 and, as in the latter case, we may assume that there is a parallel series of morphological curves: differing bird beak sizes corresponding to different preferred seed sizes; differing hydrobiid body sizes corresponding to different ingested particle sizes; and so on.

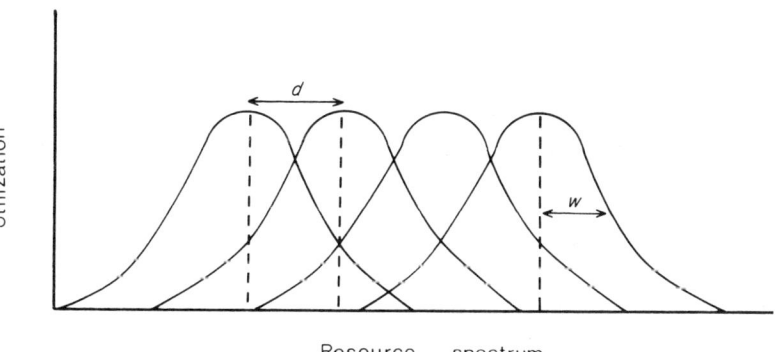

Figure 14.1. Partially overlapping resource-utilization curves in a guild of four competing species. d, intermodal separation; w, standard deviation.

The main questions that need to be asked of such systems as the one depicted in Figure 14.1 are as follows. Is some degree of separation between overlapping resource-utilization curves necessary for the species concerned to reach a state of stable coexistence? If so, what is the limiting degree of interspecific similarity in resource use that is compatible with coexistence? We also need to know whether the answers to these questions vary between species and between different environments.

May and MacArthur (1972) address all of these issues, and start by effectively assuming that the answer to the first question is 'yes'; that is, that some degree of resource heterogeneity and divergence in resource utilization is a prerequisite for stable coexistence. This assumption stems from the fact that these authors equate the degree of niche overlap between two species with the competition coefficients (α_{ij} s) of the Lotka-Volterra model. This model (see Volterra, 1926) has been stated in a variety of forms, one of which is

$$\frac{dN_i}{dt} = r_i N_i \frac{(K_i - N_i - \sum_{j=1}^{m} \alpha_{ij} N_j)}{K_i}$$

where the rate of change of poulation size of any species i, dN_i, is a function of the intrinsic rate of natural increase, r (which, under non-limiting conditions = ln R_0/generation time), the population size itself, the carrying capacity, K, and the population sizes, N_j, of m other competing species together with their inhibitory effect per individual on the growth rate of species i (α_{ij}).

The conditions for coexistence under this model are that interspecific competition is less severe than intraspecific; that is, if all species have equal carrying capacities, all α_{ij} s < 1. Thus if overlap in the resource-utilization curves of two species is equated with their αs, with total overlap of the two curves indicating that $\alpha_{12} = \alpha_{21} = 1$, the requirement of the Lotka-Volterra model that interspecific competition must be less severe than its intraspecific equivalent can be interpreted as a requirement that some degree of separation of utilization patterns is necessary if a state of stable coexistence is to be attained.

Having assumed that some degree of separation of resource utilization curves is necessary for coexistence, May and MacArthur (1972) go on to work out what the maximum degree of overlap, or *limiting similarity*, is in environments subject to different degrees of stochastic variation. The prediction that emerges from their work is that for mildly and moderately variable environments the criterion for stable coexistence is $d/w > 1$. Only in environments with a very high degree of stochastic variation does the critical value of d/w rise above 1, though it rises increasingly fast as the stochastic variation becomes more and more pronounced.

We need to consider now how this critical value of d/w determines the maximum number of competing species that can be maintained in a given environment. If we take the case of species partitioning the limiting resources according to size, this being the situation to which the model is most readily applicable, then the argument is as follows. In any particular environment, the range of sizes of food item, i.e. the length of the resource spectrum, is limited. There must also be some limit w_{min} to niche breadth (though May and MacArthur (1972) do not explicitly deal with this) and, if all species have $w_i = w_{min}$, then there is a limit to the number of competing species in our guild that can coexist which, if the length of the resource spectrum is designated L, is approximately L/w_{min}. The 'approximately' is necessary since what actually happens at the ends of the resource spectrum has not been specified. L/w_{min} is of course a pure number, as it should be, since the units of L and w_{min} (both of which will be in mm or some equivalent) cancel each other out.

Although May and MacArthur's (1972) model has been developed in various ways by a number of authors including May (1974), McMurtie (1976) and Pianka (1981), and although a prediction roughly equivalent to theirs has been made for coexistence in environments containing discrete rather than continuous resources (Stewart and Levin, 1973; Lawlor and Maynard Smith, 1976), species-packing theory remains largely untested. Even the initial assumption that some degree of resource heterogeneity and divergence in resource usage is necessary for stable coexistence has been questioned theoretically (Armstrong and McGehee, 1976) and experimentally (Levin, 1972). One demonstration that would support the efficacy, if not the necessity, of divergent resource utilization in promoting stable coexistence would be the demonstration of stable coexistence in the laboratory, when heterogeneous environments were set up, of a pair of species which, in a more homogeneous environment, do not coexist. One of the few cases in which coexistence and competitive exclusion have both been demonstrated in the laboratory (Gause, 1934, 1935) is difficult to interpret because the species pairs (of *Paramecium*) were different in the two experiments. Most laboratory experiments on interspecific competition have in fact just used a single resource (though the definition of a resource is problematical) and have resulted in competitive exclusion. One of the best examples of this is the exhaustive series of experiments of Park (1948, 1954) on the flour beetles *Tribolium castaneum* and *T. confusum*, and similar results have been obtained with *Drosophila* (see, for example, Moore, 1952a,b). MacArthur (1972) asked: "Why haven't *Drosophila* geneticists, who are now becoming interested in ecology, tried putting very heterogeneous media in population cages to see if many species of *Drosophila* can be made to coexist?" I have attempted to do this, with largely negative results (Arthur, 1980a,b). No-one, however, has set up an experimental system designed to test thoroughly the quantitative prediction of a critical value of $d/w = 1$, and the practical difficulties of doing so are considerable.

In conclusion, then, species-packing theory may or may not provide an *accurate* picture of coexistence in nature, but it does give us a *general* and *quantitative* way of picturing the manner in which competition between species limits the number of them that can coexist in a particular environment. There is little doubt that alternative formulations that may be attempted in future will again predict a maximum number of coexisting competitors in terms of L and d/w, or L and α_{ij}. Such formulations, however, may need to take into account a greater prevalence of highly asymmetrical competition than was previously envisaged (Lawton and Hassell, 1981) and also mechanisms of coexistence which do not depend on resource heterogeneity (see, for example, Atkinson and Shorrocks, 1981).

The ecological limit

We now turn to the relatively unworked area of the relationship between limiting similarity of competitors in a particular environment and limits to the evolutionary proliferation of congeneric species. The first shift of emphasis that needs to be made here is from ecology to morphology. The maximum number of coexisting competitors is, as we have seen, approximately L/w_{min}. This number is formulated in terms of resource-utilization curves, of which w is the standard deviation, and not yet in terms of the distribution curves of the corresponding morphological characters. Although the units of measurement may be the same, as in the case of beak length and seed length, the chances of the standard deviations of the two types of curve being the same are remote. If we assume that, for any particular guild, there is a fixed relationship between the standard deviation, t, of the morphological curve, and that of its ecological analogue, then the maximum number of coexisting competitors, X, can be expressed as $X = L/ct$, where c is a constant. Different guilds will, of course, have different values of c even if this parameter is indeed reasonably constant within guilds.

The other, and more difficult, shift of emphasis that needs to be made here is from single communities to large numbers of communities. The only situation in which this shift does not have to be made is where a group of competing species is endemic to a particular community. In this case, once the limit L/ct has been reached, evolution in this particular group switches from expansive to steady-state. In the vast majority of cases, however, species' ranges extend through more than one community, and indeed they often extend through more than one biogeographical realm (see Elton, 1958). The extent to which the limits to evolutionary proliferation are more complex than those to ecological overlap now becomes dependent on the degree to which some species are restricted in migration. If all species in a particular guild can readily migrate between all communities of a particular kind, then evolutionary proliferation is unaltered, compared with the simpler situation of endemic species, except in as much as it is influenced by different competitive outcomes in different places. However, if one or more species in the guild cannot reach all 'replicates' of a particular community by migration, further expansive evolution can occur after *some* communities have reached the state $S = L/ct$. The general principle here is that the more that individual species are geographically restricted, the more opportunity there is for additional species to fill their ecological niche in different areas. This principle in fact applies even in cases of 'single-species guilds' — i.e. situations in which competition is absent — and the ecologically similar but evolutionarily and spatially distinct pairs or groups of species known as ecological equivalents are an often-cited phenomenon which confirms the validity of the principle. Thus retricted migration not only favours speciation by allowing allopatric divergence and reproductive isolation in smaller areas, as noted in chapter 5; it also favours speciation by allowing the existence of ecological equivalents. That is, restricted migration provides both the *means* and the *opportunity* for speciation.

14.3 COADAPTATION AND DEVELOPMENTAL LIMITS

We turn now to the question of how mega-evolutionary changes are limited by what may loosely be termed developmental constraints. These are usually thought of as constraints imposed due to the necessity for compatibility of different organismic components both in their development and in their joint functioning. Thus any discussion of developmental limits to long-term evolution necessarily involves a discussion of the concept of coadaptation, and before proceeding further we must examine the different meanings that have been attached to this word.

The main problem here is that 'coadaptation' has been used to describe both genes and phenotypes, and is in fact more often used to describe the former. Examples of situations where the emphasis is put on genic aspects of coadaptation are Clarke and Sheppard's (1960) supergenes in *Papilio*, and Dobzhansky's (1970) coadapted gene complexes in chromosomal inversions of *Drosophila*. Although discussion of the work of these and other authors is usually couched in terms of coadapted genes, it is more accurate to talk of coadapted phenotypes. Adaptation is, after all, a phenotypic term describing the usefulness of particular structures in particular environments. Coadaptation, by extension of this, ought to refer to the joint usefulness of two or more structures.

We can loosely refer to the genes producing the coadapted structures as themselves being coadapted simply because they produce the structures concerned. However, this usage should not be confused with a second, related aspect of genic coadaptation which is as follows. If two genes jointly produce a favoured 'combined phenotype' then selection may, under certain conditions, tighten their linkage (see Fisher, 1930; Turner, 1967; Wills, 1981) including, in some cases, incorporating them into crossover-suppressing inversions. Such genetic systems will result in a more frequent co-occurrence of the favoured combinations of alleles and so a more frequent occurrence of coadapted phenotypes. The genes may now be said to be coadapted by virtue of their relative chromosomal positions which themselves are a consequence of selection for coadapted phenotypes.

For the purpose of understanding developmental limits to morphological evolution we need to concentrate on phenotypes, and so I will use 'coadaptation' primarily as a phenotypic term. Two or more *D*-genes may loosely be described as coadapted in the sense that they produce a coadapted phenotype, but the question of chromosomal location of the interacting genes will not be discussed further here.

I now wish to outline a scheme for considering coadaptational aspects of an evolutionary process where quantitative, neo-Darwinian modification is widespread, but in which major new body plans originate saltationally by

n-selection. For the purposes of this scheme it is useful to classify D-genes into two categories: very early-acting ones with major phenotypic effects; and all the later-acting D-genes with lesser effects, including the polygenes, which may be thought of conglomerately as modifiers of the early-acting genes. I will use the conventional genetic terms of 'major gene' and 'modifier' in this restricted sense. Of course, D-genes have a spectrum of magnitudes of phenotypic effect and do not in practice fall into two neat categories; but assuming such categories simplifies the argument.

The main proposal to be made here is this: modifiers are subjected to w-selection and the resultant evolutionary changes tend to increase all aspects of adaptation, including coadaptation; whereas major D-genes are usually altered in evolution by n-selection, and the evolutionary changes involved in this case tend to reduce the level of coadaptation. The fact that n-selection in species-poor environments may result in the establishment of new forms which are inefficient and not perfectly pre-adapted for any ecological role was noted in chapters 11 and 12. What is necessary to emphasize now is that, compared with the form from which it arose, a new form may be less well coadapted. In fact, this reduction in the degree of coadaptation is inevitable, assuming that the ancestral form had been subjected to prolonged w-selection on its modifiers, which will always be the case. The idea of some sort of 'generalized' ancestor not highly coadapted and adapted to a particular ecological niche seems nonsensical; and in any ancestral form whose modifiers have been progressively refined so that they modify a major gene's effect in a particular way for a particular ecological purpose, a major mutation will result in an imbalance between the gene concerned and its modifiers except, perhaps, in the special case of mirror-image mutations as in *Lymnaea*. However, providing the new phenotype can pass the test of n-selection, its modifiers will be subjected to strong w-selection which will result in a gradual restoration of a high degree of coadaptation. This process is illustrated diagrammatically in Figure 14.2. In this scheme, back-mutation of the major gene would again produce a reduction in the degree of coadaptation, and this may be what we see in the *Drosophila* homoeotics. It should be noted that the suggestion that saltational changes are de-coadapting is, along with the suggestion that they are rare and largely confined to early multicellular evolution, a major difference between the theory advanced here and the saltational theories of earlier authors such as Goldschmidt (1940).

In the scheme described above, a particular allele of a major D-gene is seen as defining both an optimum allele and a limited range of viable alleles for each of its modifiers, though these optima and ranges may also be interdependent. Thus any particular set of major D-genes controlling early development limits, to an unknown extent, the number of directions in which modifications of the basic theme can proceed without further saltational change. These limits are imposed by what may be called coadaptational constraints, developmental constraints, or selective constraints, depending on your point of view.

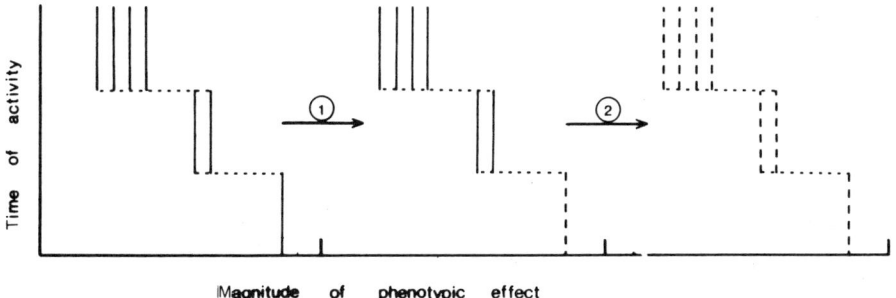

Figure 14.2. Evolutionary alterations of D-genes controlling the morphogenetic tree. Vertical lines of the same type (solid or dashed) represent coadapted genes. Step 1 represents saltational change involving mutation of an early-acting D-gene and establishment by n-selection of the variant so produced. Step 2 shows the result of mutations in, and w-selection on, modifier genes. See also Figure 10.12.

Interestingly, some neo-Darwinists tend to be hostile to the idea of developmental constraints but receptive to the idea of selective ones.

Since n-selection is, like w-selection, concerned with the present and not the future, some saltations will inevitably lead into situations where the range of permissible sets of modifiers is very restricted. Such forms may be said to be 'developmentally over-specialized'. In all cases there must be some limit to the number of permissible modifications, though this limit would be expected to vary greatly among different body plans. Also, some saltations may not only produce considerable phenotypic change in themselves but, in addition, a marked separation of the permitted patterns of modification of ancestral and derived forms. In such cases we would expect large gaps between major branches of the evolutionary tree.

In conclusion, then, the partially saltational but primarily quantitative form of evolution proposed in this book involves gradual improvement of the degree of coadaptation in most lineages most of the time but occasional reductions in coadaptation caused by saltational changes. These changes limit the number of directions in which evolution can proceed without further saltations of an equivalent magnitude. The existence of a limited series of distinct body plans, as represented by different phyla and classes, within which early developmental processes are the most constant theme, is clearly compatible with this view of evolution.

14.4 SUMMARY

Attention is focused on two distinct kinds of evolutionary limit: ecological limits to speciation within genera and families; and developmental limits to the proliferation of different body plans in mega-evolution. As regards ecological limits, the relevance of species-packing theory is noted and its prediction of a limiting similarity of coexisting competitors discussed. It is pointed out that the degree to which individuals of particular species are able to migrate between communities has to be known before the predictions of species-packing theory can be applied to the question of when evolution shifts from an expansive to a steady-state pattern. In section 14.3, developmental limits to long-term evolution are discussed after the meaning of the term coadaptation has been clarified. It is stressed that saltational changes proceeding through n-selection tend to reduce coadaptation, and gradual changes under w-selection to increase it; and that each saltational change limits the number of quantitative modifications that can be produced.

Chapter 15

Summary, Conclusions and Prospect

"Synthesis has become both more necessary and more difficult as evolutionary studies have become more diffuse and more specialized. Knowing more and more about less and less may mean that relationships are lost and the grand pattern and great processes of life are overlooked."

G. G. Simpson, 1944.

In this final chapter I will first briefly state the general conclusions on micro-and macroevolution that can be drawn from the studies discussed in chapters 3–8, and will then go on to consider the extent to which these conclusions provide an adequate explanation of morphological evolution as a whole. I will concentrate on whether they are sufficient to explain the origins of the major body plans which characterize the various higher taxa, since this is one of the most important issues yet also one where the adequacy of the microevolutionary conclusions is highly questionable.

Regarding micro- and macroevolution, then, the following conclusions, most of which are already widely accepted, can be drawn:

1. Considerable quantitative morphological variation exists within and between populations of all species of multicellular organisms.
2. This variation is usually similar in kind, and sometimes not much less marked in magnitude, than interspecific morphological variation within a genus.
3. Quantitative intraspecific variation in outbreeding populations is rarely if ever completely determined either by genes or by the environment. Almost always the heritability is somewhere between 0 and 1; and the genetic component of the variation is due to a large number of genes whose individual phenotypic effects are small.
4. Differential survival and/or reproduction of morphological variants, as effected in artificial selection experiments, is capable of causing a pronounced change in the mean value of the morphological character

concerned, and often of other correlated characters, in a handful of generations.
5. Different natural populations of a given species usually differ significantly in the mean value of one or more morphological characters.
6. Although in some cases these interpopulation differences are largely due to direct environmental effects, a strong genetic component has been demonstrated in others.
7. In a few cases, selection experiments have shown that each variant is best adapted to its own environment, which strongly suggests that the interpopulation differences in morphology originated through directional selection acting in different directions in different populations.
8. Occasionally, geographically separated populations are found which exhibit partial reproductive incompatibility with individuals from the main range of the same species.
9. This can occur with negligible change at the morphological and chromosomal levels. Studies of partially isolated populations and of closely related species suggest that morphological divergence and speciation have no close interrelationship except where the morphology of the reproductive system itself is concerned.
10. Where partial reproductive isolation occurs in allopatry, selection for reinforcement of barriers to interbreeding may occur in neo-sympatry, and this selection may cause reproductive character displacement at the morphological level, though the commonness of this process is uncertain.
11. Alternative forms of speciation, such as sympatric rather than allopatric, or chromosomally rather than genically based, have been put forward, but their importance is not yet clear.
12. After total reproductive isolation has been achieved the pair of species now formed may, if they compete, continued to diverge morphologically due to selection for ecological displacement.
13. The vast majority of new species come into being through the gradual process of quantitative morphological divergence outlined above, together with associated genetic, physiological and behavioural divergence.
14. The challenging of this assertion, by some supporters of the theory of punctuated equilibrium, is based on a dubious interpretation of data which are both very restricted spatially and purely phenotypic.

Our question regarding morphological evolution as a whole can now be put in the following form: assuming points 1–14 are correct, is the overall cyclical process they describe sufficient to produce the major events of mega-evolution, particularly the establishment of new body plans? The extreme neo-Darwinian view (see section 11.2) is that the answer to this question is a simple 'yes'. Extreme saltational theories, including those of de Vries (1905), Goldschmidt (1940) and Eldredge and Gould (1972), do not attempt to answer the above question because they go further and dispute the efficacy of gradual, quantitative change as a common form of evolution above the intraspecific level. However, in this book I have specifically tried to tackle the question in the precise form given above. Thus in chapters 3–8 some of the studies leading to the above conclusions were examined, while in the remaining chapters, especially 11 and 12, the universality of gradual, quantitative change (but not its predominance) is questioned. In particular, it is asserted that there are certain rare but important events in evolution, where new body plans are formed, which are inexplicable except in terms of saltational change. I will now attempt to summarize the main lines of thought which culminated in this assertion, starting with a single example of an evolutionary change for which a gradualistic explanation is untenable.

In *Lymnaea peregra* and *Partula suturalis*, some natural populations are polymorphic for the direction of shell coiling; i.e. some shells in these populations are sinistral, others dextral. In both species the genetic basis of this polymorphism is a single locus with two alleles and not a polygenic system operating in conjunction with a threshold. The gene for shell coiling takes effect very early in development, at the first cleavage division, and it exhibits a delayed pattern of inheritance, with the genotype of the mother determining the phenotype of her progeny.

Within the broader taxonomic group in which the phenomena of sinistrality and dextrality have meaning — the gastropods — most families consist of a number of species and genera all of which are uniformly dextral. However, some families, such as the Vertiginidae, have occasional sinistral members; and others, such as the Clausiliidae, are entirely sinistral. Interspecific differences in the direction of coiling are traceable back to the first cleavage division in the same way as are intraspecific differences, so the parallel between intra- and interspecific variation here is not just one relating to adult morphology but rather one concerning the whole developmental process from egg to adult. Although we cannot reject the hypothesis that some of the interspecific variation in the gastropods is genetically different from that within *L. peregra* and *P. suturalis* and consists of a polygenic/threshold system, this seems rather unlikely. Thus we have here an evolutionary change, which is in some cases intimately involved in the origin of new families, which is saltational at the phenotypic level and which appears to be due to the mutation of a single gene with a major effect on morphogenesis.

This single example of a saltational evolutionary change is clearly sufficient to warrant rejection of the extreme neo-Darwinian view that all evolutionary change occurs through gradual quantitative modification of existing phenotypes. However, on the rare occasions where authors have stated neo-Darwinism in this way, including Mayr's (1963) book from which the quotations representative of the extreme neo-Darwinian position were taken (section 11.2), the authors concerned would probably admit in retrospect to over-generalizing. Indeed, Mayr's footnote admitting saltational changes in colouration shows this to be true in his case. Thus there is little point in citing a single example which disposes only of the most extreme version of a theory when most of the adherents of that theory would not support that particular version anyway. This point is similar to the one Lewontin (1974) was making when, in discussing the selectionist/neutralist controversy over molecular evolution, he commented that, "The neoclassical theory cannot be refuted by erecting a neutralist strawman and refuting that." The same goes for neo-Darwinian strawmen, and thus what we need to decide is whether the case of shell coiling in gastropods is a unique one which has little general relevance or whether it exemplifies a whole category of evolutionary change whose importance has not yet been fully recognized.

The shell-coiling polymorphism is indeed unusual in three important ways. First, it occurs in species which are capable of self-fertilization, a process which might be suspected of spreading an initially unfit saltational mutation by 'clonal' production of populations. Second, because the major, early-acting gene that mutates produces a phenotype which is a mirror image of the original one, the broad body plan produced by that gene is reasonably congruent with the effects of later-acting genes which are superimposed on it; that is, the level of coadaptation of the overall phenotype remains high despite a saltational change having occurred. Finally, and perhaps as a consequence of the second point, the gene determining the direction of shell coiling is found to be polymorphic in natural populations of at least two distantly related species. This is in contrast to other genes which have been postulated as being involved in saltational morphological change, such as the homoeotics of *Drosophila*, which are never found to be truly polymorphic in nature.

Do these unusual features of the gene for shell coiling mean that we can write it off as being responsible for a form of saltational change which has no general importance? The answer to this question is, in my opinion, a categorical 'no' for the following reason. There are very many characters, in all sorts of organisms, which can only vary in a saltational manner. These include binary variables, such as the direction of coiling itself, and integer variables like the numbers of legs and body segments. Unless there is some stage in evolution where these characters are so variable within species that they have distributions which can be represented by their means and variances — and this seems most unlikely — then interspecific changes must necessarily occur saltationally at the phenotypic level.

As regards the genetic basis of these saltational evolutionary changes, there are two main possibilities: mutation and establishment of a single gene with major morphogenetic effect; or mutations in, and alteration of allele frequencies at, many polygenic loci which together result in a mutant phenotype once some underlying quantitative variable such as the concentration of a morphogen passes a threshold value. The latter process is usually associated with neo-Darwinism because it is essentially a mechanism whereby the 'gradual, accumulative' philosophy is retained at a cryptic level even in cases where the visible change is saltational.

There are several reasons for suspecting that many (but certainly not all) of the saltational phenotypic changes that have occurred in evolution are underlain by mutations of single loci where what is obviously a minor genetic change causes a 'developmental revolution' without any assistance, at least initially, from the rest of the genome. First of all, there is the rather negative reason that the fitness-depression inherent in most saltational variants, and so often emphasized by neo-Darwinists as a problem of saltational evolutionary theories, is a feature of the phenotype itself and not of any particular mechanism underlying its production. This is clear because many mutant phenotypes, such as bithorax *Drosophila* and sinistral snails, can be produced as phenocopies (by early exposure to ether and unknown causes, respectively). Bithorax phenocopies are clearly less fit than wild-type flies; and the decrease in reproductive fitness experienced by a single sinistral snail in a population of dextrals as a result of the difficulty of cross-fertilization is due to morphological differences rather than genic ones. Thus although the fitness-depression caused by saltational phenotypic changes has to be acknowledged, this problem is not circumvented by assuming a polygenic basis to the mutant forms.

So far the argument has led us to consider single-gene mutation as, perhaps, an equal candidate to switch mechanisms operating through polygenes, as the genetic basis of saltational phenotypic change. We now turn to the line of argument which asserts that single genes with major effects on morphogenesis *must* have evolved on a number of occasions, with consequent phenotypic saltations. This argument is centred on the relationship between the magnitude of a gene's morphogenetic effect and its time of earliest activity during development. Most, if not all, biologists would agree that, in general, the earlier a morphogenetic gene begins to take effect, the larger the eventual change in adult morphology that mutations in it can produce. The form of the adult, then, can be seen as a composite result of the effects of a large number of genes whose varying degrees of individual effect correspond roughly to their time of onset of activity. This proposition, which is represented diagrammatically in Figure 15.1, is supported by the observation that mutations with radical effects on morphogenesis, including those producing sinistral gastropods and homoeotically transformed fruitflies, take effect at a very early stage in development. Now we know that early developmental stages are more similar to each other, given any particular pairwise comparison between taxa, than later stages, and thus it seems that the genes controlling early development are highly constrained in evolution and that they evolve much more slowly than those taking effect later in development and having more minor morphological consequences. However, it is clear that even the very earliest stages of development have undergone modification during evolution; otherwise we would not have the different patterns of cleavage and gastrulation that characterize the different phyla and classes, and, in insects, the different patterns of compartmentalization that characterize the different orders. If the pattern shown in Figure 15.1 is fundamentally correct, then a single gene that mutates and produces even a very small change in the pattern of early morphogenesis will produce a much more radical change in the morphology of the adult than it produced in the morphology of the early embryo, simply

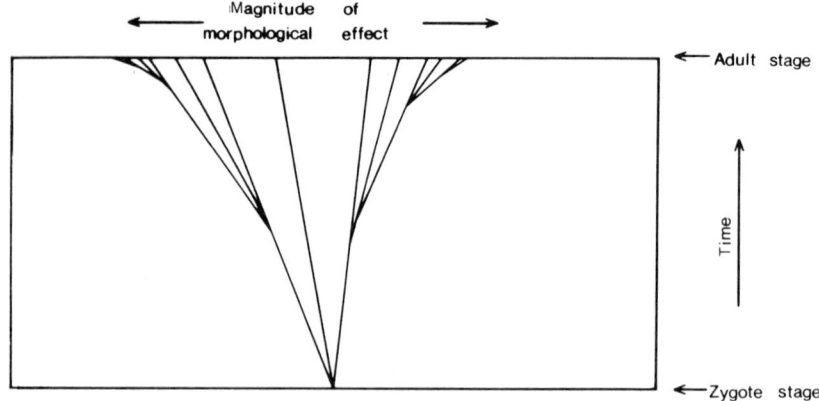

Figure 15.1. Simplified illustration of the relationship between a D-gene's magnitude of morphological effect and its time of initial activity during development. Each triangle represents a different D-gene. Note that in complex developmental systems with different life-stages, this diagram may apply separately to each stage (see section 9.4 and Figure 9.5).

because of the magnifying effect of development on early-occurring modifications. Thus the fact that early developmental stages differ between major taxa indicates that saltational changes in adult morphology caused by single mutations must have occurred in evolution. This is quite apart from the fact that mutations affecting early development may even be saltational at that level if, for example, patterns of cleavage intermediate between radial and spiral are unstable.

Having reached a position where it seems that saltational phenotypic changes produced by mutations of single major genes must have occurred from time to time in very long-term evolution, we need to enquire into the mechanisms by which the mutant forms have become established. This has been the stumbling block for all previous saltational theories of evolution because, while many diverse mutations are known, in a variety of species, which are capable of causing a particular individual to possess a radically new phenotype, a mechanism is also needed to establish the new form in the population if the mutation is to have any evolutionary significance. If saltational variants are usually associated with considerable fitness-depression, how can they be selectively favoured, or even neutral and subject to genetic drift?

There are two possible ways out of this apparent dead-end. First, restricting ourselves to the possible spread of saltational variants through conventional selection, it is advantageous not to consider two types of variation — quantitative and saltational — but rather a continuum along which the magnitude of a gene's morphogenetic effect varies, with genes such as the homoeotics at one end and the unidentified 'polygenes' at the other (see quotation from *On the Origin of Species* at the start of chapter 12). The question then becomes one of how the probability of a new mutation being selectively advantageous varies along this continuum. This representation assumes that each *locus* has some intrinsic magnitude of effect around which different mutations of that locus cluster fairly closely. As noted in chapter 12, there are grounds for believing this to be true in relation to functionally distinct alleles of loci affecting morphogenesis; but it is certainly not true in some other contexts, such as the case of different mutations affecting the structure of the haemoglobin molecule.

Two theoretical arguments (section 12.2) lead to the conclusion that the probability of mutations being selectively advantageous declines as their magnitude of phenotypic effect increases. Fisher (1930) considered the magnitude of effect of a mutation in situations where the number of characters is fixed, and concluded that increasing effect meant decreasing chances of success. Fisher's argument is outlined diagrammatically in Figure 12.1 for the special case of two characters. The alternative approach (Arthur, 1982d; see also section 12.2), is to consider the number of distinct phenotypic effects rather than the magnitude of each. The conclusion is, at least under certain simplifying assumptions, equivalent to that reached by Fisher; that is, genes with greater effect — this time expressed as the number of separate effects — have a lower probability of giving rise to selectively advantageous mutations than those with fewer effects. These two approaches are also in a sense synergistic; Fisher noted that the fall-off in the chances of a mutation being selectively advantageous with increasing magnitude of phenotypic effect becomes more rapid as the number of characters affected increases.

Since early-acting genes with large morphogenetic effects will thus evolve more slowly, but when they do evolve will lead to a greater phenotypic change than later-acting genes with smaller effects, the pattern of evolution of a morphological charcter over long periods of evolutionary time will be rather erratic. A computer simulation of this varying rate of evolution of a morphological character is shown in Figure 12.2. Although the bursts of rapid change in this figure caused by the occasional selectively driven evolution of a gene with *relatively* major morphogenetic effect can be considered as saltations, we need to ask whether this model provides an adequate explanation of really major saltations such as altered patterns of development in whole segments or indeed total re-structuring of the organism. That is: How far along our continuum of 'magnitude of effect' can we go — in the direction of increasing magnitude — before the probability of a mutation being selectively advantageous becomes effectively zero, and we reach a point where the mutations have such a drastic effect that conventional forms of selection cannot cause their spread to fixation?

Although no clear answer can be given to this question, it is my view that the evolution of most very early-acting morphogenetic genes, such as the homoeotics, cannot be explained in a conventional manner because their phenotypic effects are so severe and in particular their negative effect on the degree of coadaptation so drastic. A change in the direction of shell coiling is one of the few major saltations that seems to be explicable in conventional selective terms (see section 11.4). This is because of the unusually high level of coadaptation possessed by the mutant as a consequence of its being a mirror image of the original type. This is clearly the exception rather than the rule for mutations of genes acting very early in development.

If conventional forms of selection are, as hypothesized here, incapable of causing the spread of most major saltational variants through a population, how then do such forms become established? The answer to this question is, I believe, that they spread through what I have called (section 11.3) n-selection. This process differs from the competitive selection, or w-selection, envisaged by Darwin in that the criterion for success is not w, the cross-product ratio measuring the fitness relative to some other form, but rather R_0, the net reproductive rate of the genotype concerned, or 'absolute fitness' as I labelled it in section 1.4. In w-selection, a new variant B will spread to fixation in a population of A if, at all gene frequencies, $w_B > w_A$. However, in n-selection, a new form establishes itself if $R_0 > 1$ when resources are not in short supply. I have given this process the name n-selection to indicate that it is a subcategory within the overall heading of 'selection' because it is, like w-selection, a process involving the weeding out of unfit genotypes on grounds of survival and reproduction. The criterion for success or failure is different, as are the population conditions under which the selection occurs (see section 11.3), but nevertheless selection is occurring since some genotypes will fail the test of n-selection and others will pass it. In the absence of wild-type *D. melanogaster*, for example, morphological mutations causing larval lethality will not establish themselves as a population, while other morphological mutants, such as vestigial-winged flies, can do so with ease.

The precise conditions under which n-selection will lead to the establishment of a new form are very important. Earlier authors (for example, Frazzetta, 1975, Stanley, 1979) have hinted at similar processes but have though of them as acting in spatially isolated populations. In my opinion a saltational variant surviving simply because it has managed to build up a population in an area temporarily uninhabited by wild-type individuals of the same species has very little chance of *long-term* survival, because it is involved in a race between a process operating in evolutionary time and one operating in ecological time. That is, its own improvement through mutation of, and w-selection on, modifier genes will be much too slow to have produced useful results by the time the isolated population is re-invaded by wild-type individuals which bring with them the kind of w-selection that will eliminate the saltational variant.

The alternative type of situation in which n-selection could operate, and the one which seems far more likely to result in the more-or-less permanent establishment of a major saltational variant, is one in which the mutation concerned takes individuals carrying it out of reproductive and resource-utilization contact with the original form. In other words, the effect of the mutation must be so drastic that reproductive isolation and non-overlap in limiting resources are simultaneously achieved. In such cases it becomes inappropriate to use w as a measure of fitness even if the two forms remain geographically and locally sympatric.

The conditions necessary for this sort of selection to be successful, then, are:

1. Occurrence of a mutation of a gene with major morphogenetic effect.
2. The production of a 'propagule' of mutant organisms, for example as a result of the mutation occurring in an early germ-line cell.
3. The interbreeding of the saltational variants and establishment of their population through a period of at least a few generations when $R_0 > 1$.
4. The gradual refinement of the new form by mutation of, and w-selection on, modifiers within the newly established population.

The first three of these four processes are, of course, very improbable and their combined occurrence is even more improbable. However, *a very improbable mechanism is precisely what we need to explain a very infrequent event.* In the rare cases where all four processes do jointly occur, establishment of a new form is achieved; however, one further qualification is necessary. Interspecific competition is a form of w-selection which can operate across large taxonomic gaps and to which a newly mutated saltational form will be very susceptible because of its low degree of coadaptation (see sections 11.3, 14.3). Thus the greater the variety of species already in existence, the lower the chances that n-selection, rather than w-selection, will occur. It may, therefore, be the case that saltational variants can only become established during the early stages of the evolution of multicellular organisms. It is interesting to note, in this context, that all phyla except the chordates seem to have originated before 500 MYBP (Frazzetta, 1975).

The view of mega-evolution adopted herein should now be clear. Most new species, throughout evolution, have arisen through w-selection on minor, quantitative variations as described in points 1–14 at the beginning of this chapter. However, at least in the early stages of the evolution of multicellular organisms, major new body plans arose through n-selection on radically altered phenotypes produced by mutations in early-acting genes with major morphogenetic effects. This process may now be prohibited because of the diverse flora and fauna produced partially by it, and at any rate is much less likely to occur than it was during earlier evolution.

The difference between this view of evolution and conventional forms of neo-Darwinism can now be seen. The extreme neo-Darwinian hypothesis that all evolutionary change takes place through gradual accumulation of very slight phenotypic modifications is refuted, though, as mentioned earlier, this is of little consequence. The claims of the more moderate version of neo-Darwinism (given in section 1.5) are regarded here as essentially correct; that is, virtually all morphological evolution occurs as a result of selection on inherited variation, and the vast majority of species are products of gradual morphological divergence under w-selection. The main differences between the view taken in this book and this moderate neo-Darwinian view are that, in the theory formulated here:

(a) an explicit mechanism is given for the rare but important saltational changes; and
(b) a pattern incorporating both common quantitative change and rare but important saltational change is put forward.

Because these are additions to, rather than revisions of, the moderate neo-Darwinian proposals, and because the suggested mechanism of saltational change also involves a form of selection acting on inherited variation, I regard my own proposals as an expanded version of neo-Darwinism rather than as something distinct from it.

It is important to concentrate on the overall pattern of morphological evolution rather than the postulated saltational changes which form a small but conspicuous part of it. In this context, the overall pattern of evolution proposed here is probably best seen as a system of three dependent variables: probability of occurrence of mutations, probability of fixation under w-selection, and probability of establishment through n-selection, which are functions of several interrelated predictor variables. The latter include the magnitude of a gene's phenotypic effect, the time at which it begins to affect morphogenesis, the number of loci in each morphogenetic category, the likelihood of a mutation having a net supplementary or reductive effect, the 'ecological proximity' of other species, and the degree to which a mutation produces a coadapted phenotype. Such a system takes into account genes, development, adult phenotypes and the environment, and thus satisfies Waddington's (1957a, 1975) demand that evolutionary theory should not concentrate on genes to the virtual exclusion of environmental and phenotypic considerations.

In order to formulate this view of evolution, I have, despite a feeling that there is already too much terminology in the biological sciences, introduced three main new terms: the morphogenetic tree, D-genes and n-selection. I have tried to keep this final chapter free at least of the first two of these so that it is reasonably intelligible to that unfortunately large proportion of readers that makes straight for the concluding chapter of a book. However, for the benefit of the more conscientious readers who got here the hard way, I would like to re-emphasize the need for these terms and the concepts underlying them. The concept of morphogenetic trees is necessary to concentrate attention on the problem of how organismic heterogeneity proliferates during development rather than on the question of how it originates (which it doesn't, since some heterogeneity is carried over from the previous generation) or how it is transmitted (to which problem the French flag analogy is addressed). The concept of D-genes is important because geneticists interested in morphological evolution have all too readily jumped into discussions of unspecified 'regulatory' genes. The use of the terms D-gene and R-gene helps to emphasize that genes which regulate the between-cell component of development, i.e. morphogenesis, need not be regulatory within the cells in which they are active. Conversely, genes that regulate the production of a protein produced by a neighbouring gene within a particular cell may have no effect whatever on morphogenesis. The confusion of genic regulation and developmental regulation has, in my opinion, severely inhibited sensible comments on morphological evolution. Finally, the process I have called n-selection needed a name to emphasize that there are other possible forms of selection than w-selection, and indeed to make it clear that what we normally call 'selection' *is* w-selection.

Some brief comments are in order on the testability of evolutionary theories in general, and specifically on possible lines of research that might help to test the theory advanced herein. The main problem is that the proposed saltational changes are rare events manifested only in very long periods of evolutionary time, and the postulated evolutionary patterns of which they are a part are consequently very long-term patterns. There is no direct way of testing the underlying mechanism of such a pattern of events extending back into distant history. It is worth noting that this applies equally to any other theory of long-term evolution. The proposal that gradual quantitative phenotypic divergence occurs *within species* as a result of directional w-selection is testable in artificial selection experiments and, though with greater difficulty, in natural populations; but the proposal that cumulative repeats of this process are responsible for all, or at least nearly all, of long-term evolution is as untestable as the proposal that occasional saltational changes are involved.

In the absence of direct tests of the theory, then, we must resort to the usual evolutionists' practice of indirect testing, and four separate lines of investigation of this sort suggest themselves. First, palaeontologists need to identify, as precisely as possible, those gaps in the early fossil record which are most strongly suggestive of saltational phenotypic change. Now that the fossil record is beginning to extend back a little into the Pre-Cambrian (Olson, 1981) this may become increasingly possible. Second, population geneticists could examine the extent to which improvement of the degree of coadaptation of aberrant phenotypes, such as homoeotic mutants of *D. melanogaster*, is possible through selection on modifiers. This could be assessed by competing both a selection line and a control line at regular intervals with an inbred line of another species of *Drosophila*, and it would help to establish the speed at which improvement of a new form produced initially by n-selection might proceed. Clearly, the faster this process, the greater the chances of indefinite survival of a saltational variant, since any such form must ultimately experience competition and, when it does so, will become extinct if its degree of coadaptation has not been increased again by w-selection subsequent to the major mutation occurring. Thirdly, as developmental genetics progresses, the precise structure of the morphogenetic tree will hopefully become apparent and the hypothetical structure suggested in Figure 10.12 will therefore be tested. This in itself is of considerably evolutionary importance since the proposals made here are, to some extent, logical consequences of this structure. Also, in as much as the details of the tree turn out to be different for different higher taxa, this should lead to predictions of which groups should exhibit saltational change most readily; and these predictions can be compared with the results of the comparative palaeontological survey suggested above.

Fourthly, evolutionary ecologists could test the amount of separation of multidimensional patterns of resource utilization of wild-type and aberrant phenotypes in heterogeneous environments in order to establish which phenotypes are able to escape from competition, and so from w-selection. These various lines of indirect testing are capable of going a long way towards ascertaining the truth or otherwise of the proposal that morphological saltations are rare but important evolutionary events.

Finally, whether or not the specific hypotheses put forward herein gain general acceptance, I hope that I have made it clear that Darwinian and developmental approaches to morphological evolution are not alternatives to each other, as some authors seem to imply. Morphological evolution, as observed at the level of the adult phenotype, is a result of a definite pattern of change of the genes controlling morphogenesis in developing individuals in natural populations. Any evolutionary theory that omits reference to the genes themselves, the morphogenetic processes they control, or the populations in which mutant individuals must ultimately establish themselves is an incomplete theory; and all theories advanced so far, including Darwin's and the so-called modern synthesis of the 1940s, fall into this category. A truly synthetic theory of morphological evolution must await a complete exposition of the genetic basis of development. I have, in the absence of such an exposition, provided a working hypothesis relating to the involvement of genes in morphogenesis and have attempted, using this hypothesis, to show the directions in which we should proceed in order to arrive, eventually, at an expanded version of neo-Darwinism which will fully merit the title 'the modern synthesis'.

REFERENCES

Abbott, I. (1980). Theories dealing with the ecology of landbirds on islands. *Adv. Ecol. Res.*, **11**, 329–371.
Abbott, I., Abbott, L. K., and Grant, P. R. (1977). Comparative ecology of Galapagos ground finches (*Geospiza* Gould): Evaluation of the importance of floristic diversity and interspecific competition. *Ecol. Monogr.*, **47**, 151–184
Anderson, W. W., and King, C. E. (1970). Age-specific selection. *Proc. Natl. Acad. Sci. USA*, **66**, 780–786.
Armstrong, R. A., and McGehee, R. (1976). Coexistence of two competitors on one resource. *J. Theor. Biol.*, **56**, 499–502.
Arthur, W. (1978). Morph-frequency and coexistence in *Cepaea*. *Heredity*, **41**, 335–346.
Arthur, W. (1980a). Interspecific competition in *Drosophila*. I. Reversal of competitive superiority due to varying concentration of ethanol. *Biol. J. Linn. Soc.*, **13**, 109–118.
Arthur, W. (1980b). Interspecific competition in *Drosophila*. II. Competitive outcome in some 2-resource environments. *Biol. J. Linn. Soc.*, **13**, 119–128.
Arthur, W. (1980c). Further associations between morph-frequency and coexistence in *Cepaea*. *Heredity*, **44**, 417–421.
Arthur, W. (1982a). The evolutionary consequences of interspecific competition. *Adv. Ecol. Res.*, **12**, 127–187.
Arthur, W. (1982b). A critical evaluation of the case for competitive selection in *Cepaea*. *Heredity*, **48**, 407–419.
Arthur, W. (1982c). Control of shell shape in *Lymnaea stagnalis*. *Heredity*, **49**, 153–161.
Arthur, W. (1982d). A developmental approach to the problem of variation in evolutionary rates. *Biol. J. Linn. Soc.*, **18**, 243–261.
Arthur, W., and Middlecote, J. (1984). Evolution of pupation site and interspecific competitive ability in *Drosophila hydei*. *Biol. J. Linn. Soc.*, in press.
Arthurs, A. M. (1965). *Probability Theory*. Routledge and Kegan Paul, London.
Ashburner, M. (1969). On the problem of genetic similarity between sibling species — puffing patterns in *Drosophila melanogaster* and *Drosophila simulans*. *Am. Nat.*, **103**, 189–191.
Atkinson, W. D. (1979). A field investigation of larval competition in domestic *Drosophila*. *J. Anim. Ecol.*, **48**, 91–102.
Atkinson, W. D., and Shorrocks, B. (1981). Competition on a divided and ephemeral resource: a simulation model. *J. Anim. Ecol.*, **50**, 461–471.
Ayala, F. J. (1966). Reversal of dominance in competing species of *Drosophila*. *Am. Nat.*, **100**, 81–83.
Ayala, F. J. (1969). Evolution of fitness. IV. Genetic evolution of interspecific competitive ability in *Drosophila*. *Genetics*, **61**; 737–747.
Ayala, F. J. (ed.) (1976). *Molecular Evolution*. Sinauer, Sunderland, Massachusetts.
Ayala, F. J. and Campbell, C. A. (1974). Frequency-dependent selection. *Ann. Rev. Ecol. Syst.*, **5**, 115–138.
Ayala, F. J. and Valentine, J. W. (1979). *Evolving. The Theory and Processes of Organic Evolution*. Benjamin/Cummings, Menlo Park, California.
Bakker, K. (1961). An analysis of factors which determine success in competition for food among larvae of *Drosophila melanogaster*. *Arch. Neerl. Zool.*, **14**, 200–281.
Bantock, C. R., and Bayley, J. A. (1973). Visual selection for shell size in *Cepaea* (Held). *J. Anim. Ecol.*, **42**, 247–261.

Bantock, C. R., Bayley, J. A., and Harvey, P. H. (1975). Simultaneous selective predation on two features of a mixed sibling species population. *Evolution,* **29**, 636–649.

Barker, J. S. F. and Cummins, L. J. (1969). Disruptive selection for sternopleural bristle numbers in *Drosophila melanogaster. Genetics,* **61**, 697–712.

Barker, J. S. F., and Robertson, A. (1966). Genetic and phenotypic parameters for the first three lactations in Friesian cows. *Anim. Prod.,* **8**, 221–240.

Barnes, R. D. (1980). *Invertebrate Zoology,* fourth edition. Saunders College, Philadelphia.

Bell, M. A. (1981). Lateral plate polymorphism and ontogeny of the complete plate morph of threespine sticklebacks (*Gasterosteus aculeatus*). *Evolution,* **35**, 67–74.

Bender, W., Akam, M., Karch, F., Beachy, P. A., Peifer, M., Spierer, P., Lewis, E. B., and Hogness, D. S. (1983). Molecular genetics of the bithorax complex in *Drosophila melanogaster. Science,* **221**, 23–29.

Berry, R. J. (1982). *Neo-Darwinism.* Edward Arnold, London.

Bishop, J. A., and Cook, L. M. (1980). Industrial melanism and the urban environment. *Adv. Ecol. Res.,* **11**, 373–404.

Blackith, R. E., and Reyment, R. A. (1971). *Multivariate Morphometrics.* Academic Press, London.

Blair, W. F. (1974). Character displacement in frogs. *Am. Zool.,* **14**, 1119–1125.

Boag, P. T. and Grant, P. R. (1978). Heritability of external morphology in Darwin's finches. *Nature,* **274**, 793–794.

Bohren, B. B., Hill, W. G. and Robertson, A. (1966). Some observations on asymmetrical correlated responses to selection. *Genet. Res.,* **7**, 44–57.

Bonner, J. T. (1974). *On Development.* Harvard University Press, Cambridge, Massachusetts.

Boucot, A. J. (1982). Ecophenotypic or genotypic? *Nature,* **296**, 609–610.

Boycott, A. E. (1938). Experiments on the artificial breeding of *Limnaea involuta, Limnaea burnetti* and other forms of *Limnaea peregra. Proc. Malacol. Soc. London,* **23**, 101–108.

Boycott, A. E., and Diver, C. (1923). On the inheritance of sinistrality in *Limnaea peregra. Proc. R. Soc. London Ser. B,* **95**, 207–213.

Boycott, A. E., Diver, C., Garstang, S. L., and Turner, F. M. (1930). The inheritance of sinistrality in *Limnaea peregra* (Mollusca, Pulmonata). *Phil. Trans. R. Soc. London Ser. B,* **219**, 51–131.

Brent, L., Chandler, P., Fierz, W., Medawar, P. B., Rayfield, L. S., and Simpson, E. (1982). Further studies on supposed lamarckian inheritance of immunological tolerance. *Nature,* **295**, 242–244.

Britten, R. J., and Davidson, E. H. (1969). Gene regulation for higher cells: a theory. *Science,* **165**, 349–357.

Britten, R. J., and Davidson, E. H. (1971). Repetitive and non-repetitive DNA sequences and a speculation on the origins of evolutionary novelty. *Q. Rev. Biol.,* **46**, 111–138.

Brown, J. H., and Davidson, D. W. (1977). Competition between seed-eating rodents and ants in desert ecosystems. *Science,* **196**, 880–882.

Brown, W. L., and Wilson, E. O. (1956). Character displacement. *Syst. Zool.,* **5**, 49–64.

Bulmer, M. G. (1974). Density-dependent selection and character displacement. *Am. Nat.,* **108**, 45–58.

Bulmer, M. G. (1980). *The Mathematical Theory of Quantitative Genetics.* Clarendon Press, Oxford.

Butlin, R. K., Read, I. L., and Day, T. H. (1982). The effects of a chromosomal inversion on adult size and male mating success in the seaweed fly, *Coelopa frigida. Heredity,* **49**, 51–62.

Cain, A. J. (1971). *Animal Species and their Evolution*, third edition, Hutchinson, London.
Cain, A. J., and Currey, J. D. (1963a). Area effects in *Cepaea*. *Phil. Trans. R. Soc. London Ser. B*, **246**, 1–81.
Cain, A. J., and Currey, J. D. (1963b). Area effects in *Cepaea* on the Larkhill Artillery Ranges, Salisbury Plain. *J. Linn. Soc. London (Zool.)*, **45**, 1–15.
Carson, H. L., Hardy, D. E., Spieth, H. T., and Stone, W. S. (1970). The evolutionary biology of the Hawaiian Drosophilidae. In: *Essays in Evolution and Genetics in Honor of Theodosius Dobzhansky*, eds. M. K. Hecht and W. C. Steere. Appleton Century Crofts, New York.
Cavalli-Sforza, L. L., and Bodmer, W. F. (1971). *The Genetics of Human Populations*. Freeman, San Francisco.
Chabora, A. J. (1968). Disruptive selection for sternopleural chaeta number in various strains of *Drosophila melanogaster*. *Am. Nat.*, **102**, 525–532.
Charlesworth, B. (1981). Review of *Macroevolution, Pattern and Process*, by S. M. Stanley. *Biol. J. Linn. Soc.*, **16**, 169–172.
Charlesworth, B., and Lande, R. (1982). Morphological stasis and developmental constraint: no problem for Neo-Darwinism. *Nature*, **296**, 610.
Child, C. M. (1941). *Patterns and Problems of Development*. Univerity of Chicago Press, Chicago.
Clarke, B. (1975). The contribution of ecological genetics to evolutionary theory: detecting the direct effects of natural selection on particular polymorphic loci. *Genetics*, **79**, (Suppl.), 101–113.
Clarke, B., Arthur, W., Horsley, D. T., and Parkin, D. T. (1978). Genetic variation and natural selection in pulmonate molluscs. In: *Pulmonates*, Vol. 2A, *Systematics, Evolution and Ecology*, eds. V. Fretter and J. Peake. Academic Press, London.
Clarke, B., Camfield, R. G., Galvin, A. M., and Pitts, C. R. (1979). Environmental factors affecting the quantity of alcohol dehydrogenase in *Drosophila melanogaster*. *Nature*, **280**, 517–518.
Clarke, B., and Murray, J. (1969). Ecological genetics and speciation in land snails of the genus *Partula*. *Biol. J. Linn. Soc.*, **1**, 31–42.
Clarke, C. A., and Sheppard, P. M. (1960). Super-genes and mimicry. *Heredity*, **14**, 175–185.
Clausen, J., and Hiesey, W. M. (1958). Experimental studies on the nature of species. IV. Genetic structure of ecological races. *Carnegie Inst. Washington Publ.*, No. 615.
Clausen, J., and Hiesey, W. M. (1960). The balance betwen coherence and variation in evolution. *Proc. Natl. Acad. Sci. USA*, **46**, 494–506.
Clausen, J., Keck, D. D., and Hiesey, W. M. (1940). Experimental studies on the nature of species. I. Effect of varied environments on western North American plants. *Carnegie Inst. Washington Publ.*, No. 520.
Clayton, G. A., Morris, J. A., and Robertson, A. (1957a). An experimental check on quantitative genetical theory. I. Short-term response to selection. *J. Genet.*, **55**, 131–151.
Clayton, G. A., Knight, G. R., Morris, J. A., and Robertson, A. (1957b). An experimental check on quantitative genetical theory. III. Correlated responses. *J. Genet.*, **55**, 171–180.
Comfort, A. (1944). *Cepaea nemoralis* in the Dartry Mountains, Co. Sligo. *J. Conchol.*, **22**, 80.
Cook, L. M. (1965). Inheritance of shell size in the snail *Arianta arbustorum*. *Evolution*, **19**, 86–94.
Cook, L. M. (1967). The genetics of *Cepaea nemoralis*. *Heredity*, **22**, 397–410.
Cook, L. M. (1971). *Coefficients of Natural Selection*. Hutchinson, London.

Cook, L. M., and Cain, A. J. (1980). Population dynamics, shell size and morph frequency in experimental populations of the snail *Cepaea nemoralis* (L.). *Biol. J. Linn. Soc.,* **14**, 259–292.

Cook, L. M., and O'Donald, P. (1971). Shell size and natural selection in *Cepaea nemoralis.* In: *Ecological Genetics and Evolution,* ed. E. R. Creed. Blackwell, Oxford.

Cook, L. M., and Peake, J. F. (1960). A study of some populations of *Cepaea nemoralis* L. from the Dartry Mountains, Co. Sligo, Ireland. *Proc. Malacol. Soc. London,* **34**, 1–11.

Cowling, D. E., and Burnet, B. (1981). Courtship songs and genetic control of their acoustic characteristics in sibling species of the *Drosophila melanogaster* subgroup. *Anim. Behav.,* **29**, 924–935.

Crampton, H. E. (1932). Studies on the variation, distribution, and evolution of the genus *Partula.* The species inhabiting Moorea. *Carnegie Inst. Washington Publ.* No. 410.

Crick, F. H. C. (1970). Diffusion in embryogenesis. *Nature,* **225**, 420–422.

Cronin, J. E., Boaz, N. T., Stringer, C. B., and Rak, Y. (1981). Tempo and mode in hominid evolution. *Nature,* **292**, 113–122.

Crothers, J. H. (1977). Some observations on the growth of the common dog-whelk, *Nucella lapillus* (Prosobranchia: Muricacea) in the laboratory. *J. Conchol.,* **29**, 157–162.

Crothers, J. H. (1980). Further observations on the growth of the dog-whelk *Nucella lapillus* (L.) in the laboratory. *J. Moll. Stud.,* **46**, 181–185.

Crow, J. F., and Kimura, M. (1970). *An Introduction to Population Genetics Theory.* Harper and Row, New York.

Crozier, R. H. (1974). Niche shape and genetic aspects of character displacement. *Am. Zool.,* **14**, 1151–1157.

Darwin, C. (1859). *On the Origin of Species by Means of Natural Selection, or the Preservation of Favoured Races in the Struggle for Life,* first edition. John Murray, London.

Darwin, F. (1887). *The Life and Letters of Charles Darwin.* John Murray, London.

Davidson, E. H., and Britten, R. J. (1979). Regulation of gene expression: possible role of repetitive sequences. *Science,* **204**, 1052–1059.

Day, T. H., Dobson, T., Hillier, P. C., Parkin, D. T., and Clarke, B. (1982). Associations of enzymic and chromosomal polymorphisms in the seaweed fly, *Coelopa frigida. Heredity,* **48**, 35–44.

Degner, E. (1952). Der Erbang der Inversion bei *Laciniaria biplicata* MTG (Gastr. Pulm.). *Mitt. Hamburg Zool. Mus. Inst.,* **51**, 3–61.

Delson, E., Eldredge, N., and Tattersall, I. (1977). Reconstruction of hominid phylogeny: A testable framework based on cladistic analysis. *J. Hum. Evol.,* **6**, 263–278.

Dempster, J. P. (1982). The ecology of the Cinnabar Moth, *Tyria jacobaeae* L. (Lepidoptera: Arctiidae). *Adv. Ecol. Res.,* **12**, 1–36.

De Souza, H. M. L., da Cunha, A. B., and dos Santos, E. P. (1970). Adaptive polymorphism of behavior evolved in laboratory populations of *Drosophila willistoni. Am. Nat.,* **104**, 175–189.

De Vries, H. (1905). *Species and Varieties, Their Origin by Mutation.* Open Court, Chicago.

Dickinson, H., and Antonovics, J. (1973). Theoretical considerations of sympatric divergency. *Am. Nat.,* **107**, 256–274.

Diver, C. (1940). The problem of closely related species living in the same area. In: *The New Systematics,* ed. J. Huxley, Clarendon Press, Oxford.

Diver, C., and Anderson-Kottö, I. (1938). Sinistrality in *Limnaea peregra* (Mollusca, Pulmonata): The problem of mixed broods. *J. Genet.,* **35**, 447–525.

Diver, C., Boycott, A. E., and Garstang, S. (1925). The inheritance of inverse symmetry in *Limnaea peregra*. *J. Genet.*, **15**, 113–200.
Dobzhansky, T. (1951). *Genetics and the Origin of Species*, third edition. Columbia University Press, New York.
Dobzhansky, T. (1970). *Genetics of the Evolutionary Process*. Columbia University Press, New York.
Dobzhansky, T., Ayala, F. J., Stebbins, G. L., and Valentine, J. W. (1977). *Evolution*. Freeman, San Francisco.
Dobzhansky, T., and Spassky, B. (1969). Artificial and natural selection for two behavioural traits in *Drosophila pseudoobscura*. *Proc. Natl. Acad. Sci. USA*, **62**, 75–80.
Dunn, L. C., and Glueckson-Waelsch, S. (1953). Genetic analysis of seven newly discovered mutant alleles at locus T in the house mouse. *Genetics*, **38**, 261–271.
Dunn, L. C., and Suckling, L. (1956). Studies of the genetic variability in wild populations of house mice. I. Analysis of seven alleles at locus T. *Genetics*, **41**, 344–345.
Ede, D. A. (1978). *An Introduction to Developmental Biology*. Blackie, Glasgow.
Eldredge, N. (1971). The allopatric model and phylogeny in Paleozoic invertebrates. *Evolution*, **25**, 156–167.
Eldredge, N., and Gould, S. J. (1972). Punctuated equilibria: an alternative to phyletic gradualism. In: *Models in Palaeobiology*, ed. T. J. M. Schopf. Freeman, San Fransisco.
Ellis, A. E. (1969). *British Snails. A Guide to the Non-marine Gastropoda of Great Britain and Ireland, Pleistocene to Recent*. Clarendon Press, Oxford.
Elton, C. S. (1927). *Animal Ecology*. Macmillan, London.
Elton, C. S. (1958). *The Ecology of Invasions by Animals and Plants*. Methuen, London.
Eoff, M. (1975). Artificial selection in *Drosophila melanogaster* for increased and decreased sexual isolation from *D. simulans*. *Am. Nat.*, **109**, 225–229.
Erk, F. C., and Sang, J. H. (1966). The comparative nutritional requirements of two sibling species *Drosophila simulans* and *D. melanogaster*. *J. Insect Physiol.*, **12**, 43–51.
Falconer, D. S. (1953). Selection for large and small size in mice. *J. Genet.*, **51**, 470–498.
Falconer, D. S. (1960). *Introduction to Quantitative Genetics*, first edition. Longman, London.
Falconer, D. S. (1973). Replicated selection for body weight in mice. *Genet. Res.*, **22**, 291–321.
Falconer, D. S. (1981). *Introduction to Quantitative Genetics*, second edition. Longman, London.
Fenchel, T., (1975a). Factors determining the distribution patterns of mud snails (Hydrobiidae). *Oecologia*, **20**, 1–17.
Fenchel, T. (1975b). Character displacement and coexistence in mud snails (Hydrobiidae). *Oecologia*, **20**, 19–32.
Fenchel, T. M., and Christiansen, F. B. (1977). Selection and interspecific competition. In: *Measuring Selection in Natural Populations*, ed. F. B. Christiansen and T. M. Fenchel. Springer-Verlag, Berlin.
Fisher, R. A. (1918). The correlations between relatives on the supposition of Mendelian inheritance. *Trans. R. Soc. Edinburgh*, **52**, 399–433.
Fisher, R. A. (1930). *The Genetical Theory of Natural Selection*. Clarendon Press, Oxford.
Ford, E. B. (1971). *Ecological Genetics*, third edition. Chapman and Hall, London.
Frankham, R., Briscoe, D. A., and Nurthen, R. K. (1978). Unequal crossing over at the rRNA locus as a source of quantitative genetic variation. *Nature*, **272**, 80–81.
Frazzetta, T. H. (1975). *Complex Adaptations in Evolving Populations*. Sinauer, Sunderland, Massachusetts.
Gale, J. S. (1980). *Population Genetics*. Blackie, Glasgow.

Garcia-Bellido, A. (1975). Genetic control of wing disc development in *Drosophila*. In: *Cell Patterning, CIBA Found. Symp.*, No. 29 (new series). Associated Scientific Publishers, Amsterdam.

Garcia-Bellido, A., Lawrence, P. A., and Morata, G. (1979). Compartments in animal development. *Sci. Am.*, **241** (1), 90–98.

Garcia-Bellido, A., Ripoll, P., and Morata, G. (1973). Developmental compartmentalization of the wing disk of *Drosophila*. *Nature New Biol.* **245**, 251–253.

Garcia-Bellido, A. & Santamaria, P. (1972). Developmental analysis of the wing disc in the mutant *engrailed* of *Drosophila melanogaster*. *Genetics*, **72**, 87–104.

Garnett, I., and Falconer, D. S. (1975). Protein variation in strains of mice differing in body size. *Genet. Res.*, **25**, 45–57.

Garrod, D. R. (1973). *Cellular Development*. Chapman and Hall, London.

Gass, I. G., Smith, P. J., and Wilson, R. C. L. (eds.) (1971). *Understanding the Earth. A Reader in the Earth Sciences*. Artemis Press, Sussex.

Gause, G. F. (1934). *The Struggle for Existence*. Williams and Wilkins, Baltimore.

Gause, G. F. (1935). Vérifications expérimentales de la théorie mathématique de la lutte pour la vie. *Actual. Sci. Ind.*, No. 277.

Gilpin, M. E. (1975). Limit cycles in competition communities. *Am. Nat.*, **109**, 51–60.

Goldschmidt, R. (1940). *The Material Basis of Evolution*. Yale University Press, New Haven, Connecticut.

Goldschmidt, R. B. (1952). Homoeotic mutants and evolution. *Acta Biotheor.*, **10**, 87–104.

Goldschmidt, R. B. (1953). Experiments with a homoeotic mutant, bearing on evolution. *J. Exp. Zool.*, **123**, 79–114.

Goldschmidt, R. B. (1955). *Theoretical Genetics*. University of California Press, Berkeley.

Goodwin, B. C., and Cohen, M. H. (1969). A phase-shift model for the spatial and temporal organization of developing systems. *J. Theor. Biol.*, **25**, 49–107.

Gorczynski, R. M., and Steele, E. J. (1980). Inheritance of acquired immunological tolerance to foreign histocompatibility antigens in mice. *Proc. Natl. Acad. Sci. USA*, **77**, 2871–2875.

Gould, S. J. (1977). *Ontongeny and Phylogeny*. Harvard University Press, Cambridge, Massachusetts.

Gould, S. J. (1980). *Ever Since Darwin. Reflections in Natural History*. Penguin Books, Harmondsworth, Middlesex.

Gould, S. J. (1982). Darwinism and the expansion of evolutionary theory. *Science*, **216**, 380–387.

Gould, S. J., and Eldredge, N. (1977). Punctuated equilibria: the tempo and mode of evolution reconsidered. *Paleobiology*, **3**, 115–151.

Grant, P. R. (1972). Convergent and divergent character displacement. *Biol. J. Linn. Soc.*, **4**, 39–68.

Grant, P. R. (1975). The classical case of character displacement. In: *Evolutionary Biology*, Vol. 8, eds. T. Dobzhansky, M. K. Hecht and W. C. Steere. Plenum Press, New York.

Grant, P. R., Grant, B. R., Smith, J. N. M., Abbott, I. J., and Abbott, L. K. (1976). Darwin's finches: Population variation and natural selection. *Proc. Natl. Acad. Sci. USA*, **73**, 257–261.

Grant, V. (1966). The selective origin of incompatibility barriers in the plant genus *Gilia*. *Am. Nat.*, **100**, 99–118.

Grant, V. (1981). *Plant Speciation*, second edition. Columbia University Press, New York.

Green, E. L. (1962). Quantitative genetics of skeletal variations in the mouse. II. Crosses between four inbred strains. *Genetics*, **47**, 1085–1096.

Gregor, J. W. (1939). Experimental taxonomy. III. Population differentiation in North

American and European sea plantains allied to *Plantago maritima* L. *New Phytol.*, **39**, 293–322.
Haldane, J. B. S. (1949). Suggestions as to quantitative measurements of rates of evolution. *Evolution*, **3**, 51–56.
Harris, H. (1966). Enzyme polymorphisms in man. *Proc. R. Soc. London Ser. B*, **164**, 298–310.
Harris, H. (1975). *The Principles of Human Biochemical Genetics*, second edition. North-Holland, Amsterdam.
Hassell, M. P. (1976). *The Dynamics of Competition and Predation*. Edward Arnold, London.
Hedrick, P. W., and McDonald, J. F. (1980). Regulatory gene adaptation: an evolutionary model. *Heredity*, **45**, 85–99.
Heslop-Harrison, J. (1964). Forty years of genecology. *Adv. Ecol. Res.*, **2**, 159–247.
Hewitt, G. M. (1973). The integration of supernumerary chromosomes into the orthopteran genome *Cold Spring Harbor Symp. Quant. Biol.*, **38**, 183–194.
Ho, M. W., and Saunders, P. T. (1979). Beyond neo-Darwinism — An epigenetic approach to evolution. *J. Theor. Biol.*, **78**, 573–591.
Horton, I. H. (1939). A comparison of the salivary gland chromosomes of *Drosophila melanogaster* and *D. simulans*. *Genetics*, **24**, 234–243.
Hubby, J. L., and Lewontin, R. C. (1966). A molecular approach to the study of genic heterozygosity in natural populations. I. The number of alleles at different loci in *Drosophila pseudoobscura*. *Genetics*, **54**, 577–594.
Hubby, J. L., and Throckmorton, L. H. (1968). Protein differences in *Drosophila*. IV. A study of sibling species. *Am. Nat.*, **102**, 193–205.
Hubendick, B. (1951). Recent Lymnaeidae. Their variation, morphology, taxonomy, nomenclature and distribution. *K. Sven. Vetenskapsakad. Hand.*, **4**(3), 1–223.
Hutchinson, G. E. (1965). *The Ecological Theater and the Evolutionary Play*. Yale University Press, New Haven, Connecticut.
Huxley, A. (1982). Address of the President at the Anniversary Meeting, 30 November 1981. *Proc. R. Soc. London Ser. B*, **214**, 137–152.
Huxley, J. S. (1942). *Evolution, The Modern Synthesis*. Allen and Unwin, London.
Jackson, J. B. C., and Buss, L. W. (1975). Allelopathy and spatial competition among coral reef invertebrates. *Proc. Natl. Acad. Sci. USA*, **72**, 5160–5163.
Jaksic, F. M. (1981). Abuse and misuse of the term "guild" in ecological studies. *Oikos*, **37**, 397–400.
Johnson, M. S. (1976). Allozymes and area effects in *Cepaea nemoralis* on the Western Berkshire Downs. *Heredity*, **36**, 105–121.
Johnson, M. S. (1982). Polymorphism for direction of coil in *Partula suturalis*: behavioural isolation and positive frequency dependent selection. *Heredity*, **49**, 145–151.
Johnson, M. S., Clarke, B., and Murray, J. (1977). Genetic variation and reproductive isolation in *Partula*. *Evolution*, **31**, 116–126.
Jones, J. S. (1981). An uncensored page of fossil history. *Nature*, **293**, 427–428.
Jones, J. S., Leith, B. H., and Rawlings, P. (1977). Polymorphism in *Cepaea*: A problem with too many solutions? *Annu. Rev. Ecol. Syst.*, **8**, 109–143.
Jones, J. S., and MacDonald, A. (1976). Race, intelligence and Siamese cats. *New Sci.*, **70**, 80–82.
Jones, J. S., and Probert, R. F. (1980). Habitat selection maintains a deleterious allele in a heterogeneous environment. *Nature*, **287**, 632–633.
Jones, J. S., Selander, R. K., and Schnell, G. D. (1980). Patterns of morphological and molecular polymorphism in the land snail *Cepaea nemoralis*. *Biol. J. Linn. Soc.*, **14**, 359–387.
Kauffman, R. (1933). Variationsstatische Untersuchungen über die "Artabwandlung" und "Artumbildung" an der oberkambrischen Trilobitengattung *Olenus* Dalm. *Abh. Geol.-Pal. Inst. Univ. Greifswald*, **10**, 1–54.

Kerney, M. P., and Cameron, R. A. D. (1979). *A Field Guide to the Land Snails of Britain and North-West Europe*. Collins, London.
Khush, G. S. (1973). *Cytogenetics of Aneuploids*. Academic Press, New York.
Kimura, M. (1968). Evolutionary rate at the molecular level. *Nature*, **217**, 624–626.
Kimura, M. (1979). The neutral theory of molecular evolution. *Sci. Am.*, **241** (5), 94–104.
King, C. E., and Anderson, W. W. (1971). Age-specific selection. II. The interaction between r and K during population growth. *Am. Nat.*, **105**, 137–156.
King, M.-C., and Wilson, A. C. (1975). Evolution at two levels: Molecular similarities and biological differences between humans and chimpanzees. *Science*, **188**, 107–116.
Krebs, C. J. (1972). *Ecology. The Experimental Analysis of Distribution and Abundance*. Harper and Row, New York.
Lack, D. (1968). *Darwin's Finches. An Essay on the General Biological Theory of Evolution*. Peter Smith, Gloucester, Massachusetts.
Lande, R. (1981). The minimum number of genes contributing to quantitative variation between and within populations. *Genetics*, **99**, 541–553.
Latter, B. D. H., and Robertson, A. (1962). The effects of inbreeding and artificial selection on reproductive fitness. *Genet. Res.*, **3**, 110–138.
Lawlor, R., and Maynard Smith, J. (1976). The coevolution and stability of competing species. *Am. Nat.*, **110**, 79–99.
Lawrence, P. A., and Brower, D. L. (1982). Myoblasts from *Drosophila* wing disks can contribute to developing muscles throughout the fly. *Nature*, **295**, 55–57.
Lawrence, P. A., and Morata, G. (1976). The compartment hypothesis. In: *Insect Development*, ed. P. A. Lawrence. Blackwell, Oxford.
Lawton, J. H., and Hassell, M. P. (1981). Asymmetrical competition in insects. *Nature*, **289**, 793–795.
Lejeune, J., Gautier, M., and Turpin, R. (1959). Les chromosomes humaines en culture de tissus. *C. R. Acad. Sci.*, **248**, 602–603.
Léon, J. A. (1974). Selection in contexts of interspecific competition. *Am. Nat.*, **108**, 739–757.
Lerner, I. M. (1950). *Population Genetics and Animal Improvement*. Cambridge University Press, Cambridge.
Levene, H. (1953). Genetic equilibrium when more than one ecological niche is available. *Am. Nat.*, **87**, 331–333.
Levin, B. R. (1969). A model for selection in systems of species competition. In: *Concepts and Models in Biomathematics*, ed. F. Heinmets. Dekker, New York.
Levin, B. R. (1971). The operation of selection in situations of interspecific competition. *Evolution*, **25**, 249–264.
Levin, B. R. (1972). Coexistence of two asexual strains on a single resource. *Science*, **175**, 1272–1274.
Levin, D. A., and Kerster, H. W. (1967). Natural selection for reproductive isolation in *Phlox*. *Evolution*, **21**, 679–687.
Lewin, B. (1980). *Gene Expression*, Vol. 2, *Eucaryotic Chromosomes*, second edition. John Wiley, New York.
Lewin, B. (1983). *Genes*. John Wiley, New York.
Lewis, E. B. (1951). Pseudoallelism and gene evolution. *Cold Spring Harbor. Symp. Quant. Biol.*, **16**, 159–174.
Lewis, E. B. (1963). Genes and developmental pathways. *Am. Zool.*, **3**, 33–56.
Lewis, E. B. (1964). Genetic control and regulation of developmental pathways. In: *Role of Chromosomes in Development*, ed. M. Locke. Academic Press, New York.
Lewis, E. B. (1978). A gene complex controlling segmentation in *Drosophila*. *Nature*, **276**, 565–570.
Lewontin, R. C. (1968). Evolution of complex genetic systems. In: *Some Mathematical Questions in Biology*, ed. M. Gerstenhaber. American Mathematical Society, Providence, Rhode Island.

Lewontin, R. C. (1974). *The Genetic Basis of Evolutionary Change*. Columbia University Press, New York.
Lewontin, R. C., and Hubby, J. L. (1966). A molecular approach to the study of genic heterozygosity in natural populations. II. Amount of variation and degree of heterozygosity in natural populations of *Drosophila pseudoobscura*. *Genetics*, **54**, 595–609.
Lindsley, D. L. and Grell, E. H. (1968). Genetic Variations of *Drosophila melanogster*. *Carnegie Inst. Washington Publ.*, No. 627.
MacArthur, J. W. (1944). Genetics of body size and related characters. I. Selecting small and large races of the laboratory mouse. *Am. Nat.*, **78**, 142–157.
MacArthur, J. W. (1949). Selection for small and large body size in the house mouse. *Genetics*, **34**, 194–209.
MacArthur, R. H. (1972). *Geographical Ecology. Patterns in the Distribution of Species*. Harper and Row, New York.
MacArthur, R. H., and Levins, R. (1967). The limiting similarity, convergence and divergence of coexisting species. *Am. Nat.*, **101**, 377–385.
MacArthur, R. H., and Wilson, E. O. (1967). *The Theory of Island Biogeography*. Princeton University Press, Princeton.
McClintock, B. (1951). Chromosome organization and gene expression. *Cold Spring Harbor Symp. Quant. Biol.*, **16**, 13–47.
McMurtie, R. (1976). On the limit to niche overlap for nonuniform niches. *Theor. Popul. Biol.*, **10**, 96–107.
McNeilly, T., and Antonovics, J. (1968). Evolution in closely adjacent plant populations. IV. Barriers to gene flow. *Heredity*, **23**, 205–218.
Machin, J. (1967). Structural adaptation for reducing water loss in three species of terrestrial snail. *J. Zool.*, **152**, 55–65.
Manning, A. (1959a). The sexual isolation between *Drosophila melanogaster* and *D. simulans*. *Anim. Behav.*, **7**, 60–65.
Manning, A. (1959b). The sexual behaviour of two sibling *Drosophila* species. *Behaviour*, **15**, 123–145.
Marchant, C. J. (1967). Evolution in *Spartina* (Gramineae). I. The history and morphology of the genus in Britain. *J. Linn. Soc. London. (Bot.)*, **60**, 1–24.
Marchant, C. J. (1968). Evolution in *Spartina* (Gramineae). 2. Chromosomes, basic relationships and the problem of *S.* × *townsendii* agg. *J. Linn. Soc. London (Bot.)*, **60**, 381–409.
Mather, K., and Jinks, J. L. (1971). *Biometrical Genetics*, second edition. Chapman and Hall, London.
Mather, K., and Jinks, J. L. (1977). *Introduction to Biometrical Genetics*. Chapman and Hall, London.
May, R. M. (1974). On the theory of niche overlap. *Theor. Popul. Biol.*, **5**, 297–332.
May, R. M., and MacArthur, R. H. (1972). Niche overlap as a function of environmental variability. *Proc. Natl. Acad. Sci. USA*, **69**, 1109–1113.
Maynard-Smith, J. (1962). Disruptive selection, polymorphism and sympatric speciation. *Nature*, **195**, 60–62.
Maynard Smith, J. (1964). Group selection and kin selection: a rejoinder. *Nature*, **201**, 1145–1147.
Maynard Smith, J. (1966). Sympatric speciation. *Am. Nat.*, **100**, 637–650.
Maynard Smith, J. (1972). *On Evolution*. Edinburgh University Press, Edinburgh.
Maynard Smith, J. (1978). *The Evolution of Sex*. Cambridge University Press, Cambridge.
Mayr, E. (1963). *Animal Species and Evolution*. Harvard University Press, Cambridge, Massachusetts.
Mayr, E. (1982). Questions concerning speciation. *Nature*, **296**, 609.
Medawar, P. B. (1967). *The Art of the Soluble*. Methuen, London.

Moore, J. A. (1952a). Competition between *Drosophila melanogaster* and *Drosophila simulans*. I. Population cage experiments. *Evolution*, **6**, 407–420.
Moore, J. A. (1952b). Competition between *Drosophila melanogaster* and *Drosophila simulans*. II. The improvement of competitive ability through selection. *Proc. Natl. Acad. Sci. USA*, **38**, 813–817.
Morata, G., and Lawrence, P. A. (1977). Homoeotic genes, compartments and cell determination in *Drosophila. Nature*, **265**, 211–216.
Morata, G., and Ripoll, P. (1975). Minutes: mutants of *Drosophila* autonomously affecting cell division rate. *Dev. Biol.*, **42**, 211–221.
Morgan, T. H., and Tyler, A. (1938). The relation between the entrance point of the spermatozoon and bilaterality of the egg in *Chaetopterus. Biol. Bull.*, **74**, 401–402.
Mozley, A. (1935). The variation of two species of *Lymnaea. Genetics*, **20**, 452–465.
Mozley, A. (1939). The variation of *Lymnaea stagnalis* (Linné). *Proc. Malacol. Soc. London*, **23**, 267–269.
Murphy, P. G. (1976). Electrophoretic evidence that selection reduces ecological overlap in marine limpets. *Nature*, **261**, 228–230.
Murray, J. (1972). *Genetic Diversity and Natural Selection*. Oliver and Boyd, Edinburgh.
Murray, J. (1975). The genetics of the Mollusca. In: *Handbook of Genetics*, Vol. 3, ed. R. C. King. Plenum Press, New York.
Murray, J., and Clarke, B. (1966). The inheritance of polymorphic shell characters in *Partula* (Gastropoda). *Genetics*, **54**, 1261–1277.
Murray, J., and Clarke, B. (1968a). Inheritance of shell size in *Partula. Heredity*, **23**, 189–198.
Murray, J., and Clarke, B. (1968b). Partial reproductive isolation in the genus *Partula* (Gastropoda) on Moorea. *Evolution*, **22**, 684–698.
Murray, J., and Clarke, B. (1976). Supergenes in polymorphic land snails. II. *Partula suturalis. Heredity*, **37**, 271–282.
Niemann-Sørensen, A., and Robertson, A. (1961). The association between blood-groups and several production charcteristics in three Danish cattle breeds. *Acta Agric. Scand.*, **11**, 163–196.
Nüsslein-Volhard, C., and Wieschaus, E. (1980). Mutations affecting segment number and polarity in *Drosophila. Nature*, **287**, 795–801.
Ochman, H., Jones, J. S., and Selander, R. K. (1983). Molecular area effects in *Cepaea. Proc. Natl. Acad. Sci. USA*, **80**, 4189–4193.
Odum, E. P. (1971). *Fundamentals of Ecology*. Saunders, Philadelphia.
Ohno, S. (1970). *Evolution by Gene Duplication*. Springer-Verlag, New York.
Oldham, C. (1931). Some scalariform examples of *Arianta arbustorum* infested by parasitic mites. *Proc. Malacol. Soc. London*, **19**, 240–242.
Olson, E. C. (1981). The problem of missing links: today and yesterday. *Q. Rev. Biol.*, **56**, 405–442.
Ouweneel, W. J. (1976). Developmental genetics of homoeosis. *Adv. Genet.*, **18**, 179–248.
Park, T. (1948). Experimental studies of interspecies competition. I. Competition between populations of the flour beetles *Tribolium confusum* Duval and *Tribolium castaneum* Herbst. *Ecol. Monogr.*, **18**, 265–307.
Park, T. (1954). Experimental studies of interspecies competition. II. Temperature, humidity, and competition in two species of *Tribolium. Physiol. Zool.*, **27**, 177–238.
Parsons, P. A. (1975). The comparative evolutionary biology of the sibling species *Drosophila melanogaster* and *D. simulans. Q. Rev. Biol.*, **50**, 151–169.
Patterson, J. T., and Stone, W. S. (1952). *Evolution in the Genus Drosophila*. Macmillan, New York.
Pelseneer, P. (1920). *Les Variations et leur Hérédité chex les Mollusques*. M. Hayez, Brussels.

Piaget, J. (1929a). Les races lacustres de la *Limnaea stagnalis* (L.). *Bull. Biol. Fr. Belg.*, **63**, 424–455.
Piaget, J. (1929b). L'adaptation de la *Limnaea stagnalis* aux milieux lacustres de la Suisse romande. *Rev. Suisse Zool.*, **36**, 263–531.
Piaget, J. (1979). *Behaviour and Evolution*, English translated edition. Routledge and Kegan Paul, London.
Pianka, E. R. (1978). *Evolutionary Ecology*. Harper and Row, New York.
Pianka, E. R. (1981). Competition and niche theory. In: *Theoretical Ecology, Principles and Applications*, second edition, ed. R. M. May. Blackwell, Oxford.
Pimentel, D., Feinberg, E. H., Wood, P. W., and Hayes, J. T. (1965). Selection, spatial distribution and the coexistence of competing fly species. *Am. Nat.*, **99**, 97–109.
Poodry, C. A. (1980). Imaginal discs: morphology and development. In: *The Genetics and Biology of Drosophila*, Vol. 2d, eds. M. Ashburner and T. R. F. Wright, Academic Press, London.
Poulson, D. F. (1950). Histogenesis, organogenesis, and differentiation in the embryo of *Drosophila melanogaster* Meigen. In: *Biology of Drosophila*, ed. M. Demerec. John Wiley, New York.
Prakash, S. (1972). Origin of reproductive isolation in the absence of apparent genic differentiation in a geographical isolate of *Drosophila pseudoobscura*. *Genetics*, **72**, 143–155.
Prakash, S., Lewontin, R. C., and Hubby, J. L. (1969). A molecular approach to the study of genic heterozygosity in natural populations. IV. Patterns of genic variation in central, marginal and isolated populations of *Drosophila pseudoobscura*. *Genetics*, **61**, 841–858.
Raff, R. A., and Kaufman, T. C. (1983). *Embryos, Genes and Evolution. The Developmental Genetic Basis of Evolutionary Change*. Macmillan, New York.
Raup, D. (1966). Geometric analysis of shell coiling: general problems. *J. Paleontol.*, **40**, 1178–1190.
Raven, C. P. (1964). Development. In: *Physiology of Mollusca*, Vol. 1, eds. K. M. Wilbur and C. M. Yonge. Academic Press, New York.
Reeve, E. C. R., and Robertson, F. W. (1953). Studies in quantitative inheritance. II. Analysis of a strain of *Drosophila melanogaster* selected for long wings. *J. Genet.*, **51**, 276–316.
Reeve, E. C. R., and Robertson, F. W. (1954). Studies in quantitative inheritance. VI. Sternite chaeta number in *Drosophila*: a metameric quantitative character. *Z. Indukt. Abstamm. Vererbungsl.*, **86**, 269–288.
Rensch, B. (1959). *Evolution Above the Species Level*. Methuen, London.
Roberts, D. B., and Graziosi, G. (1977). Protein synthesis in the early *Drosophila* embryo; analysis of the protein species synthesised. *J. Embryol. Exp. Morphol.*, **41**, 101–110.
Roberts, D. F., Billewicz, W. Z. and McGregor, I. A. (1978). Heritability of stature in a West African population. *Ann. Hum. Genet.*, **42**, 15–24.
Robertson, A. (1970). A note on disruptive selection experiments in *Drosophila*. *Am. Nat.*, **104**, 561–569.
Robertson, F. W. (1955). Selection response and the properties of genetic variation. *Cold Spring Harbor Symp. Quant. Biol.*, **20**, 166–177.
Robertson, F. W. (1957). Studies in quantitative inheritance. XI. Genetic and environmental correlation between body size and egg production in *Drosophila melanogaster*. *J. Genet.*, **55**, 428–443.
Root, R. B. (1967). The niche exploitation pattern of the blue-gray gnatcatcher. *Ecol. Monogr.*, **37**, 317–350.
Rose, S. (1966). *The Chemistry of Life*. Penguin Books, Harmondsworth, Middlesex.
Roszkowski, W. (1914). Contribution à l'étude des Limnées du Lac Leman. *Rev. Suisse Zool.*, **22**, 457–539.
Roughgarden, J. (1972). Evolution of niche width. *Am. Nat.*, **106**, 683–718.

Roughgarden, J. (1976). Resource partitioning among competing species — a coevolutionary approach. *Theor. Popul. Biol.,* **9**, 388–424.
Sandler, L., Hiraizumi, Y., and Sandler, I. (1959). Meiotic drive in natural populations of *Drosophila melanogaster*. I. The cytogenetic basis of segregation distortion. *Genetics,* **44**, 233–250.
Sarich, V., and Cronin, J. E. (1977). Generation length and rates of hominoid molecular evolution. *Nature,* **269**, 354.
Saunders, P. T., and Ho, M. W. (1976). On the increase in complexity in evolution. *J. Theor. Biol.,* **63**, 375–384.
Scharloo, W., den Boer, M., and Hoogmoed, M. S. (1967). Disruptive selection on sternopleural chaeta number in *Drosophila melanogaster. Genet. Res.,* **9**, 115–118.
Scheller, R. H., Constantini, F. D., Kozlowski, M. R., Britten, R. J., and Davidson, E. H. (1978). Specific representation of cloned repetitive DNA sequences in sea urchin RNAs. *Cell,* **15**, 189–203.
Schindel, D. E. (1982). The gaps in the fossil record. *Nature,* **297**, 282–284.
Schopf, T. J. M., and Oehler, D. Z. (1976). How old are eukaryotes? *Science,* **193**, 47–49.
Shorrocks, B. (1972). *Invertebrate Types: Drosophila*. Ginn and Company, London.
Siegel, S. (1956). *Non-parametric Statistics for the Behavioural Sciences*. McGraw-Hill Kogakusha, Tokyo.
Simpson, G. G. (1944). *Tempo and Mode in Evolution*. Columbia University Press, New York.
Simpson, G. G. (1953). *The Major Features of Evolution*. Columbia University Press, New York.
Singh, R. S., Lewontin, R. C., and Felton, A. A. (1976). Genetic heterogeneity within electrophoretic "alleles" of xanthine dehydrogenase in *Drosophila pseudoobscura. Genetics.* **84**, 602–629.
Sinnott, E. W., Dunn, L. C., and Dobzhansky, T. (1958). *Principles of Genetics*, fifth edition. McGraw-Hill Kogakusha, Tokyo.
Slatkin, M. (1980). Ecological character displacement. *Ecology,* **61**, 163–177.
Snodgrass, R. E. (1902). The relation of the food to the size and shape of the bill in the Galapagos genus *Geospiza. Auk,* **19**, 367–381.
Sokal, R. R., and Rohlf, F. J. (1969). *Biometry*. Freeman, San Francisco.
Sonnenblick, B. P. (1950). The early embryology of *Drosophila melanogaster*. In: *Biology of Drosophila*, ed. M. Demerec. John Wiley, New York.
Stanley, S. M. (1975). A theory of evolution above the species level. *Proc. Natl. Acad. Sci. USA,* **72**, 646–650.
Stanley, S. M. (1979). *Macroevolution. Pattern and Process*. Freeman, San Francisco.
Stein, G. H. W. (1951). Populationsanalytische Untersuchungen am europäischen Maulwurf. II. Uber zeitliche Grössenschwankungen. *Zool. Jahrb. (Syst.),* **79**, 567–590.
Stelfox, A. W. (1945). A large race of *Cepaea nemoralis* L. (and other Mollusca) at high altitudes in the Galtee Mountains, Co. Tipperary South. *J. Conchol.,* **22**, 168.
Stewart, A. D., and Hunt, D. M. (1982). *The Genetic Basis of Development*. Blackie, Glasgow.
Stewart, F. M., and Levin, B. R. (1973). Partitioning of resources and the outcome of interspecific competition: a model and some general considerations. *Am. Nat.,* **107**, 171–198.
Struhl, G. (1981). A gene product required for correct initiation of segmental determination in *Drosophila. Nature,* **293**, 36–41.
Struhl, G. (1982). Decapentaplegic — hopes held out. *Nature,* **298**, 13–14.
Sturtevant, A. H. (1923). Inheritance of direction of coiling in *Limnaea. Science,* **58**, 269–270.

Tan, W. Y., and Chang, W. C. (1972). Convolution approach to the genetic analysis of quantitative characters of self-fertilized populations. *Biometrics,* **28**, 1073–1090.
Tank, P. W., and Holder, N. (1981). Pattern regulation in the regenerating limbs of urodele amphibians. *Q. Rev. Biol.,* **56**, 113–142.
Taylor, J. W. (1914). *Monograph of the Land and Freshwater Mollusca of the British Isles,* Vol. 3. Taylor Brothers, Leeds.
Templeton, A. R. (1977). Analysis of head shape differences between two interfertile species of Hawaiian *Drosophila. Evolution,* **31**, 630–641.
Thoday, J. M. (1972). Review lecture: Disruptive selection. *Proc. R. Soc. London Ser. B,* **182**, 109–143
Thoday, J. M., and Gibson, J. B. (1962). Isolation by disruptive selection. *Nature,* **193**, 1164–1166.
Thompson, D'A. W. (1942). *On Growth and Form,* second edition. Cambridge University Press, Cambridge.
Thompson, J. N., Ashburner, M., and Woodruff, R. C. (1977). Presumptive control mutation for alcohol dehydrogenase in *Drosophila melanogaster. Nature,* **270**, 363.
Turesson, G. (1922a). The species and the variety as ecological units. *Hereditas,* **3**, 100–113.
Turesson, G. (1922b). The genotypical response of the plant species to the habitat. *Hereditas,* **3**, 211–350.
Turesson, G. (1923). The scope and import of genecology. *Hereditas,* **4**, 171–176.
Turesson, G. (1925). The plant species in relation to habitat and climate. *Hereditas,* **6**, 147–236.
Turing, A. M. (1952). The chemical basis of morphogenesis. *Phil. Trans. R. Soc. London Ser. B,* **237**, 37–72.
Turner, J. R. G. (1967). Why does the genotype not congeal? *Evolution,* **21**, 645–656.
Turner, J. R. G. (1977). Butterfly mimicry: the genetical evolution of an adaptation. In: *Evolutionary Biology,* Vol. 10, eds. M. K. Hecht, W. C. Steere and B. Wallace. Plenum Press, New York.
Val, F. C. (1977). Genetic analysis of the morphological differences between two interfertile species of Hawaiian *Drosophila. Evolution,* **31**, 611–629.
Vaurie, C. (1950). Notes on some Asiatic nuthatches and creepers. *Am. Mus. Novit.,* **1472**, 1–29.
Vaurie, C. (1951). Adaptive difference between two sympatric species of nuthatches (*Sitta*). *Proc. Xth Int. Ornithol. Congr. Uppsala,* 163–166.
Volterra, V. (1926). Variations and fluctuations of the number of individuals of animal species living together. Translation from Italian in: *Animal Ecology,* R. N. Chapman. McGraw-Hill, New York, 1931.
Von Neumann, J. (1966). *Theory of Self-reproducing Automata,* ed. A. W. Burks. University of Illinois Press, Urbana, Illinois.
Waddington, C. H. (1942). Canalization of development and the inheritance of acquired characters. *Nature,* **150**, 563–565.
Waddington, C. H. (1953). The genetic assimilation of an acquired character. *Evolution,* **7**: 118–126.
Waddington, C. H. (1956). Genetic assimilation of the bithorax phenotype. *Evolution,* **10**, 1–13.
Waddington, C. H. (1957a). *The Strategy of the Genes.* Allen and Unwin, London.
Waddington, C. H. (1957b). The genetic basis of the 'assimilated bithorax' stock. *J. Genet.,* **55**, 241–245.
Waddington, C. H. (1969). Computer simulation of a molluscan pigmentation pattern. *J. Theor. Biol.,* **25**, 219–225.

Waddington, C. H. (1975). *The Evolution of an Evolutionist*. Edinburgh University Press, Edinburgh.
Wallace, A. R. (1889). *Darwinism: An Exposition of the Theory of Natural Selection*. Macmillan, London.
Wallace, B. (1968). Polymorphism, population size, and genetic load. In: *Population Biology and Evolution*, ed. R. C. Lewontin. Syracuse University Press, Syracuse, New York.
Wallace, B. (1981). *Basic Population Genetics*. Columbia University Press, New York.
Walter, M. R., Buick, R., and Dunlop, J. S. R. (1980). Stromatolite, 3,400 – 3,500 Myr old from the North Pole area, Western Australia. *Nature*, **284**, 443–445.
Watson, J. D. (1976). *Molecular Biology of the Gene*, third edition. Benjamin, Menlo Park, California.
Westoll, T. S. (1949). On the evolution of the Dipnoi. In: *Genetics Paleontology and Evolution*, eds. G. L. Jepsen, E. Mayr and G. G. Simpson. Princeton University Press, Princeton, New Jersey.
Whalen, M. D. (1978). Reproductive character displacement and floral diversity in *Solanum* section *Androceras*. *Syst. Bot.*, **3**, 77–86.
White, M. J. D. (1978). *Modes of Speciation*. Freeman, San Francisco.
Wieschaus, E., and Gehring, W. (1976). Clonal analysis of primordial disc cells in the early embryo of *Drosophila melanogaster*. *Dev. Biol.* **50**, 249–263.
Williamson, P., Cameron, R. A. D., and Carter, M. A. (1976). Population density affecting adult shell size of the snail *Cepaea nemoralis* L. *Nature*, **263**, 496–497.
Williamson, P. G. (1981a). Palaeontological documentation of speciation in Cenozoic molluscs from Turkana Basin. *Nature*, **293**, 437–443.
Williamson, P. G. (1981b). Morphological stasis and developmental constraint: real problems for neo-Darwinism. *Nature*, **294**, 214–215.
Williamson, P. G. (1982). Reply. *Nature*, **296**, 611–612.
Wills, C. (1981). *Genetic Variability*. Clarendon Press, Oxford.
Wilson, E. O., and Bossert, W. H. (1971). *A Primer of Population Biology*. Sinauer, Sunderland, Massachusetts.
Wolda, H. (1963). Natural populations of the polymorphic land snail *Cepaea nemoralis* (L.). *Arch. Neerl. Zool.*, **15**, 381–471.
Wolda, H. (1967). The effect of temperature on reproduction in some morphs of the land snail *Cepaea nemoralis*. *Evolution*, **21**, 117–129.
Wolda, H., and Kreulen, D. A. (1973). Ecology of some experimental populations of the landsnail *Cepaea nemoralis* (L.). II. Production and survival of eggs and juveniles. *Neth. J. Zool.*, **23**, 168–188.
Wolpert, L. (1968). The French flag problem: a contribution to the discussion on pattern development and regulation. In: *Towards a Theoretical Biology*, 1. *Prolegomena*, ed. C. H. Waddington. Edinburgh University Press, Edinburgh.
Wolpert, L. (1969). Positional information and the spatial pattern of cellular differentiation. *J. Theor. Biol.*, **25**, 1–47.
Wolpert, L. (1978). Pattern formation in biological development. *Sci. Am.*, **239**(4), 124–137.
Wright, S. (1931). Evolution in Mendelian populations. *Genetics*, **16**, 97–159.
Wright, S. (1950). Genetical structure of populations. *Nature*, **166**, 247–249.
Wright, S. (1952). The genetics of quantitative variability. In: *Quantitative Inheritance*, eds. E. C. R. Reeve and C. H. Waddington. Agricultural Research Council, HMSO, London.
Wright, S. (1968). *Evolution and the Genetics of Populations*, Vol. 1, *Genetic and Biometric Foundations*. University of Chicago Press, Chicago.
Wynne-Edwards, V. C. (1962). *Animal Dispersion in Relation to Social Behaviour*. Oliver and Boyd, London.

Yunis, J. J., and Prakash, O. (1982). The origin of man: a chromosomal pictorial legacy. *Science,* **215**, 1525–1530.
Zouros, E. (1982). On the role of chromosomal inversions in speciation. *Evolution,* **36**, 414–416.

Author Index

Abbott, I., 79, 81–83
Abbott, L.K., 79, 81–83
Akam, M., 170
Anderson, W.W., 184
Anderson-Kottö, I., 154
Antonovics, J., 69, 90
Armstrong, R.A., 233
Arthur, W., 3, 26, 44, 67, 70, 71, 73–75, 78, 83, 88, 89, 93, 108–112, 124, 132, 149, 150, 167, 185, 204, 215, 233, 245
Arthurs, A.M., 207
Ashburner, M., 97, 150
Atkinson, W.D., 77, 233
Ayala, F.J., 3, 7, 16, 67, 218

Bakker, K., 18, 182
Bantock, C.R., 78, 79
Barker, J.S.F., 41, 63
Barnes, R.D., 194, 197
Bayley, J.A., 78, 79
Beachy, P.A., 170
Bell, M.A., 179
Bender, W., 170
Berry, R.J., 178, 180
Billewicz, W.Z., 41
Bishop, J.A., 215
Blackith, R.E., 26
Blair, W.F., 103
Boag, P.T., 41, 79–81
Boaz, N.T., 121, 126–128
Bodmer, W.F., 99
Bohren, B.B., 59
Bonner, J.T., 26
Bossert, W.H., 8
Boucot, A.J., 124
Boycott, A.E., 29, 70, 152, 154
Brent, L., 10
Britten, R.J., 137, 169, 227
Brower, D.L., 157
Brown, J.H., 229
Brown, W.L., 103, 110

Buick, R., 5
Bulmer, M.G., 4, 109
Burnet, B., 91
Buss, L.W., 18
Butlin, R.K., 13

Cain, A.J., 79, 92, 93
Cameron, R.A.D., 75, 76, 94, 197
Camfield, R.G., 150
Campbell, C.A., 16
Carson, H.L., 96
Carter, M.A., 75, 76
Cavalli-Sforza, L.L., 99
Chabora, A.J., 63, 64
Chandler, P., 10
Chang, W.C., 99
Charlesworth, B., 126, 130, 134
Child, C.M., 138
Christiansen, F.B., 109
Clarke, B., xii, 13, 41, 44, 70, 78, 88, 89, 101, 102, 132, 150, 154, 155, 198, 199, 215
Clarke, C.A., 235
Clausen, J., 84–86
Clayton, G.A., 41, 55, 59, 60
Cohen, M.H., 138
Comfort, A., 77
Constantini, F.D., 169
Cook, L.M., 16, 40, 41, 77–79, 215
Cowling, D.E., 91
Crampton, H.E., 153, 154
Crick, F.H.C., 138, 141, 142, 144
Cronin, J.E., 121, 126–128
Crothers, J.H., 26
Crow, J.F., 2
Crozier, R.H., 109, 112, 177
Cummins, L.J., 63
Currey, J.D., 92

Da Cunha, A.B., 67
Darwin, C., xi, 7, 14, 19, 22, 31, 51,

105, 114, 116, 118, 134, 173, 180,
    182, 204, 214, 229, 230, 250
Darwin, F., 119, 134, 173
Davidson, D.W., 229
Davidson, E.H., 137, 169, 227
Day, T.H., 13
Degner, E., 155
Delson, E., 126
Dempster, J.P., 18
Den Boer, M., 63, 64
De Souza, H.M.L., 67
De Vries, H., 7, 134, 178, 179, 218, 240
Dickinson, H., 90
Diver, C., 29, 70, 114, 152, 154
Dobson, T., 13
Dobzhansky, T., 7, 13, 41, 57, 58, 120,
    189, 235
Dos Santos, E.P., 67
Dunlop, J.S.R., 5
Dunn, L.C., 10, 189

Ede, D.A., 156
Eldredge, N., 88, 116, 117, 119–122,
    125, 126, 129, 134, 178, 200, 218, 240
Ellis, A.E., 94, 152, 169
Elton, C.S., 70, 231, 234
Eoff, M., 103
Erk, F.C., 91

Falconer, D.S., 4, 31, 34–37, 40–42, 45,
    46, 48, 49, 53, 55–57, 62, 65, 66, 168
Feinberg, E.H., 67
Felton, A.A., 46, 96
Fenchel, T., 109–112
Fierz, W., 10
Fisher, R.A., 14, 32, 139, 150, 181, 187,
    189, 194, 208, 210, 211, 215, 220,
    235, 245
Ford, E.B., 182
Frazzetta, T.H., 191, 246, 247

Gale, J.S., 5
Galvin, A.M., 150
Garcia-Bellido, A., 138, 148, 157,
    159–165, 167, 200
Garnett, I., 66
Garrod, D.R., 136
Garstang, S.L., 29, 70, 152, 154
Gass, I.G., 6
Gause, G.F., 233
Gautier, M., 94
Gehring, W., 159
Gibson, J.B., 62–64, 88, 90
Gilpin, M.E., 17

Glueckson-Waelsch, S., 10
Goldschmidt, R.B., 4, 88, 147, 148,
    178, 179, 193, 200–203, 218, 236, 240
Goodwin, B.C., 138
Gorczynski, R.M., 10
Gould, S.J., 19, 88, 116, 117, 119, 121,
    122, 125, 126, 129, 131, 134, 178,
    200, 205, 218, 223, 225, 240
Grant, B.R., 79
Grant, P.R., 41, 79–83, 103, 108, 110
Grant, V., 69, 90, 102
Graziosi, G., 156
Green, E.L., 46
Gregor, J.W., 84
Grell, E.H., 147, 179, 224

Haldane, J.B.S., 204, 206
Hardy, D.E., 96
Harris, H., 3, 209
Harvey, P.H., 78
Hassell, M.P., 107, 233
Hayes, J.T., 67
Hedrick, P.W., 131, 137
Heslop-Harrison, J., 84
Hewitt, G.M., 95
Hiesey, W.M., 84–86
Hill, W.G., 59
Hillier, P.C., 13
Hiraizumi, Y., 11
Ho, M.W., 221, 226
Hogness, D.S., 170
Holder, N., 136
Hoogmoed, M.S., 63, 64
Horsley, D.T., 44, 78, 88, 89, 132, 215
Horton, I.H., 95
Hubby, J.L., 3, 96, 97, 101
Hubendick, B., 72, 73, 93
Hunt, D.M., 165
Hutchinson, G.E., 175, 187, 231
Huxley, A., 118, 119, 121, 173
Huxley, J.S., 4, 7, 173, 220
Huxley, T.H., 119, 134, 173

Jackson, J.B.C., 17
Jaksic, F.M., 231
Jinks, J.L., 4, 28, 36, 37
Johnson, M.S., 92, 102, 198, 199
Jones, J.S., 5, 43, 92, 117, 123, 125,
    127, 132, 134, 191

Karch, F., 170
Kauffman, R., 119
Kaufman, T.C., 167
Keck, D.D., 84

Kerney, M.P., 94, 197
Kerster, H.W., 103
Khush, G.S., 95
Kimura, M., 2, 3, 11
King, C.E., 184
King, M.-C., 137
Knight, G.R., 59, 60
Kozlowski, M.R., 169
Krebs, C.J., 15, 183
Kreulen, D.A., 78

Lack, D., 79–82
Lande, R., 45, 100, 116, 126, 134
Latter, B.D.H., 12
Lawlor, R., 109, 233
Lawrence, P.A., 138, 148, 157, 159–161, 163–165, 173, 200
Lawton, J.H., 233
Leith, B.H., 127, 132
Lejeune, J., 94
Léon, J.A., 109
Lerner, I.M., 4
Levene, H., 186
Levin, B.R., 109, 185, 233
Levin, D.A., 103
Levins, R., 107, 109
Lewin, B., 119, 227
Lewis, E.B., 13, 25, 62, 161, 163, 170, 200
Lewontin, R.C., 3, 17, 46, 89, 96, 97, 101, 106, 107, 131,, 242
Lindsley, D.L., 147, 179, 224

MacArthur, J.W., 53–55
MacArthur, R.H., 107, 109, 183, 184, 191, 218, 230, 232, 233
MacDonald, A., 43
McClintock, B., 170
McDonald, J.F., 131, 137
McGehee, R., 233
McGregor, I.A., 41
McMurtie, R., 233
McNeilly T., 69, 90
Machin, J., 26
Manning, A., 91
Marchant, C.J., 90
Mather, K., 4, 28, 36, 37
May, R.M., 109, 230, 232, 233
Maynard Smith, J., 17, 19, 62, 68, 90, 109, 132, 192, 205, 220, 222, 225, 233
Mayr, E., 68, 70, 88, 89, 120, 124, 174, 175, 201, 215, 241
Medawar, P.B., 10, 139, 227
Mendel, G., xi, 139

Middlecote, J., 67
Moore, J.A., 93, 233
Morata, G., 138, 148, 157, 159–161, 163–165, 173, 200
Morgan, T.H., 140
Morris, J.A., 41, 55, 59, 60
Mozley, A., 31, 71–73
Murphy, P.G., 112
Murray, J., 41, 88, 89, 101–103, 132, 154, 155, 198, 199

Niemann-Sørensen, A., 66
Nüsslein-Volhard, C., 183

Ochman, H., 92
O'Donald, P., 78
Odum, E.P., 107
Oehler, D.Z., 5, 192
Ohno, S., 225
Oldham, C., 44
Olson, E.C., 5, 7, 249
Ouweneel, W.J., 10, 121, 156, 157

Park, T., 233
Parkin, D.T., 13, 44, 78, 88, 89, 132, 215
Parsons, P.A., 91
Patterson, J.T., 93
Peake, J.F., 77
Peifer, M., 170
Pelseneer, P., 155
Piaget, J., 26, 71–74
Pianka, E.R., 2, 109, 233
Pimentel, D., 67
Pitts, C.R., 150
Poodry, C.A., 157, 158
Poulson, D.F., 157, 176
Prakash, O., 91
Prakash, S., 100, 101
Probert, R.F., 191

Raff, R.A., 167
Rak, Y., 121, 126–128
Raup, D., 26
Raven, C.P., 153
Rawlings, P., 127, 132
Rayfield, L.S., 10
Read, I.L., 13
Reeve, E.C.R., 36, 59
Rensch, B., 7, 225
Reyment, R.A., 26
Ripoll, P., 157, 159
Roberts, D.B., 156
Roberts, D.F., 41

Robertson, A., 12, 41, 55, 59, 60, 63, 66
Robertson, F.W., 36, 41, 57–59
Rohlf, F.J., 29
Root, R.B., 231
Rose, S., 222
Roszkowski, W., 71
Roughgarden, J., 108, 109

Sandler, I., 11
Sandler, L., 11
Sang, J.H., 91
Santamaria, P., 162
Sarich, V., 126
Saunders, P.T., 221, 226
Scharloo, W., 63, 64
Scheller, R.H., 169
Schindel, D.E., 123
Schnell, G.D., 5, 92
Schopf, T.J.M., 5, 192
Selander, R.K., 5, 92
Sheppard, P.M., 235
Shorrocks, B., 93, 233
Siegel, S., 102
Simpson, E., 10
Simpson, G.G., 4, 88, 172, 178, 179, 204, 206, 239
Singh, R.S., 46, 96
Sinnott, E.W., 189
Slatkin, M., 109
Smith, J.N.M., 79
Smith, P.J., 6
Snodgrass, R.E., 79, 82
Sokal, R.R., 29
Sonnenblick, B.P., 136, 140, 189
Spassky, B., 41, 57, 58
Spierer, P., 170
Spieth, H.T., 96
Stanley, S.M., 2, 19, 116, 119, 187, 204, 229, 246
Stebbins, G.L., 7
Steele, E.J., 10
Stein, G.H.W., 70
Stelfox, A.W., 77
Stewart, A.D., 165
Stewart, F.M., 233
Stone, W.S., 93, 96
Stringer, C.B., 121, 126–128
Struhl, G., 135, 162, 164, 167
Sturtevant, A.H., 152, 153
Suckling, L., 10

Tan, W.Y., 99
Tank, P.W., 136

Tattersall, I., 126
Taylor, J.W., 113
Templeton, A.R., 98, 99
Thoday, J.M., 62–64, 88, 90
Thompson, D'A.W., 21–24, 134, 173, 179, 213
Thompson, J.N., 150
Throckmorton, L.H., 96, 97
Turesson, G., 83, 87
Turing, A.M., 140
Turner, F.M., 29, 70, 152, 154
Turner, J.R.G., 215, 235
Turpin, R., 94
Tyler, A., 140

Val, F.C., 98–100
Valentine, J.W., 7, 218
Vaurie, C., 110
Volterra, V., 232
Von Neumann, J., 221

Waddington, C.H., v, 1, 4, 19, 24, 49, 61, 62, 74, 131–133, 135, 165, 176, 228, 248
Wallace, A.R., 14, 102, 180
Wallace, B., 184
Walter, M.R., 5
Watson, J.D., 140
Westoll, T.S., 26, 206
Whalen, M.D., 103
White, M.J.D., 88, 91, 92, 93, 95, 107, 120
Wieschaus, E., 159, 183
Williamson, P., 75, 76
Williamson, P.G., 4, 9, 26, 116, 117, 121–125, 131
Wills, C., 235
Wilson, A.C., 137
Wilson, E.O., 8, 103, 110, 183, 184
Wilson, R.C.L., 6
Wolda, H., 78
Wolpert, L., 136, 138, 141, 144
Wood, P.W., 67
Woodruff, R.C., 150
Wright, S., 45, 151, 201
Wynne-Edwards, V.C., 18

Yunis, J.J., 91

Zouros, E., 90

Subject Index

Adaptive
  landscape, 151
  radiation, 151
Age structure, 75
Agrostis tenuis, 69, 90
Animal phyla, time of origin, 191
Anodonta, 93
Area effect, 92
Arianta arbustorum
  shell shape, 44
  shell size, 40, 41
Artificial selection, 51–66, 68, 111
  by truncation, 52, 65
  intensity of, 49, 53
  relaxation of, 52, 56–59
  response to, 47, 48, 52–60
  selective differential, 47, 48, 52, 53, 55, 66
Asexual reproduction, and *n* selection, 184, 188, 192
Assortative mating, 10, 45
Australopithecus
  afarensis, 126, 127
  africanus, 126–128
  boisei, 126, 127
  robustus, 126, 127

Behaviour
  altruistic, 18
  courtship, 91, 192
  feeding, 2, 108
  territorial, 18
Bellamya unicolor, 112, 113
Biogeographical realm, 234
Bithorax complex, 25, 161–165, 170, 181, 200
Body plan, 172, 173, 181, 193, 199, 202, 235, 239, 240, 242
Breeding, animal and plant, 51

Cell
  differentiation, 135–137, 140, 162
  division, 135–137
  growth, 135, 136
  lineage, 159–161
  migration, 135, 136
  types, 136, 137, 221, 222
Cepaea hortensis
  banding, 32
  range, 113
  shell size, 32, 75–77
Cepaea nemoralis
  banding, 132, 133
  range, 113
  shell colour, 27, 132
  shell size, 75–79
Character displacement
  competitive, 82, 83, 105–115
  reproductive, 103, 106, 107, 198, 199
Chromosome
  B, 95
  inversion, 9, 90, 95, 235
  repatterning, 201
Cladogenesis, 117, 121, 129
Cleavage
  radial, 244
  spiral, 153, 244
Cline, 84
Clone, in development, 159–161
Coadaptation, 184, 187, 193, 201, 226, 235–237, 242, 246, 249
Coefficient of variation, 29
Coelopa frigida, 13
Coexistence of competing species, 230–233
Compartments, 138, 159–165, 173, 199–203
Competition, 81, 105–115, 180, 182, 185–187, 215, 216, 229–233
Competitive ability, 67, 185

Complexity
  definition, 220–222
  developmental increase, 140, 141
  evolutionary increase, 7, 8, 220–227
Correlation of characters, 22
  in response to selection, 58–60
Cross-product ratio (w), 15, 17, 180, 182, 246

Degree of genetic determination (DGD), 36, 37, 46
Developmental
  constraints, 125, 131, 134, 195, 235, 237
  principles, 132, 135, 139, 165
  revolutions, xii, 243
DNA, 169
Dominance, 33, 36, 37, 188, 189
Down's syndrome, 94
Drosophila
  heteroneura, 98–100
  melanogaster
    alcohol dehydrogenase, 150
    bristle number, 12, 38, 41, 55, 59–64
    cross-veinless, 61
    development, 156–165
    egg production, 41
    identification, 93
    reproductive isolation, 62, 63, 91
    rRNA locus, 66
    segregation distorter, 11
    thorax length, 59
    wing length, 59
  persimilis, 101, 113
  pseudoobscura
    phototaxis, 41, 57, 58
    range, 101, 113
    reproductive isolation, 100, 101
    xanthine dehydrogenase, 46, 96
  silvestris, 98–100
  simulans, 91, 93
Duplication
  of genes, 9, 225
  of structures, 225

Ecocline, 84
Ecological equivalents, 234
Ecophenotypic variation, 9, 70, 71, 77, 123–125, 129
Entropy, 227
Environmental heterogeneity, 191, 250
Epistasis, 33, 36, 37
Escherichia coli, 137

Eukaryotes, time of origin, 5
Evolutionary
  cycle, 69, 80, 105, 174
  divergence, 105–108, 113
  stable strategy (ESS), 17
  tree, 117–119, 126–128, 177, 193–195, 216–219
Expansive evolution, 228, 234
Experimental selection, 66, 67
Extinction, 177
  pseudo, 229

Fitness
  decline in, 17
  depression, 173, 175–177, 243
  measurement of, 14–17, 182–186
  non-transitive, 17
Fixation, 16
  rate, 205–212
Fossil record, 5–7, 116–130, 249
Founder effect, 10
French flag model, 141–143, 162

Gasterosteus aculeatus, 179
Gastropod
  direction of coiling, 148, 151–156, 195–199
  shell shape, 25
Gastrulation, 136, 140
Gene
  D, 149, 150, 156, 160, 165–170, 175–177, 188, 193–195, 199, 200, 208, 214, 234–237, 248
  duplication, 9, 225
  frequency (q), 35, 64–66
  R, 150, 165
  regulatory, 66, 121, 125, 131, 132, 137, 148, 149
  S, 150, 165
  selector, 160–163, 167, 173, 200–203
  structural, 137, 149, 150
Genetic
  assimilation, 49, 60–62, 74, 75
  code, 132
  drift, 10–12, 189, 208
Genome size, 226, 227
Genotype frequency, 9, 10, 131
Genotype–environment correlation, 35
Genotype–environment interaction, 35
Genotypic value, 33–35
Geological time scale, 6
Geospiza
  conirostris, 81
  difficilis, 81

fortis, 41, 79, 81, 82
fuliginosa, 79, 81, 82
magnirostris, 79, 82
Gradient, of morphogen, 138, 141, 147
Growth, isometric, 25
Guild, 231, 234

Habitat choice, 191, 192
Hardy–Weinberg equilibrium, 8, 33
Helix
 aspersa, 94
 pomatia, 94
Heritability (h^2), 31, 37–50, 53, 55, 79–81
 of differences between populations, 43–45, 81, 124, 125
Higher taxa, origin of, 7, 134, 151, 172–203, 239
Histoblast, 158
Homo
 erectus, 126–128
 habilis, 126–128
 sapiens, 126–128
Homoeotic mutations, 148, 150, 156–165, 175, 181, 193, 200–203
Homosequential species, 90, 91, 94–96, 98
Hopeful monster, 148, 193
Human
 evolution, 126–128
 height, 41
Hybridization of species, 62, 92, 94, 97–100, 198, 199
Hydrobia
 neglecta, 110, 112
 ulvae, 110–112
 ventrosa, 110–112

Imaginal discs, 157–159, 176
Imbalance
 between morphology and environment, 181
 coadaptational, 181, 199
Inbreeding, 52, 54
Isolating mechanisms, 91, 92

Kurtosis, 9, 27–29

Laciniaria biplicata, 155
Lamarckian evolution, 10, 19, 49
Life and fertility tables, 14, 15
Life cycle, 135, 140, 145, 146, 166, 168
Limiting similarity, 107, 232, 233
Lotka–Volterra model, 232

Lymnaea
 emarginata, 71, 73
 palustris, 71, 72
 peregra
  direction of coiling, 29, 70, 151–154, 198, 241
  shell shape, 71–73
 stagnalis, 71–74, 124

Macroevolution
 conclusions on, 239, 240
 definition of, 178
Maternal inheritance, 140, 151–156, 162, 190
Mega-evolution, definition of, 179
Meiotic drive, 10, 11
Melanoides tuberculata, 125
Mice
 body weight, 53–55
 t locus, 10
 vertebrae, 46
Microevolution
 conclusions on, 239, 240
 definition of, 178
Migration, 10, 11
 and speciation, 234, 235
Minute technique, 159
Modern synthesis, xi, 250
Molecular evolution, 2, 3, 119, 137
Morphogen, 132, 133, 141–143, 147, 150, 162
 network, 143, 144–146
 physico-chemical nature of, 142, 143
 serial, 143–145
 terminal, 143
Morphogenesis, 135–146, 149, 215
Morphogenetic tree, 142–146, 165–169, 248, 249
 and coadaptation, 236, 237
 and mega-evolutionary tree, 216–219
Morphological character
 complexes, 25, 206, 207
 definition of, 5
 measurement of, 21–30
Morphological evolution
 definition of, 8–10, 27
Muller's ratchet, 225
Multi-species systems
 character shifts, 114
 coexistence, 230–233
Multivariate morphometric analysis, 26, 121–126

Mutation
   and n-selection, 188–190
   germ cell, 12, 188, 189
   germ cell precursor, 189, 190
   macro, 7, 179
   magnitude of effect, 12, 13, 121, 150, 151, 167, 174, 208–212, 245
   rate, 12, 205, 209
   reductive, 224, 225
   somatic, 12
   status quo, 224, 225
   supplementary, 224, 225

Natural selection
   climatic, 78, 86
   directional, 69
   frequency-dependent, 16
   group, 18
   kin, 18, 19
   n, 182–193, 202, 203, 208, 215, 219, 226, 235–237, 246–248
   normalizing, 119, 125
   r, 183, 184
   requirements for, 68, 182
   species, 19, 204, 205
   visual, 78
   w, 180–182, 184, 191–193, 198, 201–203, 208, 215, 219, 229, 235–237, 246–248
Neo-Darwinism, xi, xii, 125, 126, 129–139, 150, 151, 172–178, 180, 203, 215, 218, 219, 242, 247–250
   definition of, 19, 20
   extreme version, 174, 175, 203, 215, 240, 241, 247
Neoteny, 225
Net reproductive rate (R_0), 14, 15, 17, 182–186, 189–192, 246, 247
Neutrality, of morphological characters, 12, 70, 182
Niche, 177, 187, 193, 201, 230, 231

Olenus, 119
Operon, 137, 149
Overspecialization, developmental, 237

Papilio, 235
Parallel variation, 75–77, 123
Paramecium, 233
Parasites, 44
Partula
   aurantia, 101
   exigua, 101, 102

suturalis
   direction of coiling, 154, 155, 198, 199, 241
   shell size, 41
   taeniata, 101, 102
Pattern formation, 135, 136, 162, 215
Peripheral populations, 78, 119, 121, 125
Phacops, 117
Phase-shift model, 138
Phenocopy, 62, 156, 176
Phenotype
   and natural selection, 132
   and neo-Darwinism, 19
   coadapted, 235
   four-dimensional, 26
Phenotypic
   cycle, 17
   value, 34, 35, 65
Phlox
   glaberrina, 103
   pilosa, 103
Phyletic gradualism, 116–119, 127
Physa fontinalis, 198
Pigmentation, 2, 5, 214, 215
Pleiotropy, 43, 188, 210
Polygenes, 25, 32, 35, 42, 43, 46, 66, 113, 129, 150, 166–169
Polymorphism
   and interspecific competition, 112, 185
   chromosomal, 94, 95
   lateral plate, 179
   of polygenes, 42
   pigmentation, 215
   shell coiling, 29, 151–154
Polyploidy, 90, 91
Population
   density, 18, 75–77
   mean, 9, 34, 35
   regulation, 18
Position effect, 201
Positional information, 137, 138, 162
Potamopyrgus jenkinsi, 110, 112
Potentilla glandulosa
   ecotypes, 83–87
   selection experiment, 86
Pre-adaptation, 193, 201
Progressiveness in evolution, 7, 220
Propagule, of mutants, 183, 188–190
Pseudoallele, 162
Pseudogene, 169, 170, 176
Puffing pattern, 97

Punctuated equilibrium, 8, 116–130
Punctuations, ecological, 120, 121, 129

Rate of evolution
 and evolutionary mechanisms, 8
 measurement of, 204–207
 variation in, 3, 116–130, 204–219
Ratio, in shape measurement, 25, 26, 70–75
Recombination, mitotic, 159
Regeneration, 136, 137
Resource utilization function, 108, 109, 231

Saltational evolution, xi, xii, 7, 131–135, 172, 173, 178–203, 217, 218, 235–237, 240–244
Selective coefficient, 15, 90
Sex, time of origin, 192
Sexual dimorphism, 29, 41, 54
Sibling species, 93–97
Simulation, 212
Sitta
 neumayer, 103
 tephronota, 103
Skew, 9, 27–29
Solanum, 103
Spartina townsendii, 90
Speciation
 allopatric, 68, 69, 88–90
 and morphological change, 93, 94, 103, 116–130
 parapatric, 88–90
 sympatric, 62–64, 88, 90
Steady state evolution, 228
Stratigraphical completeness, 123
Supergene, 235
Systemic mutation, 201

Talpa europaea, 70
Testability of evolutionary theories, 180, 249, 250
Thermodynamics, second law of, 227
Threshold characters, 45–49, 61, 62
 incidence, 47
 liability, 47, 62
Transformation, 22–24, 134
Transplant experiment, 44, 72–74
Transposable elements, 10, 170
Tribolium
 castaneum, 233
 confusum, 233

Variance, 9, 28, 29, 35–40, 55
Variation
 discrete, 27, 32
 meristic, 29, 32, 38, 41
 quantitative (continuous), 27–29, 32, 41
Vertebrates, and n selection, 192

Wallace effect, 88, 102, 103

DATE DUE

PRINTED IN U.S.A.